I0493564

# Radio Engineering and Antennas

## First Edition

## Iqbal Singh Josan, P.E.

*Copyright © US Purtek LLC, 2014 All Rights Reserved. No part of this publication may be reproduced without the prior written permission of US Purtek LLC, uspurtek.com*

To all Wireless Communications Professionals for their sincerity
and dedication to the profession

# Contents

# Preface

The book 'Radio Engineering and Antennas' is intended as a ready reference, study guide and a one-stop source for wireless communications professionals, practicing telecommunication engineers, technology professionals, engineering graduates and students.

The guiding principle in writing this book is, to provide a simplified understanding of concepts related to radio engineering and antennas, with a special emphasis on their practical application to the wireless communications standards that are practiced currently around the world, such as WiFi, WiMax, GSM, CDMA, and LTE. The general flow of various topics is to begin with a review of the basics, and then move on to discussion of current applications of wireless technologies through examples and illustrations.

This book serves as an excellent companion to learning webinars offered on the web site www.uspurtek.com. These webinars are conducted via live and interactive online sessions by experienced instructors and are based on the contents of this book.

The book and the webinars can be used in conjunction to study for the 'Radio Engineering and Antennas' section of the IEEE WCET (Wireless Communication Engineering Technologies) certification exam, which is required to earn the IEEE WCP (Wireless Communications Professional) credential.

A list of acronyms, bibliography and web sites, is included at the end of the book as a quick reference for additional information.

# 1. Introduction

In any wireless communications system, the wireless communications link or the radio link is the most important component. In fact the RF (Radio Frequency) link is a critical limiting factor to the overall performance of a wireless system. **Figure 1.a** shows a radio channel between two microwave towers that have dish like antennas. In order to get maximum performance out of this wireless communications link, we must study RF propagation characteristics of the radio link that exists between antenna systems on each of these towers, and their impact on the overall design and performance of the antenna systems.

Similar considerations can be extended to RF link between a smart phone and eNodeB in a LTE network, or a NodeB in a UMTS network, or a BS (Base Station) in a GSM network. It does not matter which wireless technologies are used, the fundamental principles remain the same.

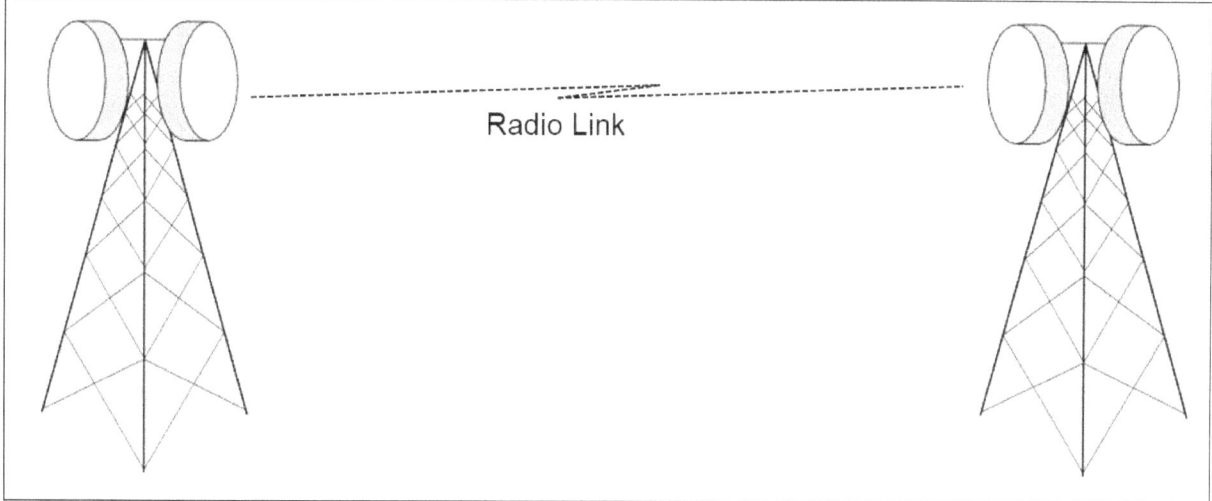

**Figure 1.a**

# 2. Key Factors for Radio Link Design

The key factors which influence the design of a radio link include, the *radio frequency* used, the *path of radio frequency propagation* and the *propagation medium*. We will now explore each of these factors in more detail.

## 2.1 Frequency

The frequencies for RF propagation are allocated to several bands which are defined by ITU (International Telecommunication Union). **Table 2.1.a** lists the frequency allocations. As you will notice, for cellular communications, Band 9, which ranges from 300-3000 MHz, is used for GSM, UMTS & LTE technologies. The frequencies within this band are allocated for commercial use by regulatory authorities in different geopolitical regions of the world.

It is essential to know which frequencies are used in a wireless system, and whether they are licensed or unlicensed. The choice of frequency will influence the design of wireless system that includes the antenna systems as well as the radio link between antennas.

| Band Number | Symbols | Frequency Range | Metric subdivision | Metric abbreviation |
|---|---|---|---|---|
| -1 | ---- | 0.03 – 0.3 Hz | Gigametric waves | B.Gm |
| 0 | ELF | 0.3 – 3 Hz | Hectomegametric waves | B.hMm |
| 1 | ---- | 3 – 30 Hz | Decamegametric waves | B.daMm |
| 2 | ---- | 30 – 300 Hz | Megametric | B.Mm |
| 3 | ULF | 300 – 3000 Hz | Hectokilometric waves | B.hkm |
| 4 | VLF | 3 – 30 kHz | Myriametric waves | B.Mam |
| 5 | LF | 30 – 300 kHz | Kilometric waves | B.km |
| 6 | MF | 300-3000 kHz | Hectometric waves | B.hm |
| 7 | HF | 3-30 MHz | Decametric waves | B.dam |
| 8 | VHF | 30 – 300 MHz | Metric waves | B.m |
| 9 | UHF | 300 – 3000 MHz | Decimetric waves | B.dm |
| 10 | SHF | 3-30 GHz | Centimetric waves | B.cm |
| 11 | EHF | 30 – 300 GHz | Millimetric waves | B.mm |
| 12 | ---- | 300 – 3000 GHz | Decimillimetric waves | B.dmm |
| 13 | ---- | 3 – 30 THz | Centimillimetric waves | B.cmm |
| 14 | ---- | 30 – 300 THz | Micrometric waves | B.μm |
| 15 | ---- | 300 – 3000 THz | Decimicrometric waves | B.dμm |

**Table 2.1.a**

## 2.2  Path of Propagation

The path traversed by radio frequency waves is defined as the *path of propagation*. The most basic is the *line of sight* (LOS) path, which exists when transmitter and receiver can "see" each other in a straight line. A familiar example of LOS path is the wireless communications between cell towers (**Figure 2.2.a**). Another example is the wireless communication between cell tower and the cell phone in a rural area with open spaces (**Figure 2.2.b**).

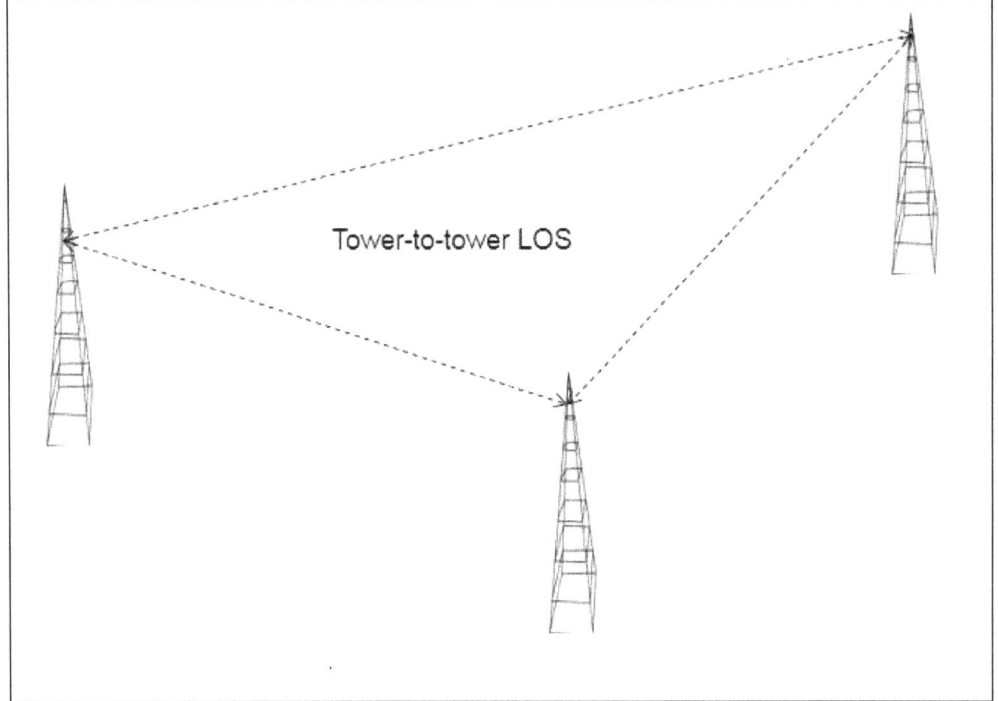

Tower-to-tower LOS

**Figure 2.2.a**

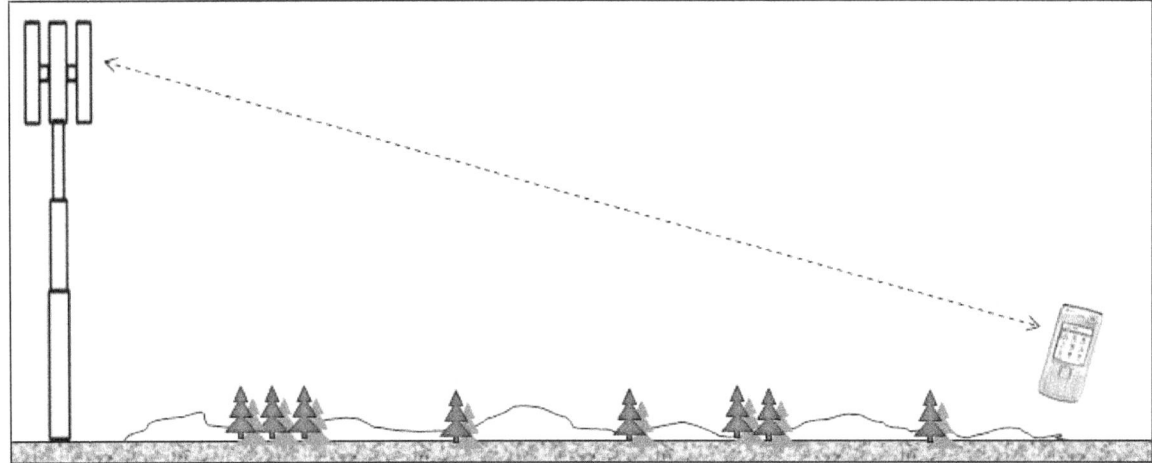

**Figure 2.2.b**

In urban and suburban areas, and in building interiors, hallways and tunnels, there is hardly any scope for LOS propagation. The handset or the receiver usually cannot "see" the base station or the access point. In these situations, the radio signals suffer one or more of the following phenomenon: *reflection, diffraction, scattering* and *absorption.*

As a result, the handset receives radio signals that have propagated over NLOS (Near Line of Sight) path, as shown in **Figure 2.2.c**. We will see in **Chapter 16** how the *path of propagation* impacts *link budget analysis* for each situation.

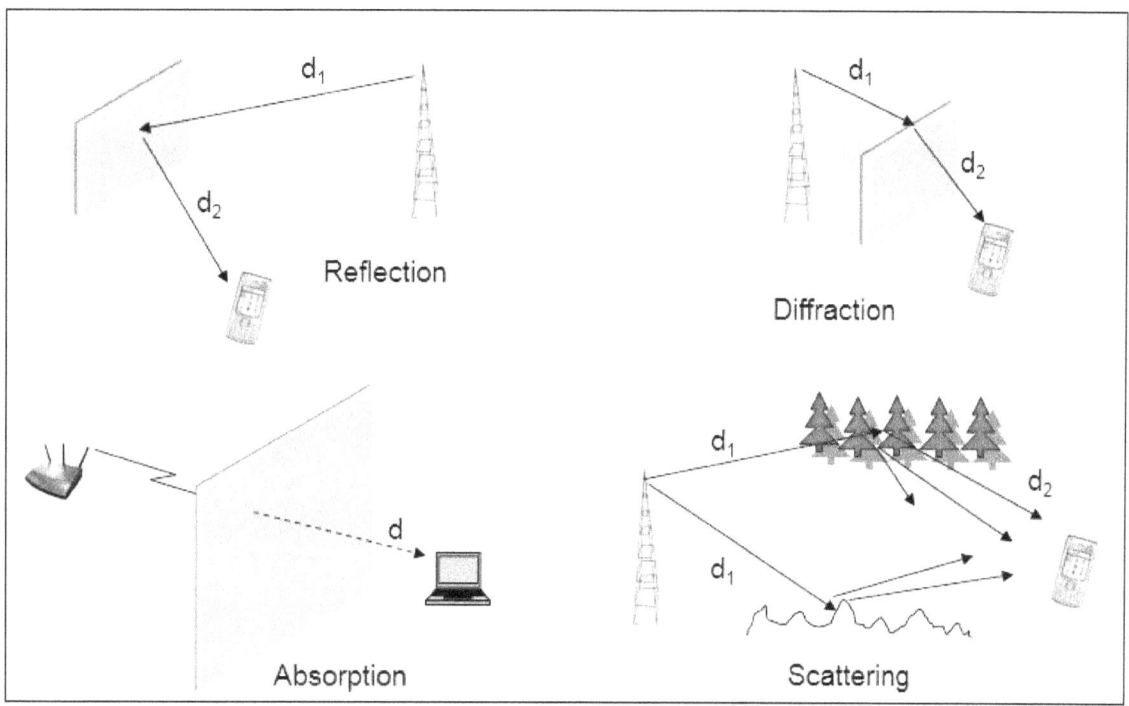

**Figure 2.2.c**

## 2.3 Propagation Medium

The medium through which radio waves propagate has an important bearing on the performance of radio link. Radio waves are essentially *electromagnetic* (EM) waves, and are subject to various types of losses in the medium through which they propagate.

The ideal propagation medium for RF is *free space*, which means, empty space, although for all practical purposes, every propagation medium is usually less than ideal. However, analysis of RF propagation in *free space* provides an excellent reference for performance comparison with all other types of propagation media.

For common situations, propagation medium consists of air, which is a combination of various gases, such as oxygen, nitrogen and carbon dioxide. The presence of gases, water vapors, fog, mist, rain drops, and dust changes the propagation characteristics of the medium (**Figure 2.3.a**).

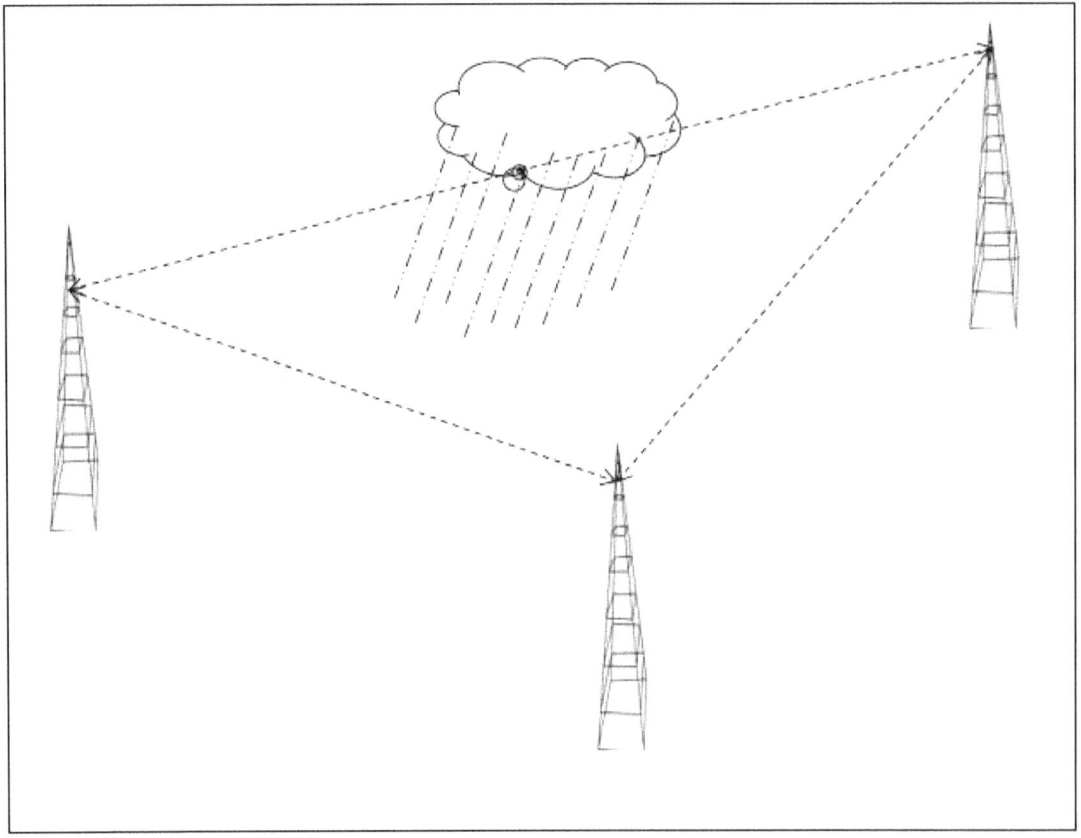

**Figure 2.3.a**

The height at which radio waves propagate from the surface of earth must be considered. The atmospheric composition changes as we rise above the surface of earth. The lower atmosphere, which is called troposphere, extends up to a height of 6 miles above earth's surface. The stratosphere extends from 6 – 30 miles and the ionosphere extends from 30 – 400 miles. The different layers of earth's atmosphere are shown in **Figure 2.3.b**. We will study in **Chapter 4**, how the propagation characteristics of radio waves change from one atmospheric sub-layer to another.

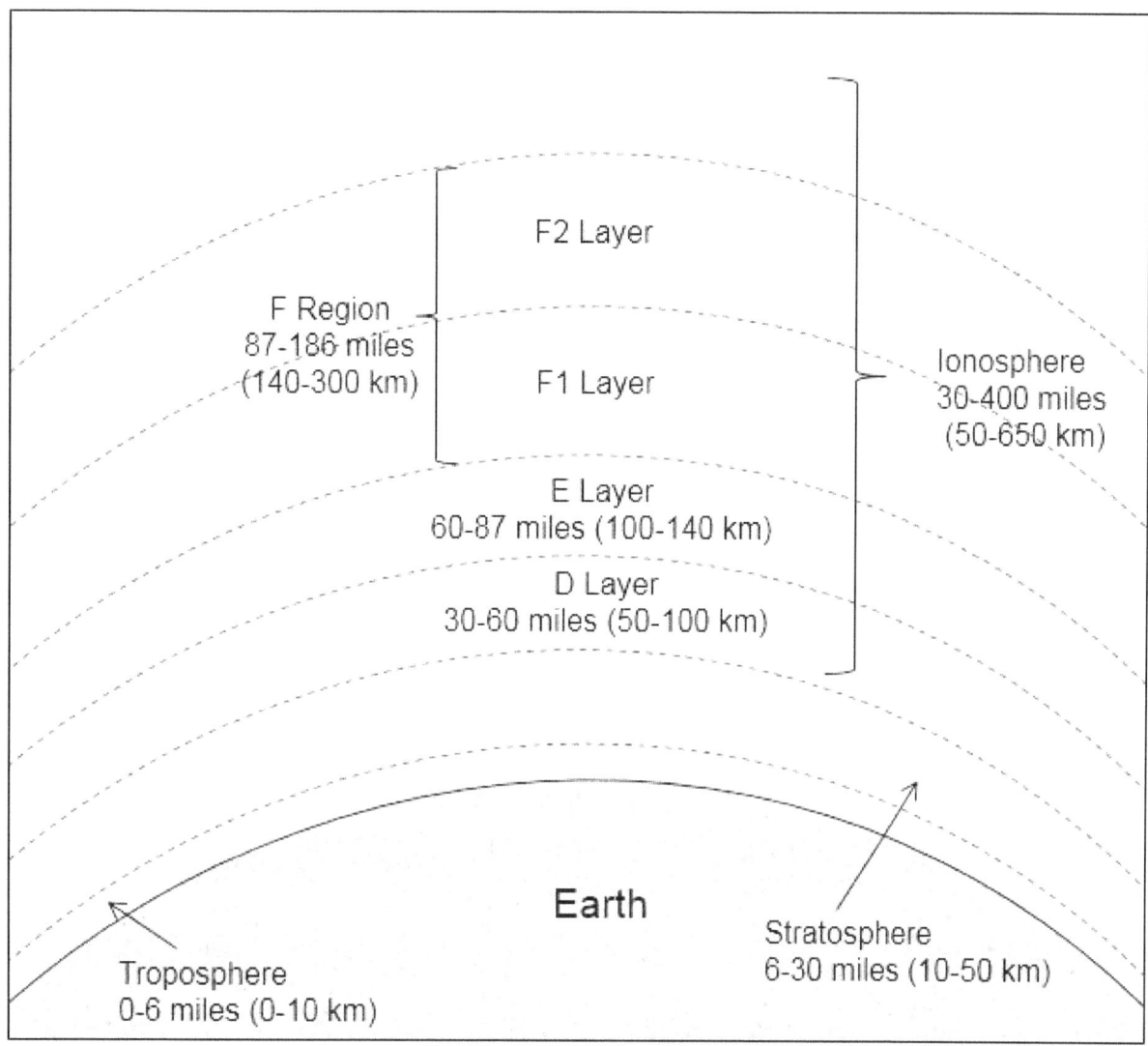

**Figure 2.3.b**

# 3.  RF Propagation Models

RF propagation models are defined as the mathematical formulations that predict the behavior of RF propagation for a radio link under a unique constraint. The constraints may include scenarios, such as propagation inside a building or outdoor propagation in a dense metropolitan area. These models are a function of *frequency*, *path of propagation* and the *propagation medium*. For each unique constraint, a unique propagation model is formulated.

RF propagation models are derived from large collections of data for a specific constraint or situation, so they are empirical in nature. They predict the most likely behavior of a radio link under a specific constraint.

## 3.1  Large Scale Variation

In a given wireless communications system, the first thing we want to know is the strength of received RF signal power in *decibel* (dB) at a distance d from the transmitting antenna (**Figure 3.1.a**). As we move the receiver away from transmitter, the signal strength is reduced. You may get more signal strength bars on your cell phone if you are at a distance of 1 mile from the base station, than, if you move away to a distance of 30 miles. This is called *large scale variation* of radio signal strength as the receiver moves away to a greater distance from the transmitter.

RF propagation models help us predict the *large scale variation* in signal strength for a radio link, and the predictions are used to plan RF signal coverage.

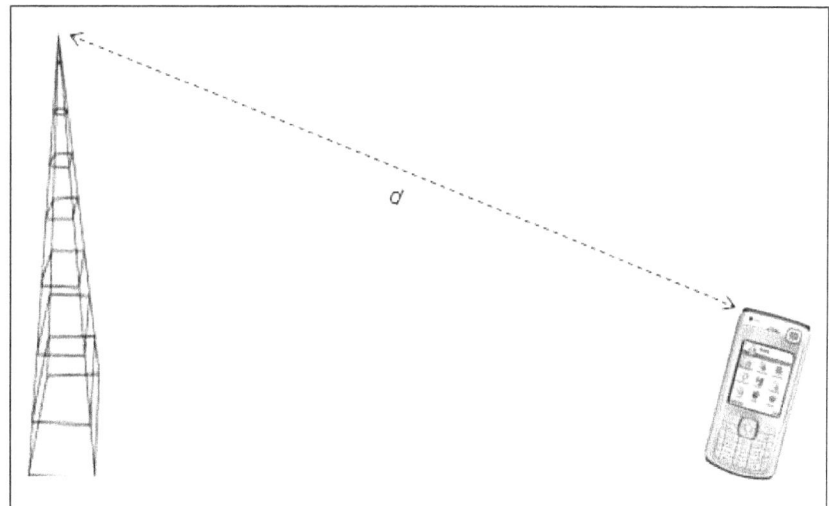

**Figure 3.1.a**

## 3.2  Small Scale Variation

It is common to have *fading* effects in a radio receiver, when the received signal loses its strength simply by moving the receiver around its usual location. For example, you will notice the number of signal strength bars change when you move in or out of a building. There could even be dead zones in certain areas. This is called *small scale variation* in received radio signal strength.

The *fading* effects are the result of *multipath propagation*. As in **Figure 3.2.a**, the radio signals are shown as propagating along three different paths from cell tower to the cell phone. Path 1 is a reflected path, where the signal bounces off a hill top. Path 2 is also a reflected path from a tall building; Path 3 is the LOS path.

The radio signals arrive at the cell phone from three paths of different *path lengths*. In addition, there may be a *phase change* in the radio wave because of reflection. These three signals add up or subtract from each other at the receiver, based on their respective phase differences that are caused by their different path lengths. In other words, the cell phone receives a vector sum of the three signals. At some locations, it may add up, because some of the received signals are in phase with each other. However, if you move the cell phone around, the signals may cancel out, because some of the signals are in opposite phase, causing *fading* effects.

It is common to observe a *fading* of 20-30 dB in a typical situation. RF propagation models help us predict the *small scale variation* in signal strength for a given situation. We will explore fading models in more detail in **Chapter 16**.

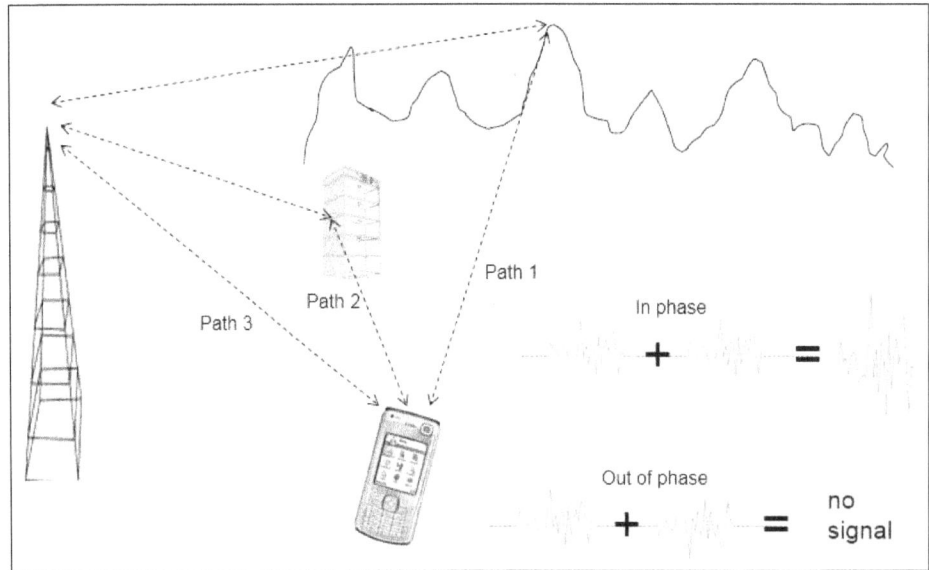

**Figure 3.2.a**

# 4.  RF Propagation Modes

Based on their frequency (or wavelength), radio waves can propagate as ground waves, sky waves, direct waves or tropospheric waves. Now we will study each of these RF propagation modes.

## 4.1  Ground Wave Propagation

Ground wave propagation follows the curvature of earth's surface. These RF waves are also called surface waves and they do not leave troposphere, which is up to a height of 6 miles above earth's surface. In fact, they are actually in contact with ground that causes the lower part of the wave front to lose energy due to currents induced in the ground. This slows down the lower part of the wave, and causes the entire wave front to tilt forward slightly, resulting in bending of the wave along the curvature of earth, as shown in **Figure 4.1.a**.

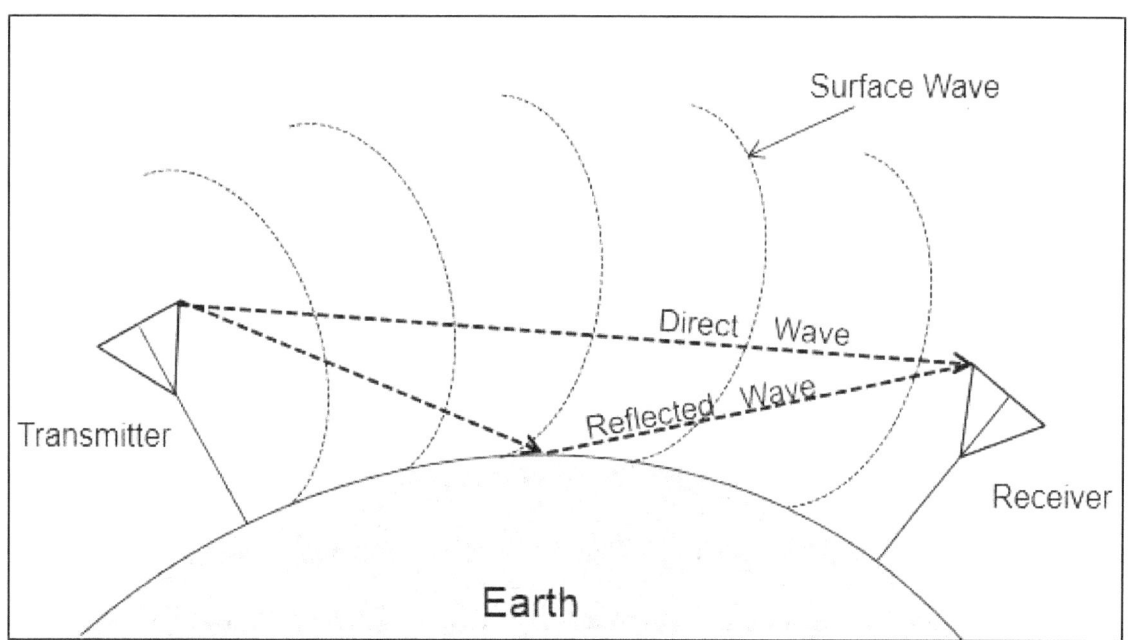

**Figure 4.1.a**

In actual practice, a ground wave receiver will also receive direct waves and reflected waves from the ground. The net result at the receiver will be a vector sum of surface waves, direct waves and reflected waves, as shown in **Figure 4.1.a**.

Ground wave propagation mode is applicable to frequencies less than 3 MHz (**Table 4.1.b**). The range for these waves is up to 100 miles for higher bands and up to 160 miles for lower bands. So, as the frequency gets higher, the range is reduced.

The conductivity of ground also influences the range of ground waves. Higher ground conductivity extends the range of ground waves. For this reason, they perform best over salt water and oceans, and have the poorest performance over earth, fresh water and ice.

Some of the common applications of ground wave propagation are radio navigation, maritime mobile services and public radio broadcasts.

| Band Number | Symbols | Frequency Range | Metric subdivision | Metric abbreviation |
|---|---|---|---|---|
| -1 | ---- | 0.03 – 0.3 Hz | Gigametric waves | B.Gm |
| 0 | ELF | 0.3 – 3 Hz | Hectomegametric waves | B.hMm |
| 1 | ---- | 3 – 30 Hz | Decamegametric waves | B.daMm |
| 2 | ---- | 30 – 300 Hz | Megametric | B.Mm |
| 3 | ULF | 300 – 3000 Hz | Hectokilometric waves | B.hkm |
| 4 | VLF | 3 – 30 kHz | Myriametric waves | B.Mam |
| 5 | LF | 30 – 300 kHz | Kilometric waves | B.km |
| 6 | MF | 300-3000 kHz | Hectometric waves | B.hm |

**Table 4.1.b**

## 4.2  Ionosphere (Sky) Wave Propagation

Ionosphere or sky wave propagation utilizes the phenomenon of refraction in the ionosphere (**Figure 4.2.a**), which is part of the upper atmosphere of earth and ranges from 30 – 400 miles above earth's surface (**Figure 2.3.b**). The ionosphere is formed by an abundance of ionized constituents of atomic oxygen, molecular oxygen, molecular nitrogen and nitric oxide. These ions are formed by solar ultraviolet and x-ray radiations and have a significant impact on radio waves passing through the ionosphere.

The electron density rises as we move from the lower sub layers, D and E to higher sub layers F1 and F2 of the ionosphere. In the dense regions of F2 sub layer, the electron density can reach a trillion electrons per cubic meter. This results in a reduction in the value of refractive index from an initial value of 1.0, as we move from lower D sub layer to the F1 and F2 sub layers

The reduction in refractive index causes refraction of radio waves, and if the *elevation angle* is smaller than the *critical angle*, the radio wave is entirely reflected from the ionosphere, and it bounces back to the surface of earth, as shown in **Figure 4.2.a**. We will define and study these angles a little later in this section.

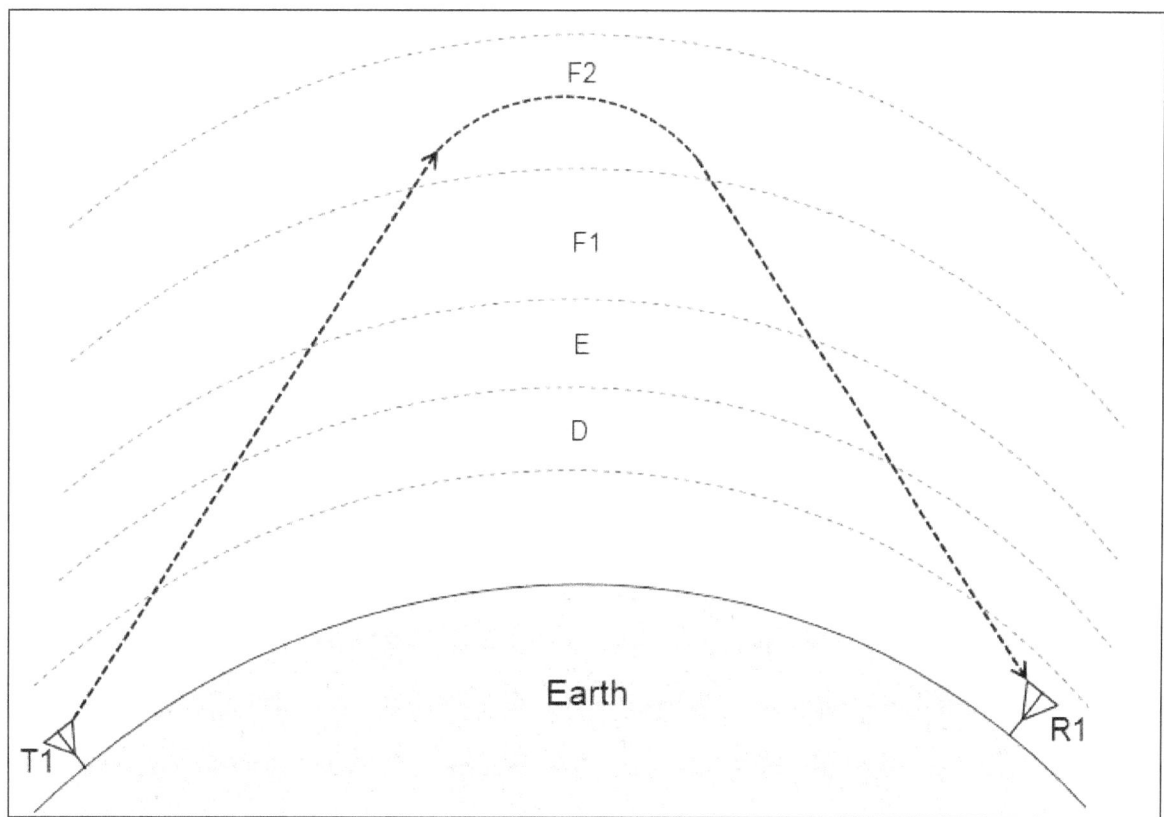

**Figure 4.2.a**

The refraction characteristics of the ionosphere are dependent on solar radiation, and vary from day to day, from region to region, and are also based on 27-day solar rotation and the 11-year solar cycle. Since the refraction characteristics are always changing, the propagation path is not always available. It is rather challenging to predict the availability of propagation path through the ionosphere. However, despite these drawbacks it is possible to achieve long distance RF propagation.

The frequencies from 3-30 MHz in the HF Band 7 are used for ionosphere wave propagation (**Table 4.2.b**).

| Band Number | Symbols | Frequency Range | Metric subdivision | Metric abbreviation |
|---|---|---|---|---|
| 7 | HF | 3-30 MHz | Decametric waves | B.dam |

**Table 4.2.b**

In order to utilize the long distance propagation benefits of ionosphere, the radio waves must be launched at an angle to the surface of earth. This angle is called the *elevation angle, take off angle, launch angle* or *wave angle*. (**Figure 4.2.b**).

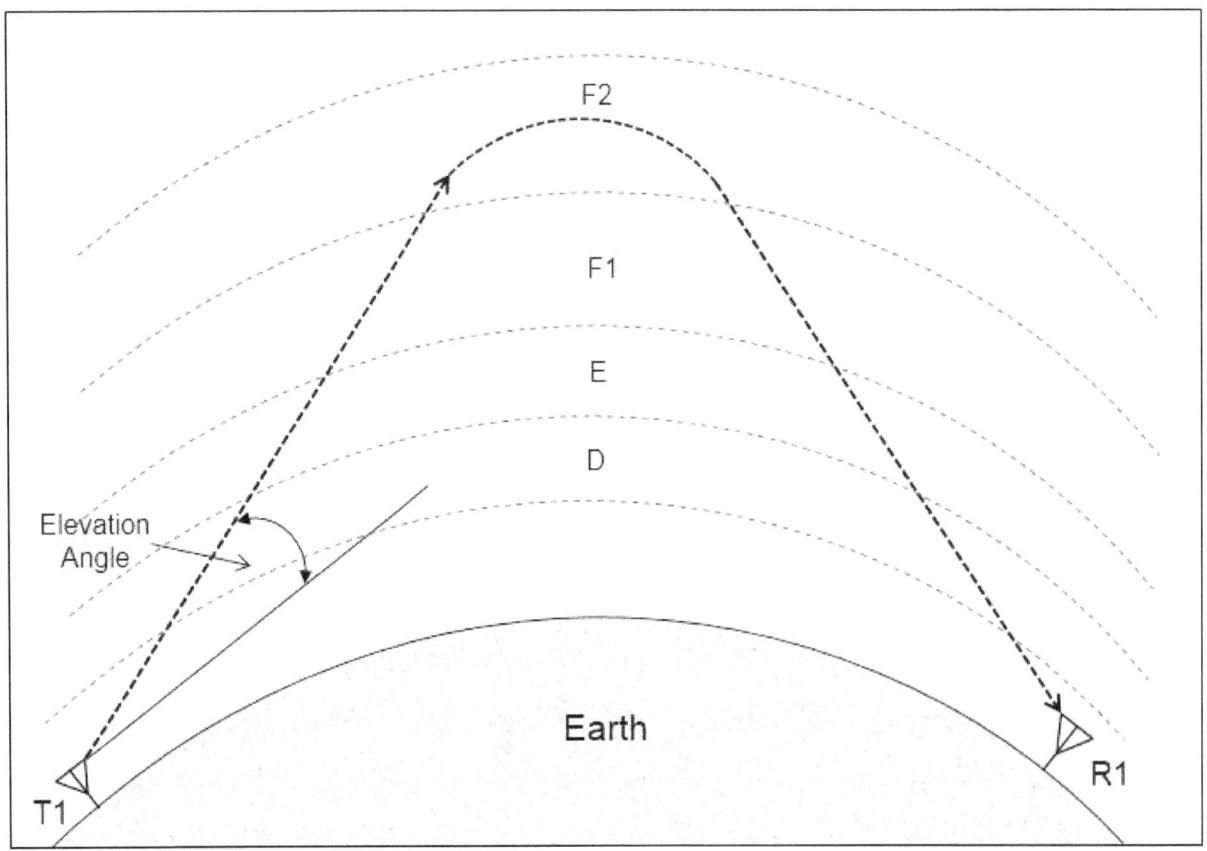

**Figure 4.2.b**

At very high *elevation angle* that is greater than the *critical angle*, the radio waves do not get reflected from the ionosphere. This is shown in **Figure 4.2.c**, where Ray 1, goes out into deep space and is not reflected. However, as the elevation angle is reduced; at *critical angle*, the radio waves begin to return to earth after reflection from the ionosphere. Ray 2 is transmitted from T1 at critical angle and returns to earth at receiver R1. The distance between T1 and R1, where the reflected radio waves begin to return is called the *skip distance*. If the *elevation angle* is further reduced, as shown for Ray 3, the reflected Ray 3 returns to earth at receiver R2, which is at a much larger distance from transmitter T1. So, to get longer range in ionosphere wave propagation, the *elevation angle* should be reduced below the *critical angle*.

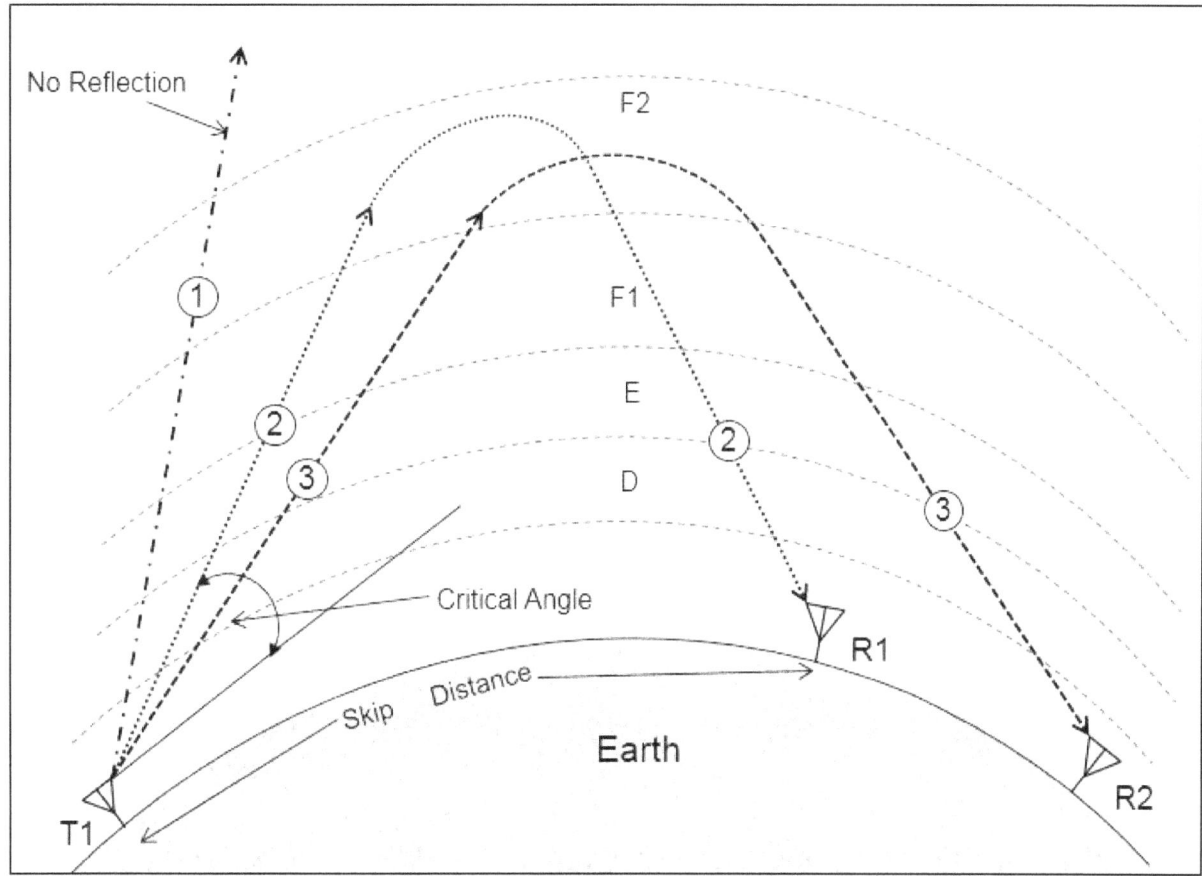

**Figure 4.2.c**

A single ionosphere hop can cover a distance of about 2500 miles (4000 km). For example, Ray 2 and Ray 3 in **Figure 4.2.c** are single hop propagation paths.

Beyond 2500 miles, most ionosphere wave propagation takes place through multiple hops. As shown in **Figure 4.2.d**, for Ray 2 and Ray 3, after the first ionosphere hop, the radio waves are reflected from the surface of earth and propagate back into the ionosphere for second ionosphere refractions, for ultimate reflection from the F2 sub layer of the ionosphere.

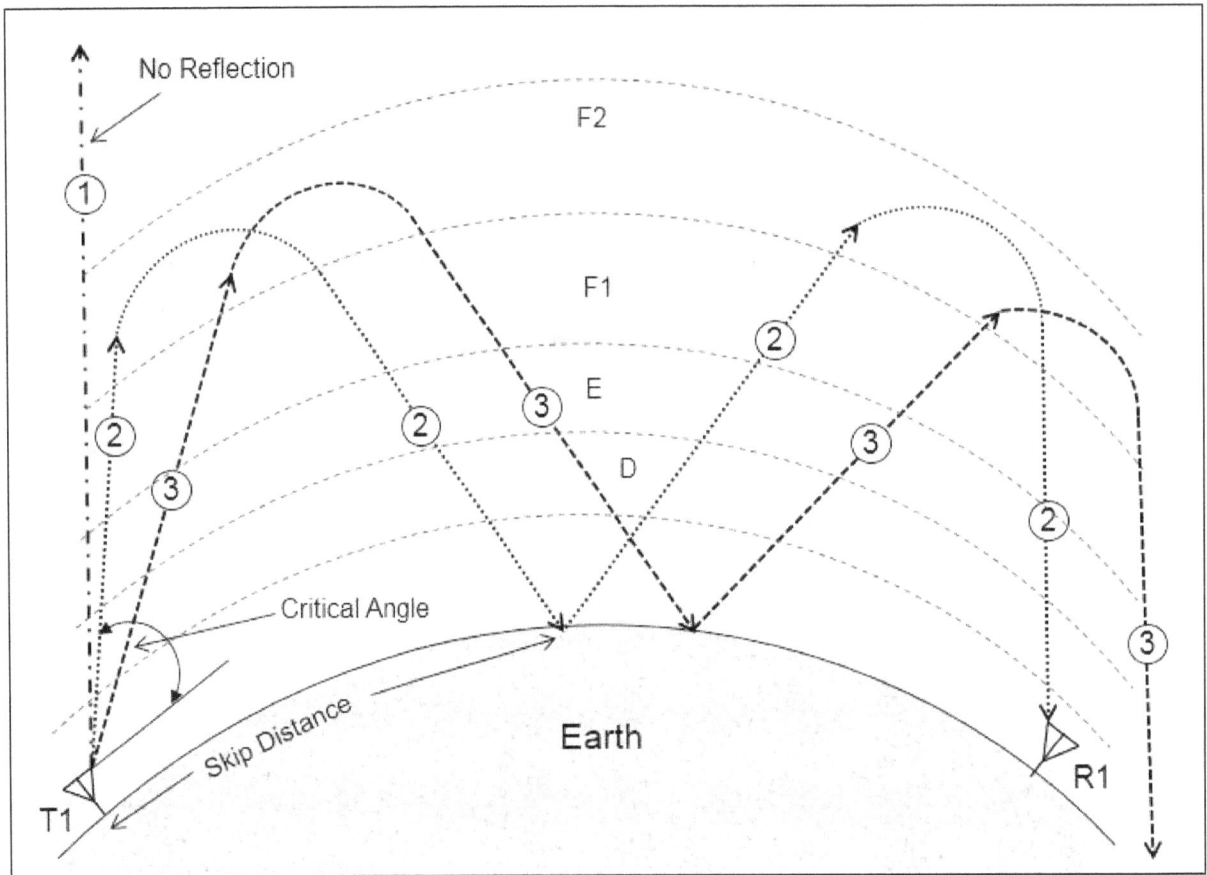

**Figure 4.2.d**

Radio waves that propagate into the ionosphere at high elevation angles can cover long distances in a single hop, as shown in **Figure 4.2.e**. These waves are called high angle *Pederson* waves. This type of long distance propagation through the ionosphere is possible when refractive index is such that the waves almost seem to "duct" through the upper layers of the ionosphere, as they encircle the earth. They appear to follow the curvature of earth and can cover long distances beyond 2500 miles in a single hop.

However, high angle long distance wave propagation is highly dependent on stable conditions in the ionosphere, and the long distance path may not always be available.

For ionosphere wave propagation, the *critical frequency* is defined as the highest frequency that returns echo from E and F sub layers of the ionosphere at vertical incidence. **Figure 4.2.f** shows a ray transmitted in the vertical direction from transmitter T1 and returning to the same location at receiver R1. The *critical frequency* is typically between 1 and 15 MHz for F sub layer. It is a function of the peak electron density in a region of the ionosphere, higher the ionization, higher will be the *critical frequency*.

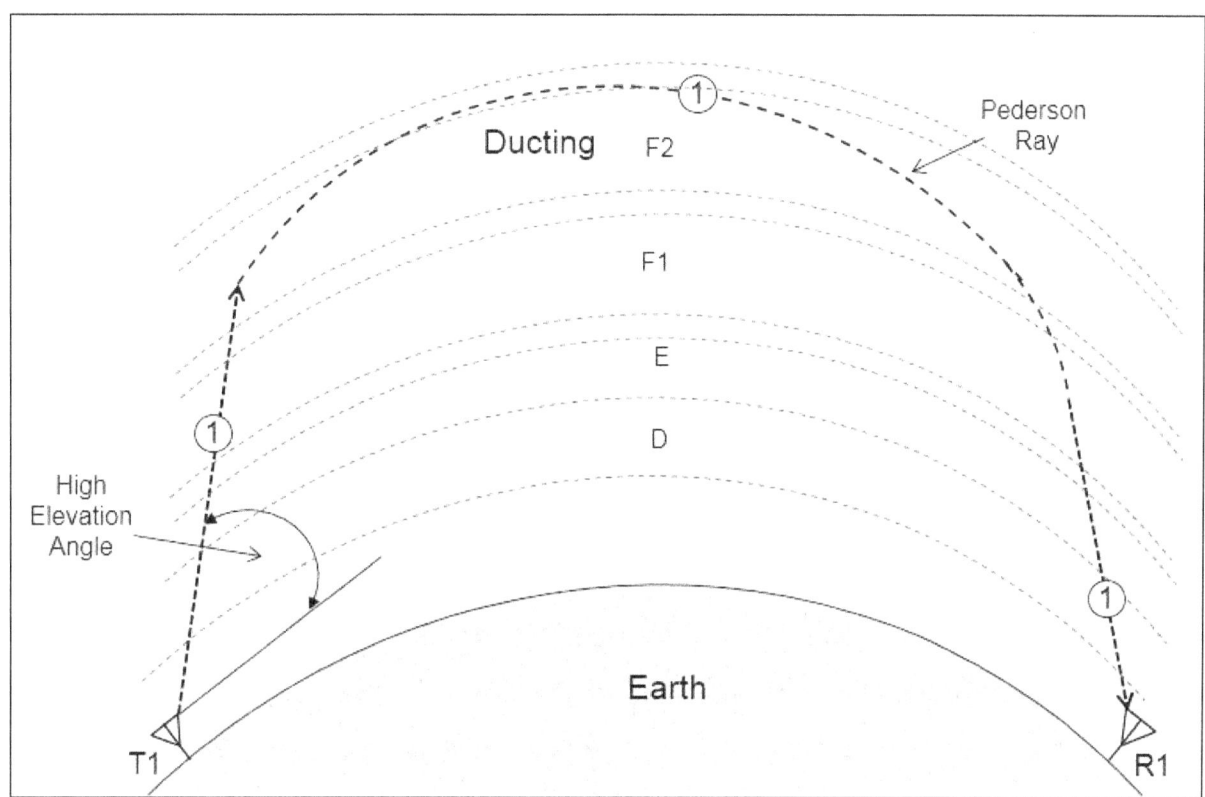

**Figure 4.2.e**

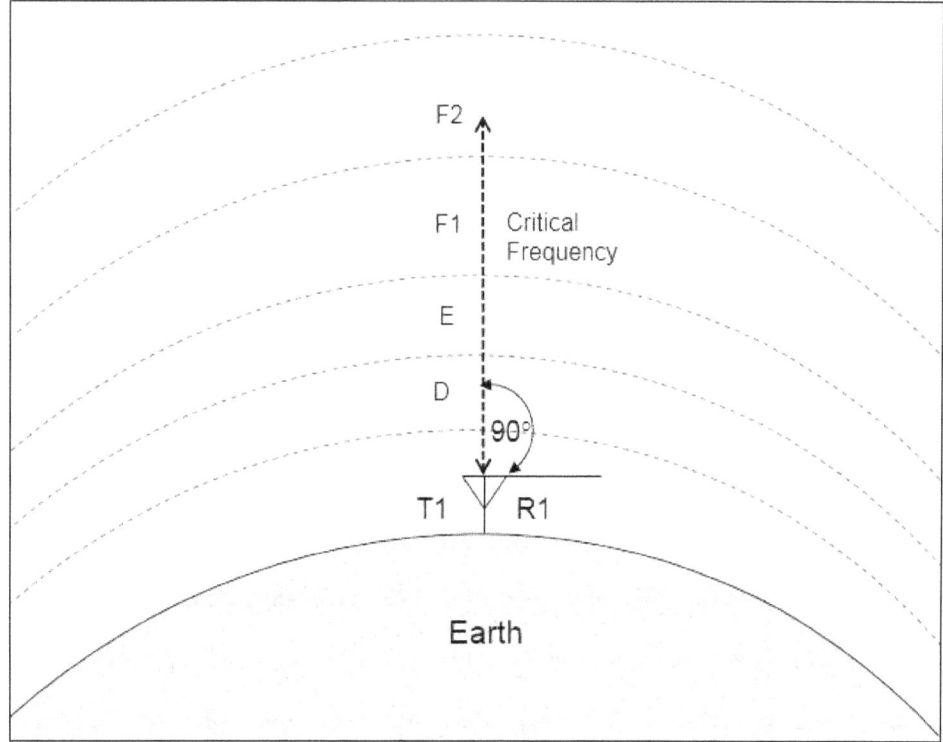

**Figure 4.2.f**

The *maximum usable frequency* (MUF) is the highest frequency supported by the ionosphere for a propagation path between a transmitter and a receiver. Lower elevation angles have higher values for MUF. **Figure 4.2.g** shows Ray1 at vertical incidence has MUF = 10 MHz for transmitter T1 and Receiver R1, which is the same as the *critical frequency*. So, stations in the neighborhood can benefit from vertically reflected signals and frequencies less than 10 MHz can be used for those stations.

Ray 3 has lower elevation angle with MUF = 15 MHz between T3 and R3. Ray 2 has the lowest elevation angle with MUF = 30 MHz between T2 and R2. Therefore, for stations that are farther apart, higher frequencies at oblique angles can be used.

The MUF is a function of propagation path, time of day, season, geo location, solar ultraviolet and x-ray radiation levels, and *sudden ionosphere disturbances* (SID).

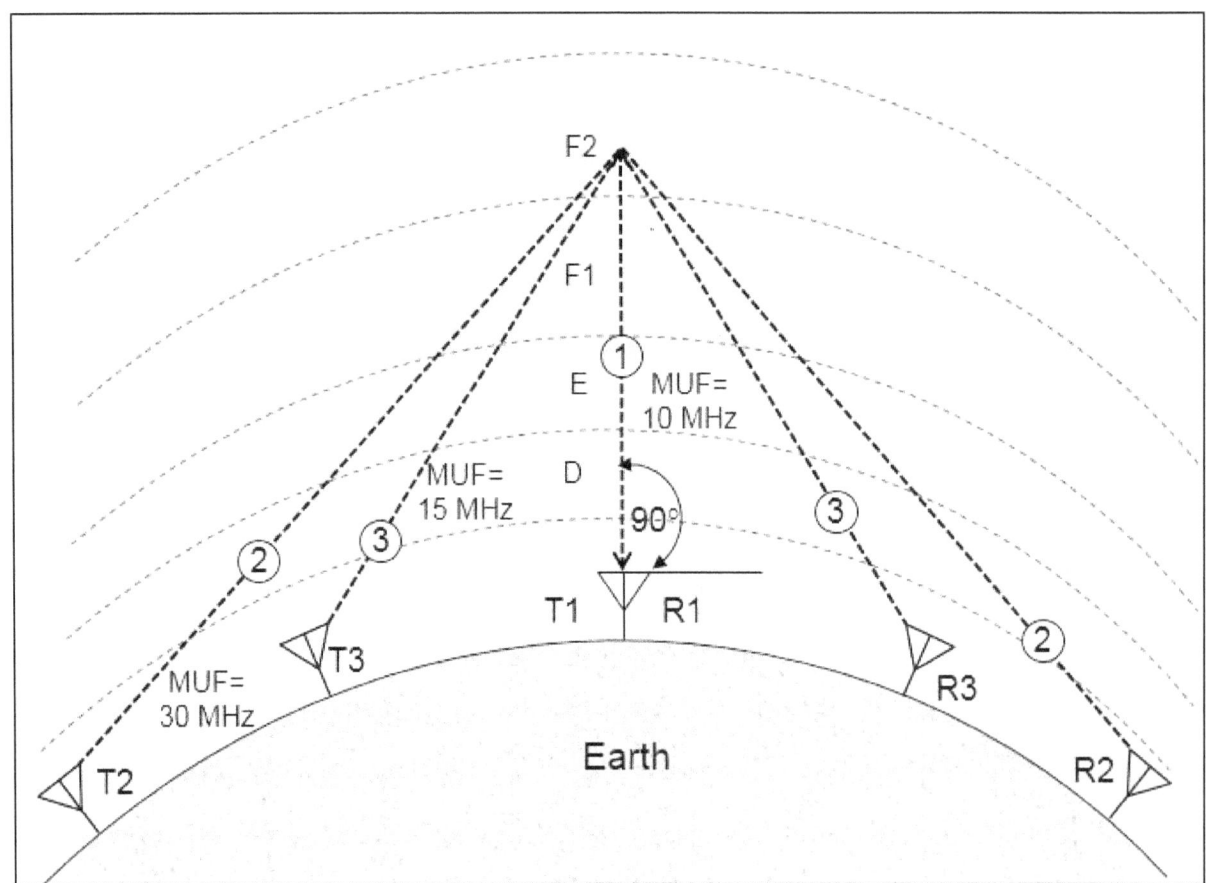

**Figure 4.2.g**

The *lowest usable frequency* (LUF) is the lowest frequency supported by the ionosphere between a transmitter and receiver. Lower elevation angles have higher values for LUF. Also, there is more absorption of signal power in the atmosphere at lower frequencies. **Figure 4.2.h** shows Ray 1 at vertical incidence has LUF = 5 MHz. Ray 3 at lower elevation angle has LUF = 7 MHz between T3 and R3. Ray 2 at lowest elevation angle has LUF = 10 MHz between T2 and T3.

It is possible to reduce LUF by using high power transmitters, highly directive antennas and receivers that are sensitive enough with lower *signal to noise ratio* (SNR).

Like MUF, LUF also changes with time of day, season and other ionosphere disturbances.

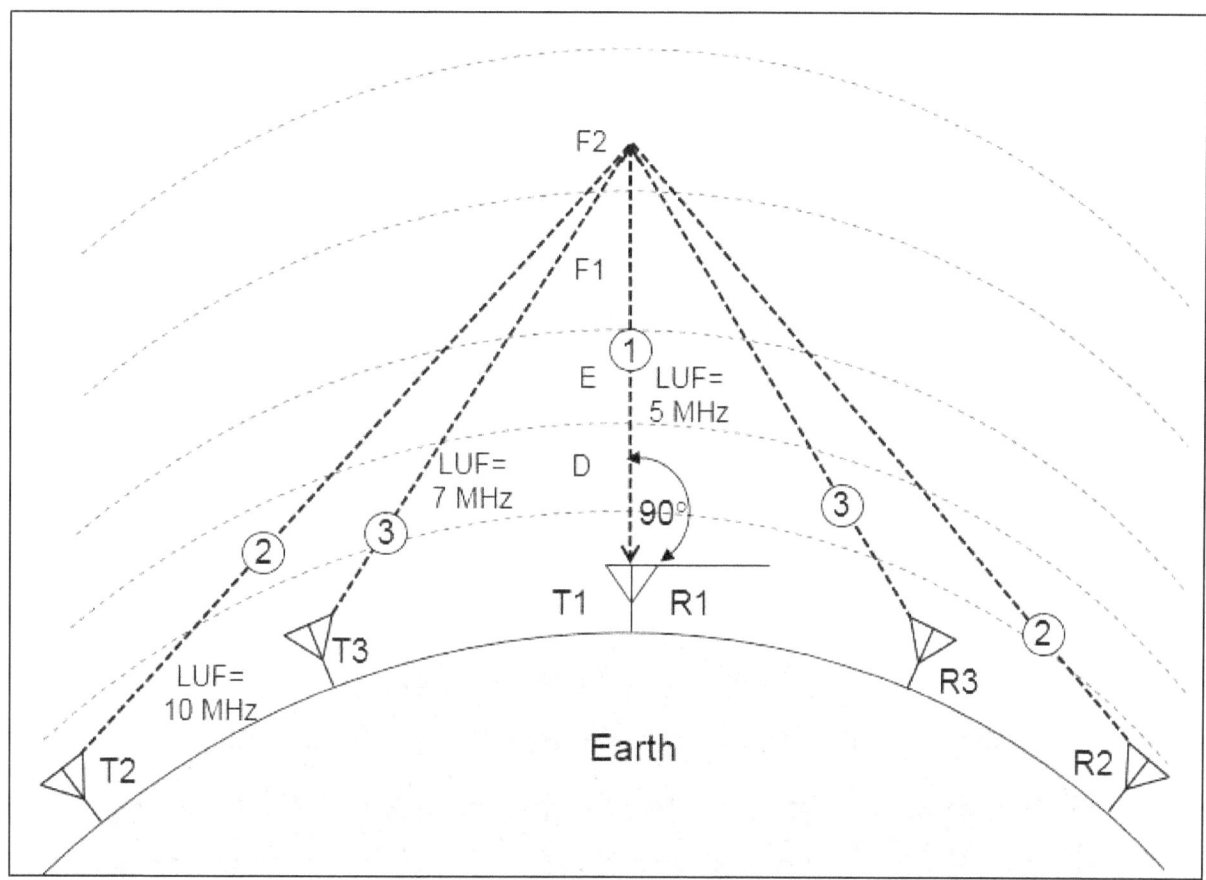

**Figure 4.2.h**

The *operating frequency* (OF) is defined as the frequency that the operator will actually use for wireless communication between two stations by using ionosphere wave propagation path. It is clear that the OF must be between MUF and LUF that is supported by the ionosphere between a transmitter and receiver. For example, in **Figure 4.2.i**, between T1 and R1, the MUF = 30 MHz and LUF = 10 MHz and the OF is determined to be 15 MHz for a specific time of day and season.

However, the challenge lies in the actual determination of the best value for OF that will provide optimum communication between two stations. Ionosphere wave propagation varies with time of day, season and physical location. The variation is usually more pronounced at sunrise and sunset and is relatively stable at midday. In general, higher frequencies perform better at midday and lower frequencies perform best after sunset. The wireless operator may have to change OF 2 to 4 times a day.

One way to predict the OF is by using *high frequency* (HF) propagation models, such as ICEPAC, VOACAP and ITU recommended REC533. These models provide best possible theoretical approximations and combined with operator experience, are a good cost effective method of prediction.

Advanced techniques use *ionosphere sounders*, which are separate pulsed radio systems that run in parallel with the HF radio links. The system consists of a sounder transmitter at one end, which uses a separate antenna. At the other end, is a sounder receiver, which may use the same antenna as the one used for HF wireless link. The receiver is synchronized with the transmitter with accurately controlled clocks. The RF pulses are transmitted from the sounder at discrete frequencies. The sounder system typically initiates a frequency scan of 10 discrete frequencies per second. The receiver scans for the range of frequencies. The ones detected by the receiver are determined to be between the range of LUF and MUF and are a potential candidate for use as OF. The sounder receiver generates a frequency display called the *ionogram*. The wireless operator uses these ionograms to schedule the operating frequencies that will be used at a specific time of day or a time of year.

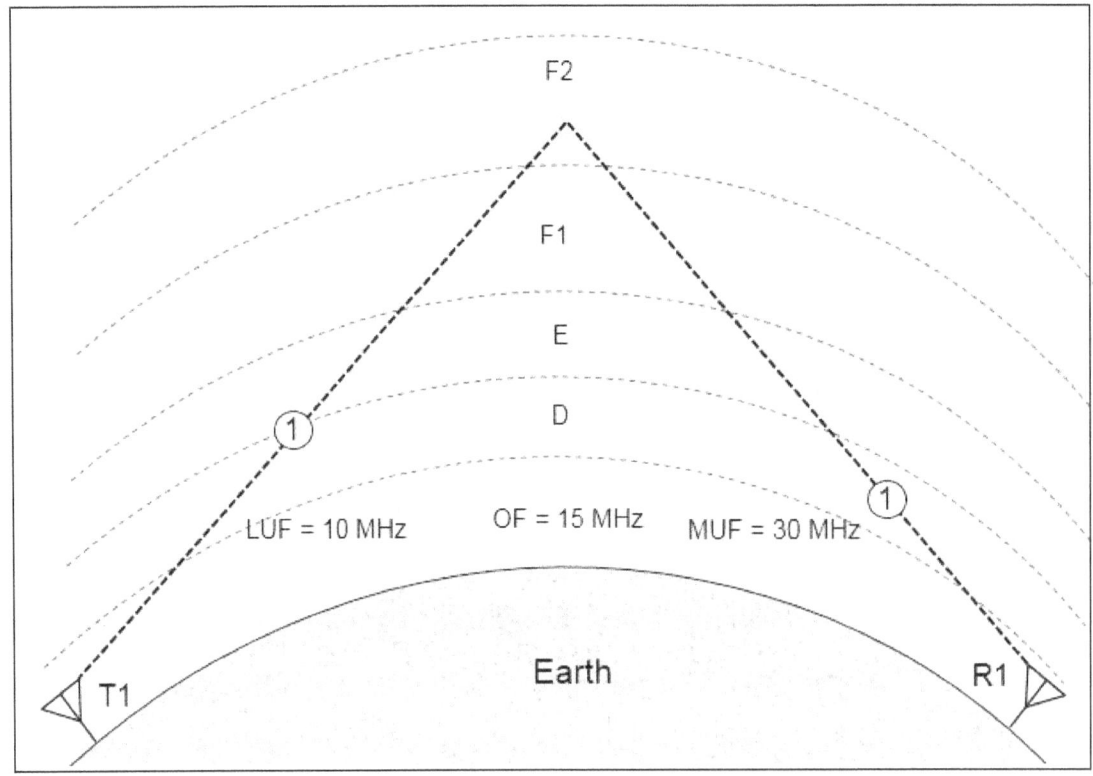

**Figure 4.2.i**

## 4.3   Direct Wave Propagation

This is the most common mode of RF wave propagation when the transmitting antenna and the receiving antenna are in a *line of sight* (LOS). However, in practical situations there will usually be obstacles such as buildings and terrain, which reflect the RF signal. The resulting RF signal at the receiving antenna is a vector sum of the direct LOS signal as well as the signals received after reflection from various obstacles.

**Figure 4.3.a**.shows that Ray 1 propagates direct from the transmitter T1 to receiver. R1. Ray 3 also propagates direct from transmitter T1 to the satellite. Ray 2 propagates from T1, suffers a reflection from the ground and finally reaches R1. Receiver R2 is not in the LOS range of transmitter T1, and therefore does not receive the RF signal.

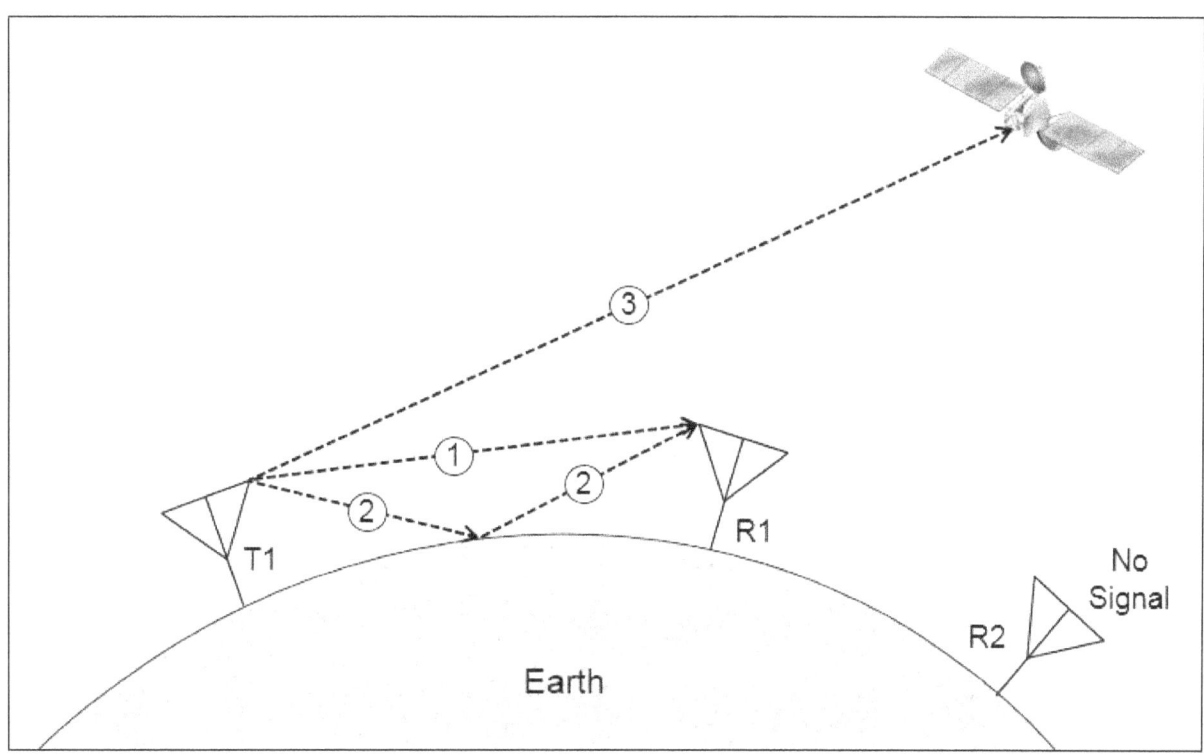

**Figure 4.3.a**

Direct (LOS) wave propagation is used with frequencies higher than 30 MHz. These include frequencies from Band 8 to Band 15, shown in **Table 4.3.b** and cover the widely used commercial UHF and VHF bands. The terrestrial range for these frequencies is limited; higher the frequency, lower is the range covered. This is because of the rapid loss of signal power as the frequency is increased. We will cover more detail about this phenomenon in **Section 5.1**.

Direct wave mode of propagation finds uses in several commercial applications, such as radio, television and satellite broadcasts, GPS navigation, mobile radio and satellite communication services, wireless local area networks (WLAN), and cellular communication networks (GSM, UMTS, CDMA, WiMax, LTE).

| Band Number | Symbols | Frequency Range | Metric subdivision | Metric abbreviation |
|---|---|---|---|---|
| 8 | VHF | 30 – 300 MHz | Metric waves | B.m |
| 9 | UHF | 300 – 3000 MHz | Decimetric waves | B.dm |
| 10 | SHF | 3-30 GHz | Centimetric waves | B.cm |
| 11 | EHF | 30 – 300 GHz | Millimetric waves | B.mm |
| 12 | ---- | 300 – 3000 GHz | Decimillimetric waves | B.dmm |
| 13 | ---- | 3 – 30 THz | Centimillimetric waves | B.cmm |
| 14 | ---- | 30 – 300 THz | Micrometric waves | B.µm |
| 15 | ---- | 300 – 3000 THz | Decimicrometric waves | B.dµm |

**Table 4.3.b**

## 4.4   Troposphere Wave Propagation

RF wave propagation in the troposphere (up to 6 miles above earth's surface) is affected by *refraction* and *scattering* phenomena.

Refraction in the troposphere is caused by changes in temperature, humidity and pressure. Its net effect is to extend the range of RF propagation beyond *line of sight* (LOS). Typically the range is extended by 15%. In other words, the radio horizon is extended 1.15 times away from the visual horizon. **Figure 4.4.a** shows Ray 3 propagating along an extended path from transmitter T1 to Receiver R2 due to refraction in the troposphere. The figure also shows a formula in the box to calculate the distance, D, to radio horizon in miles, if the height, H, of transmitting antenna is known in ft.

Scattering is caused by the presence of fog, snow, rain or dust in the troposphere. As shown in **Figure 4.4.b**, the presence of a *common scattering volume* in the troposphere enables long range RF wave propagation. The *common scattering volume* is the common enclosed area where the 2 beams from the transceiver (transmitter + receiver) antennas, T1/R1 and T2/R2 intercept. The fog, dust, snow or rain present in this enclosed volume scatters RF energy in all directions. A small portion of this RF energy is scattered in the direction of earth's surface, making it possible to propagate beyond horizon up to a maximum distance of 500 miles.

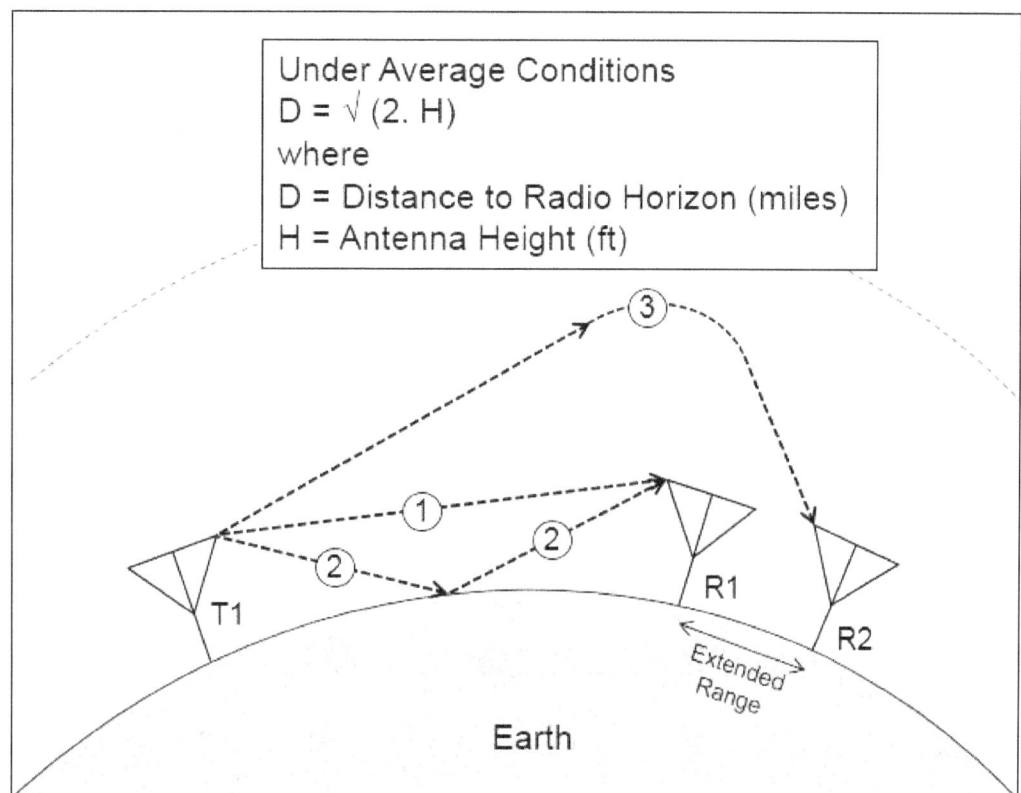

Under Average Conditions
$$D = \sqrt{(2 \cdot H)}$$
where
D = Distance to Radio Horizon (miles)
H = Antenna Height (ft)

**Figure 4.4.a**

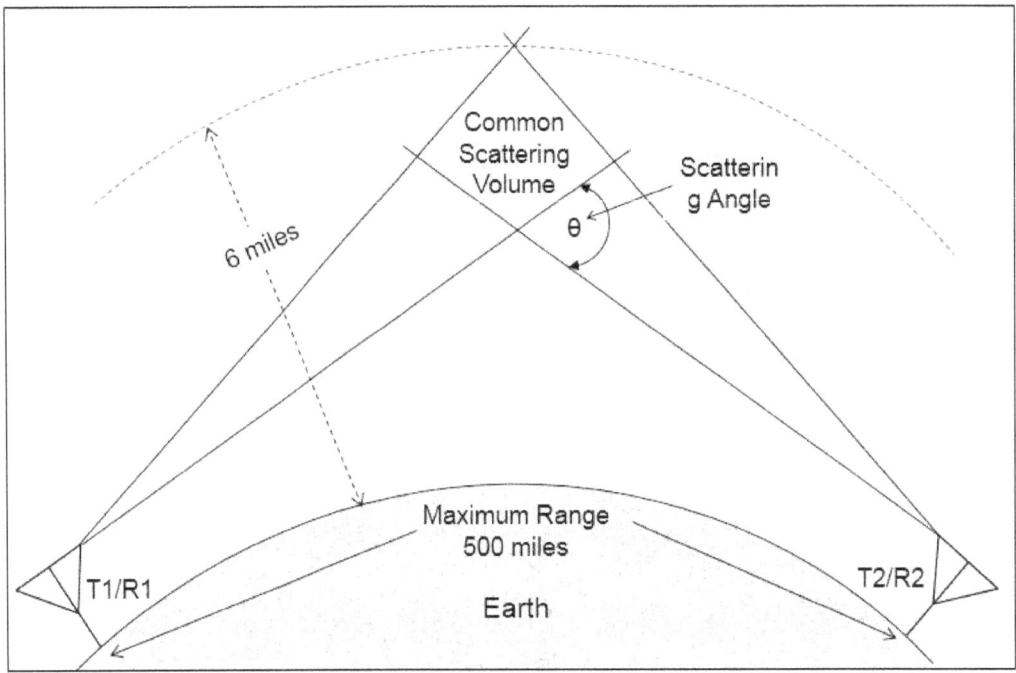

**Figure 4.4.b**

The received signal strength is directly proportional to the scattering volume. Higher scattering volume scatters more RF energy in the direction of earth's surface.

The attenuation of received signal is proportional to the scattering angle shown in the **Figure 4.4.b**. Lower scattering angles cause lower overall signal attenuation for the same scattering volume.

The troposphere wave propagation is used for frequencies higher than 30 MHz, which extend from Band 8 to Band 15, as shown in **Table 4.3.b**. The most common applications include long range point to point RF backhaul communication links.

# 5.   RF Propagation Mechanisms

In this chapter, we will explore various mechanisms that come into play when RF waves propagate in free space, through the atmosphere or when they encounter obstacles in their path of propagation, such as buildings, undulating terrain and hilltops. In particular we will study phenomena of free space loss, reflection, diffraction, refraction, scattering and atmospheric losses. It is important to understand these mechanisms to build our background for estimation of link budget in **Chapter 10.**

## 5.1  Free Space Loss

RF waves are essentially *electromagnetic* (EM) waves, and like all EM waves, they lose power when they propagate through free space. In order to estimate how much power is lost for a given distance and to arrive at a practically usable formula, it is assumed that the transmitting and receiving antennas are in *line of sight* (LOS) of each other. It is also assumed that the two antennas are *isotropic*. (For a definition of isotropic antennas, see **Section 6.7**)

**Figure 5.1.a** shows how RF power decays with distance, d and frequency f. *Friis equation* in the box represents the received signal power as a function of transmitted power, transmitter gain, receiver gain, wavelength and distance.

It should be obvious from *Friis equation* that received power is inversely proportional to the square of distance, d. Also, because, wavelength $\lambda = c / f$, where $c$ is the velocity of light and $f$ is the frequency, it is clear that received power is also inversely proportional to the square of frequency, f.

In practical terms, it means that RF power suffers a reduction of 6 dB when the distance is doubled or when the frequency is doubled. This is called 6 dB/octave reduction in RF power. Another way to put it is 20 dB/decade reduction, which is 20 dB reduction in RF power when the distance or frequency is increased by 10 times.

In order to simplify RF power calculations, the unit *decibel* (dB) is used, which is defined as $10 \log_{10} (P/P_{ref})$, where P is the measured power value and $P_{ref}$ is the reference power value. If $P_{ref}$ is 1 Watt, then the resulting decibel value is designated dBw, or simply as dB. If $P_{ref}$ is 1 milliWatt, then the resulting decibel value is designated as dBm.

For Free Space Loss calculations in dB, the simplified formula in **Figure 5.1.b** should be used.

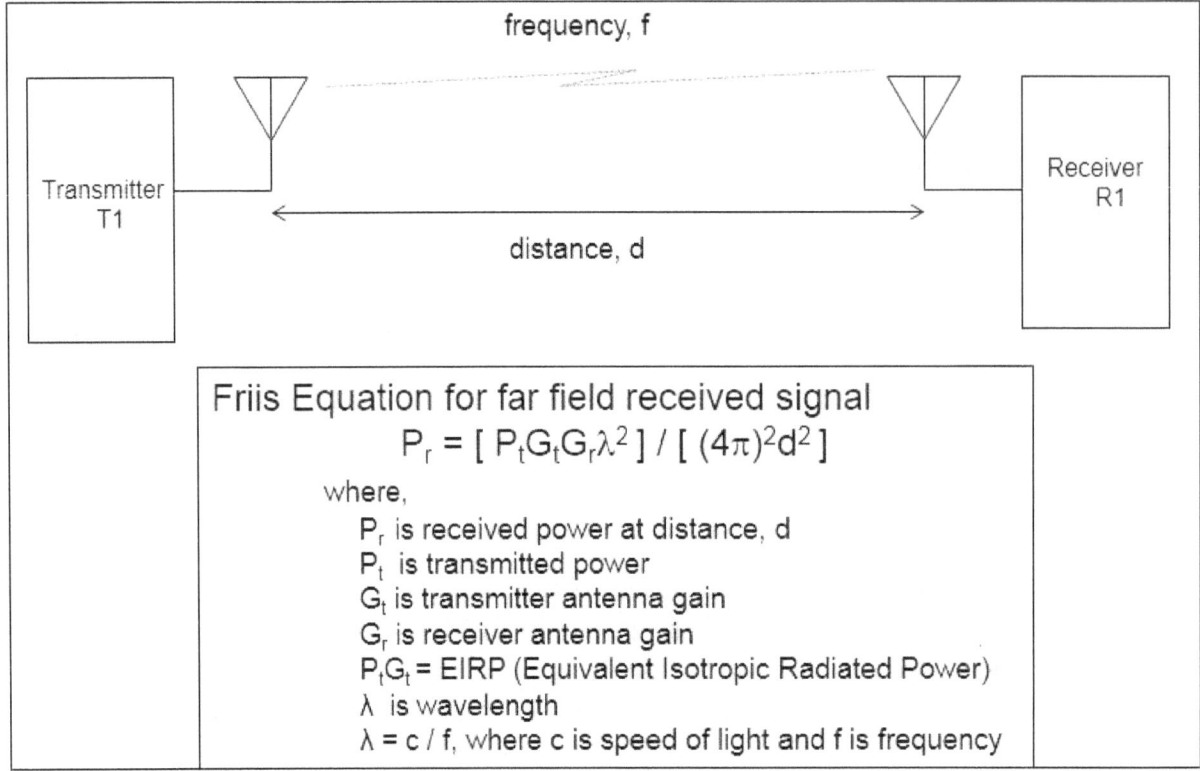

**Figure 5.1.a**

Free Space Power Loss
$L_{free}$ (dB) = 32.45 + 20 log f (MHz) + 20 log d (km)

**Figure 5.1.b**

## 5.1.1.  Microwave Link Example

Let us now consider an example for calculation of free space loss for a microwave link, as shown in **Figure 5.1.1.a**.

In Case 1, the two microwave towers are separated by a distance of 30 km and the frequency used is 1 GHz. For this case, the free space loss is calculated as 122 dB. This means that the RF signal suffers a power loss of 122 dB when it propagates over a distance of 30 km between the two microwave towers at a frequency of 1 GHz.

In Case 2, the distance is doubled to 60 km, and as a consequence the power loss rises by 6 dB to 128 dB, as shown in the calculation. This follows the principle of 6 dB/octave reduction in power when distance is doubled.

In Case 3, the frequency is doubled to 2 GHz, whereas the distance remains at the original value of 30 km. Once again, the power loss rises by 6 dB to 128 dB, as shown in the calculation. This also follows the principle of 6 dB/octave reduction in power when frequency is doubled.

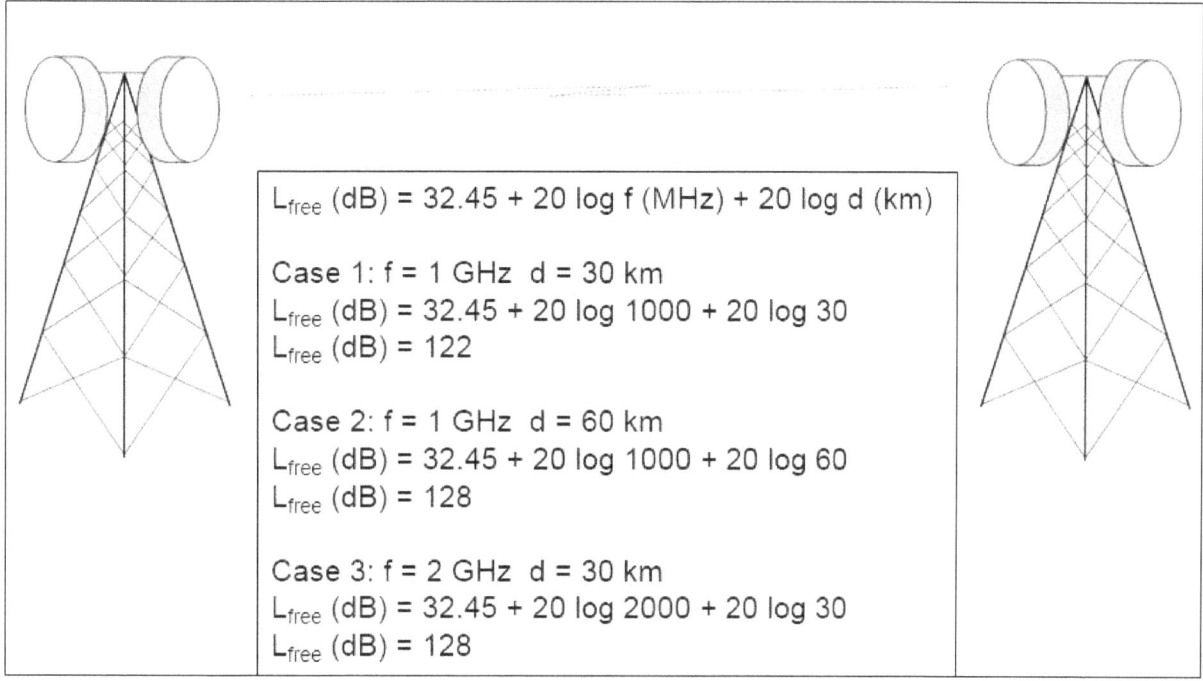

$$L_{free} (dB) = 32.45 + 20 \log f (MHz) + 20 \log d (km)$$

Case 1: f = 1 GHz  d = 30 km
$$L_{free} (dB) = 32.45 + 20 \log 1000 + 20 \log 30$$
$$L_{free} (dB) = 122$$

Case 2: f = 1 GHz  d = 60 km
$$L_{free} (dB) = 32.45 + 20 \log 1000 + 20 \log 60$$
$$L_{free} (dB) = 128$$

Case 3: f = 2 GHz  d = 30 km
$$L_{free} (dB) = 32.45 + 20 \log 2000 + 20 \log 30$$
$$L_{free} (dB) = 128$$

**Figure 5.1.1.a**

## 5.2  Reflection

Radio waves are reflected by terrain and other physical obstacles. In order to derive a practical formula to model the phenomenon of reflection for RF waves, it is assumed that the path of propagation between the transmitter and receiver consists of a LOS path and a reflected path. It is also assumed that the distance, d, between the transmitter and receiver is very large compared to the combined heights of the transmitting and the receiving antennas at both ends of the radio link, as shown in **Figure 5.2.a**

With the above assumptions, the behavior of RF power loss gets modified. The equation in **Figure 5.2.a** shows that the received power with reflection is inversely proportional to the 4th power of distance, d. However, the received power is directly proportional to the square of antenna height

In practical terms, it means that with reflection, RF power suffers a reduction of 12 dB/octave or a reduction of 12 dB when the distance is doubled. Also, there is an increase in received power by 6 dB/octave with the height of transmitter or receiver antenna, or an increase in power by 6 dB when antenna height is doubled.

For reflection calculations of RF power loss in dB, the simplified formula in **Figure 5.2.b** should be used.

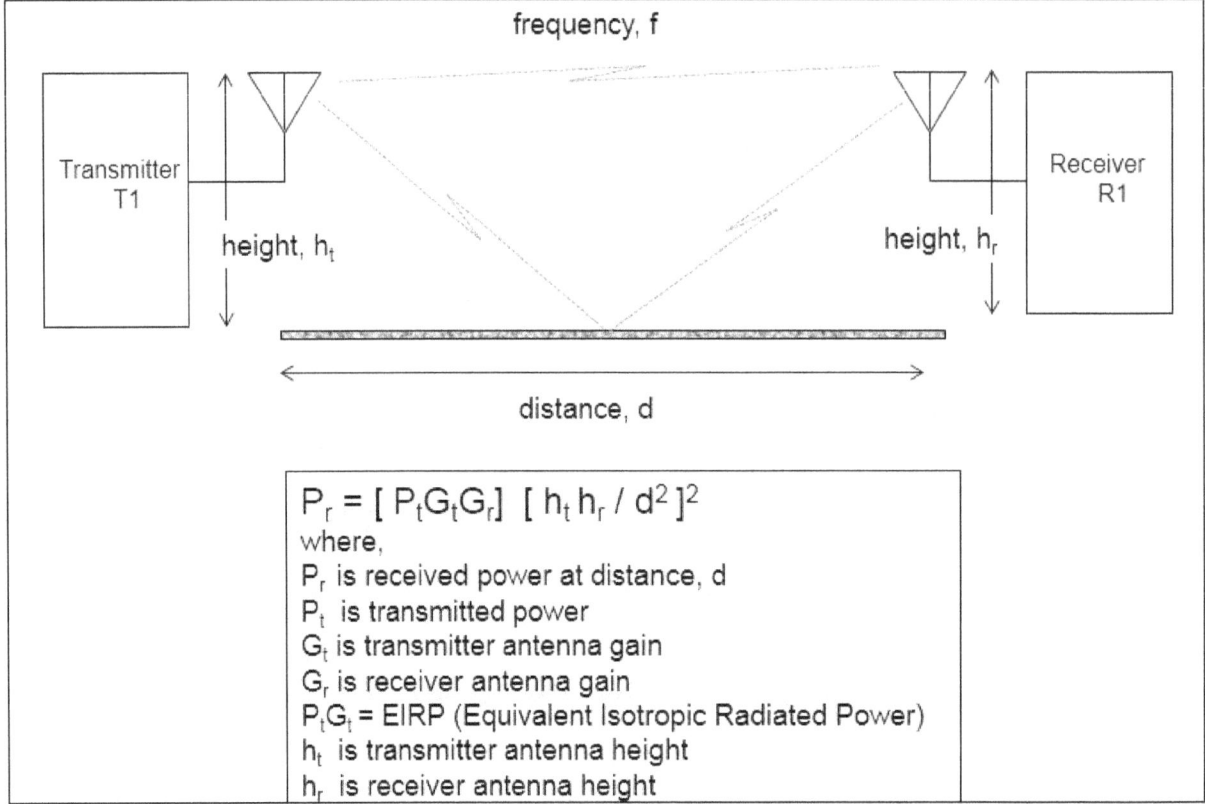

$$P_r = [ P_t G_t G_r ] \ [ h_t h_r / d^2 ]^2$$
where,
$P_r$ is received power at distance, d
$P_t$ is transmitted power
$G_t$ is transmitter antenna gain
$G_r$ is receiver antenna gain
$P_t G_t$ = EIRP (Equivalent Isotropic Radiated Power)
$h_t$ is transmitter antenna height
$h_r$ is receiver antenna height

**Figure 5.2.a**

Reflection Power Loss
$L \ (dB) = -10 \log (G_t) - 10 \log (G_r) - 20 \log (h_t) - 20 \log (h_r) + 40 \log (d)$

**Figure 5.2.b**

## 5.2.1    Mobile Link Example

Let us now consider an example for calculation of reflection power loss for a mobile RF link as shown in **Figure 5.2.1.a**.

In Case 1, for transmitter gain of 24 dB, receiver gain of 20 dB, transmitter height of 40 m, receiver height of 1 m, and a distance of 10 km, the power loss is calculated to be 84 dB.

In Case 2, the distance is doubled to 20 km and as a consequence the calculated power loss rises to 96 dB. This is 12 dB more power loss, or 12 dB/octave reduction in received power.

In Case 3, the transmitter gain is doubled to 48 dB and the calculated power loss is reduced to 60 dB. This is 24 dB less power loss due to an addition of 24 dB transmitter gain.

In Case 4, the transmitter antenna height is doubled to 80 m and the calculated power loss is reduced to 78 dB. This is 6 dB less power loss due to doubling of antenna height, or 6B/octave increase in received power due to doubling of antenna height.

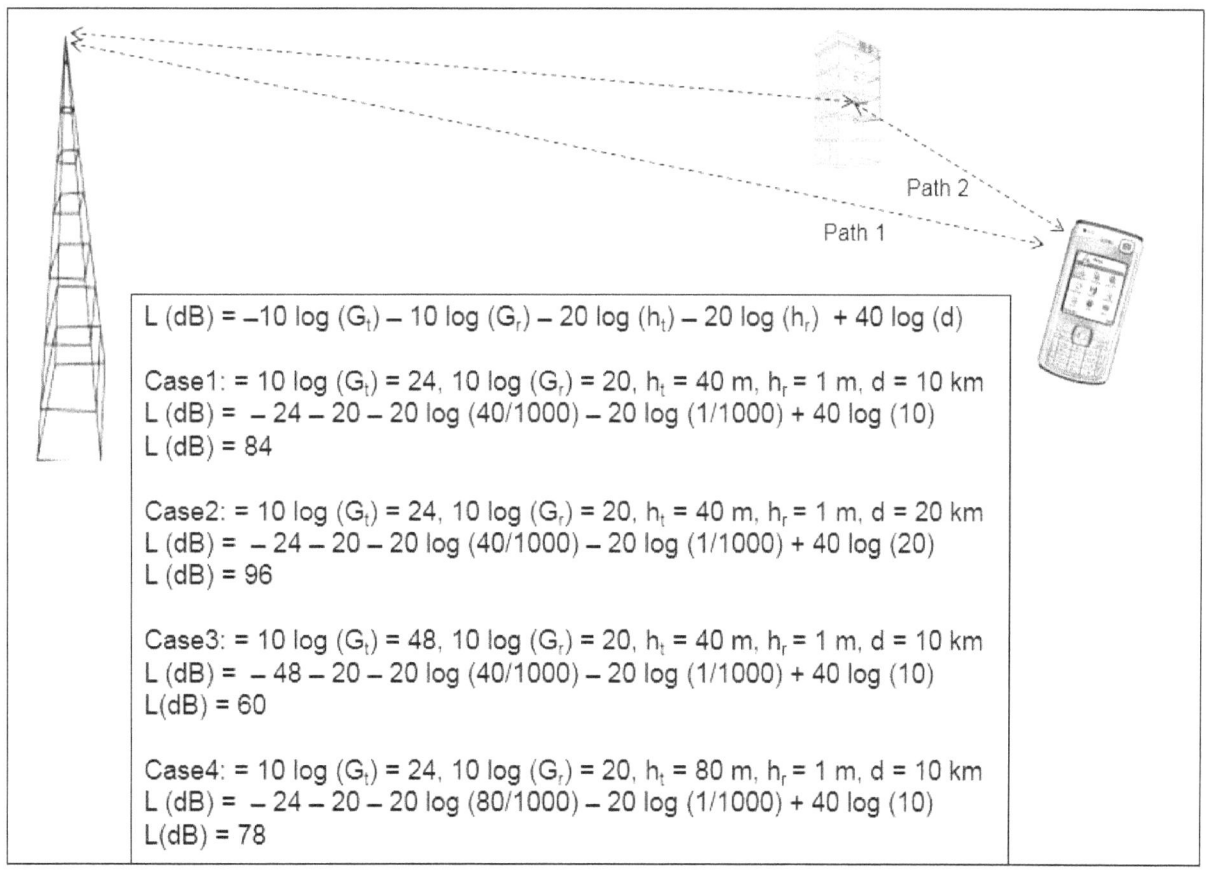

**Figure 5.2.1.a**

## 5.3  Diffraction

RF waves can propagate behind solid objects with sharp edges, such as mountains and hill tops. This phenomenon is called knife edge *diffraction*, and is common with light waves, sound waves or any other wave propagation.

**Figure 5.3.a** shows a barrier between the transmitter and receiver. From common sense, we know that the LOS path is blocked and there should be no signal received behind the barrier. However, because of the sharp edge at the top of the barrier, the radio waves suffer *diffraction* at the knife edge, and are received in the region behind the barrier, which is called *shadow zone*.

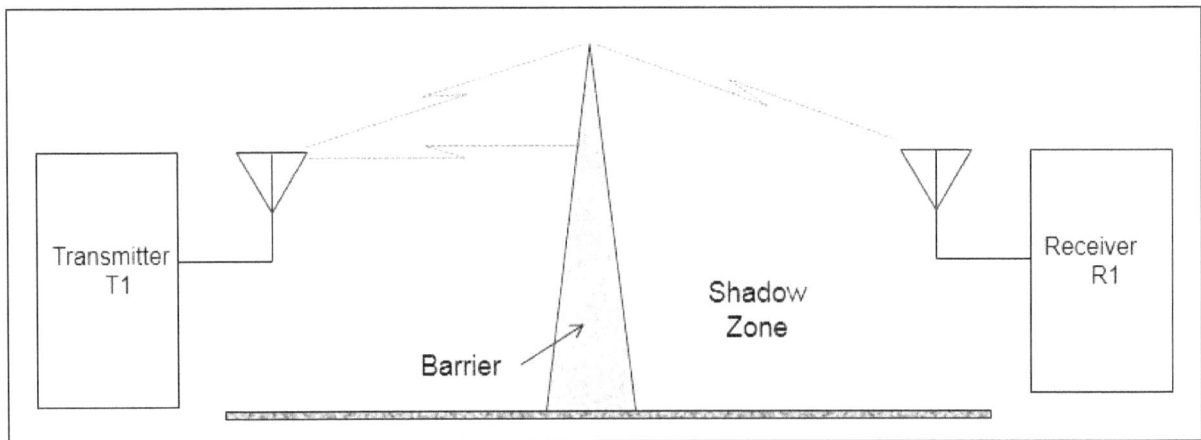

**Figure 5.3.a**

Let us now consider three different scenarios based on the height of the barrier. In Case 1, as shown in **Figure 5.3.b**, the top edge of the barrier has a clearance greater than one wavelength, λ, from the LOS path between the transmitter and receiver, In this situation, there are no additional losses due to diffraction, and only free space LOS losses are observed.

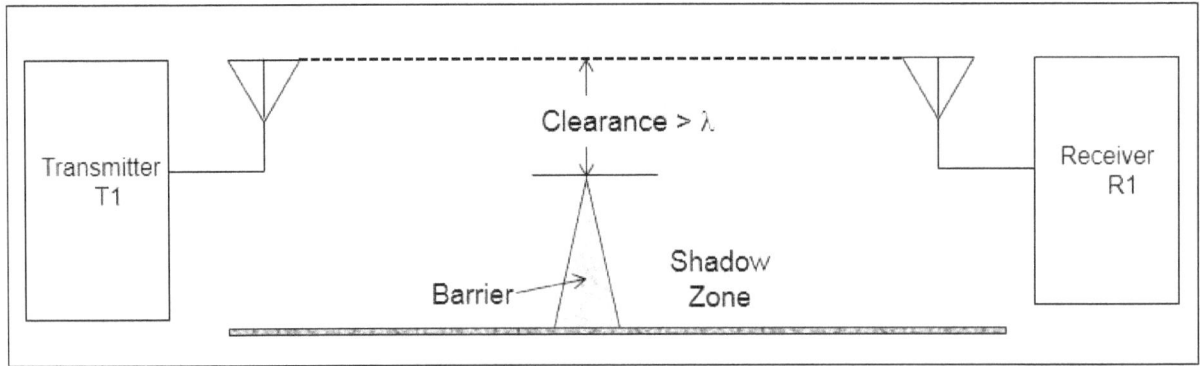

**Figure 5.3.b**

In Case 2, as shown in Figure **5.3.c**, the top edge of the barrier has a clearance less than one wavelength, λ, from the LOS path between the transmitter and receiver. In this situation, there are additional losses due to diffraction, which are in addition to the free space losses.

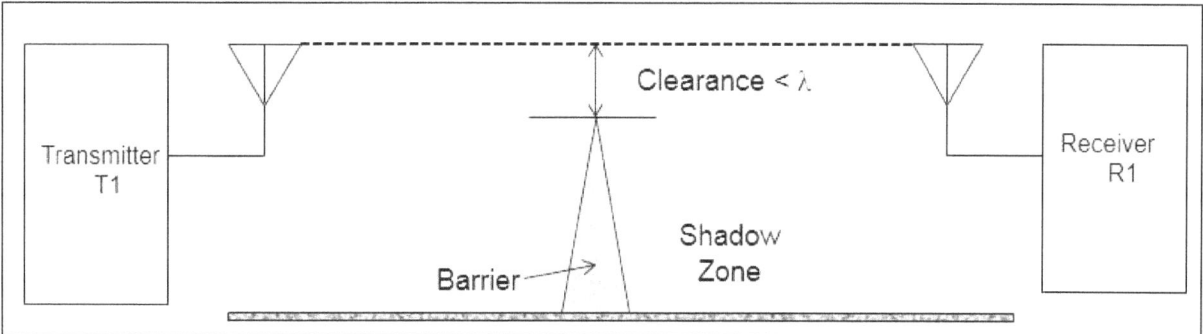

**Figure 5.3.c**

In Case 3, as shown in **Figure 5.3.d**, the LOS path is completely blocked by the barrier, and the RF waves at the top edge of the barrier suffer diffraction, and bend around the barrier, causing signal to be received in the shadow zone.

A mathematical representation of the diffraction gain is shown in **Figure 5.3.d** and it represents the gain in received signal power due to diffraction when the LOS path is completely blocked. As you can see, this equation has integration functions, and it requires numerical methods for computation. In practice, numerical or graphical techniques are used to compute diffraction gain or any additional losses introduced due to diffraction.

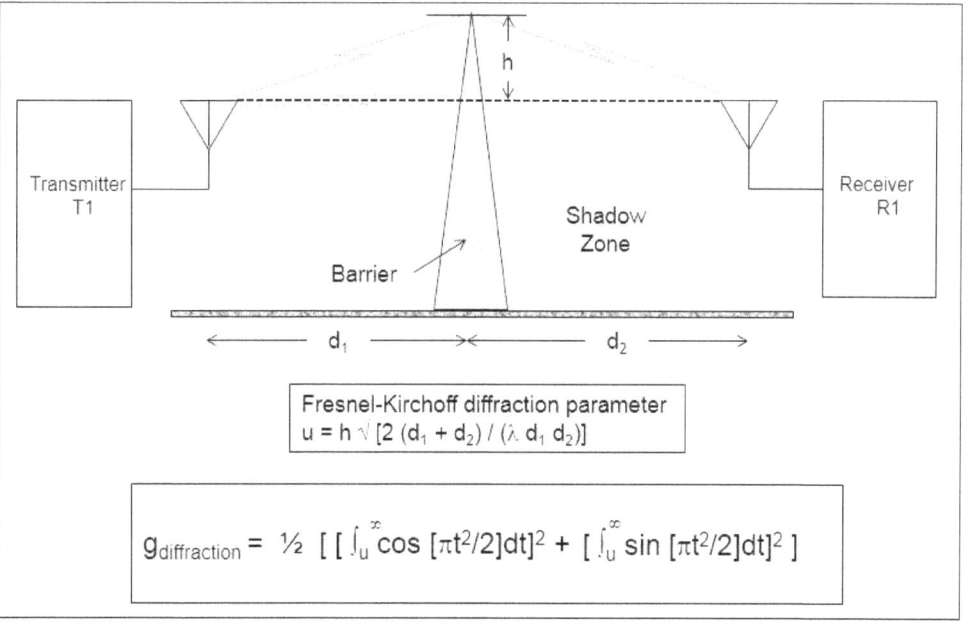

Fresnel-Kirchoff diffraction parameter
$u = h \sqrt{[2 (d_1 + d_2) / (\lambda\, d_1\, d_2)]}$

$$g_{diffraction} = \tfrac{1}{2}\, [\, [\, \int_u^\infty \cos\,[\pi t^2/2]\,dt]^2 + [\, \int_u^\infty \sin\,[\pi t^2/2]\,dt]^2\, ]$$

**Figure 5.3.d**

## 5.4  Refraction

RF waves suffer *refraction* when there is a change in the *refractive index*, n, of propagation medium. Refractive index is defined as the ratio of the velocity of a wave in vacuum to its velocity in the propagation medium. Refraction causes bending of the path of propagation when the RF wave propagates from a medium with higher refractive index to a medium with lower refractive index.

In troposphere, *refraction* is common because the refractive index decreases with increase in height from the surface of earth. This results in a curvilinear path of propagation for RF waves, as shown in **Figure 5.4.a**. The range covered will depend on the difference between the wave's curvature and earth's curvature. To account for this difference in curvature, *effective earth radius* is defined as shown in **Figure 5.4.a**. *k-factor* is defined as the ratio of *effective earth radius* and *real earth radius*.

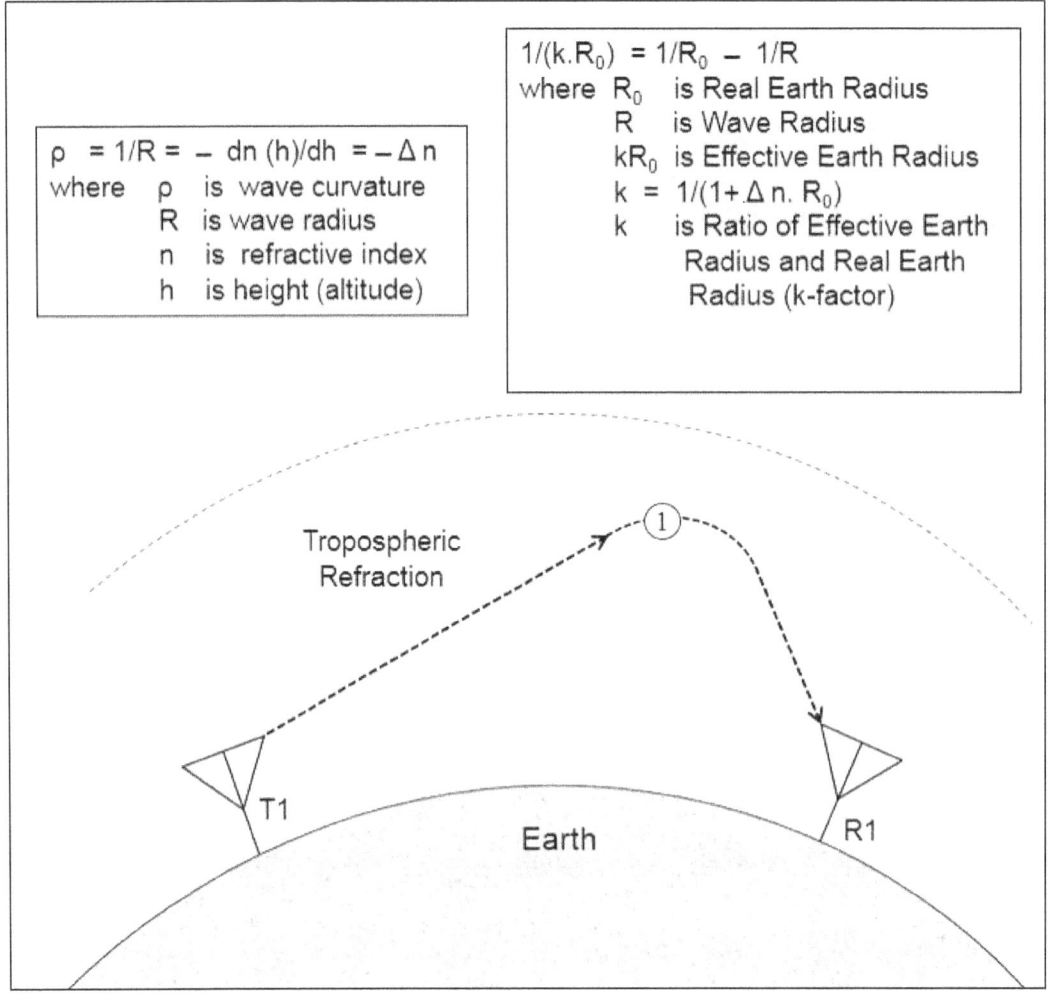

**Figure 5.4.a**

The *k-factor* can be used to estimate the effective range of RF wave propagation due to refraction in the troposphere. We will consider two scenarios for *k-factor*.

In Case 1, as shown in **Figure 5.4.b**, the value of $k$ is greater than 1. Therefore, *effective earth radius* is bigger than *real earth radius*, and hence the range is longer.

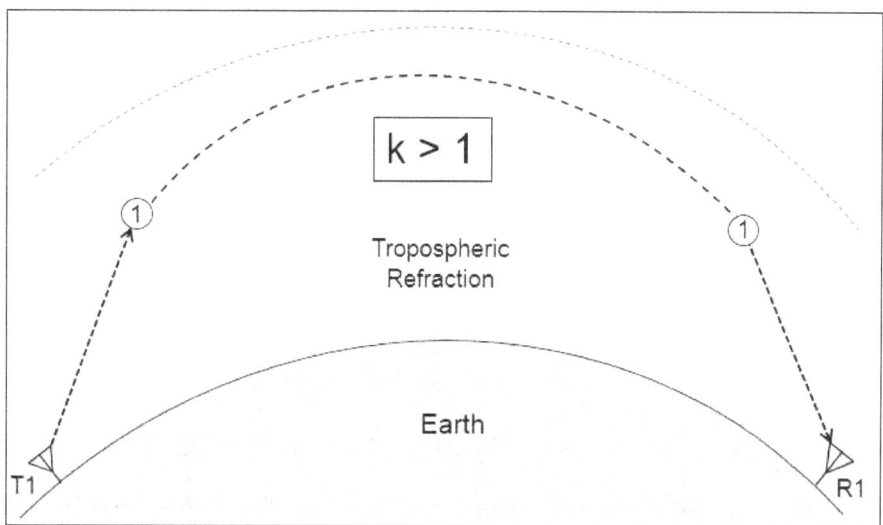

**Figure 5.4.b**

In Case 2, as shown in **Figure 5.4.c**, the value of *k* is between 0 and 1. Therefore, *effective earth radius* is smaller than *real earth radius*, and hence the range is shorter.

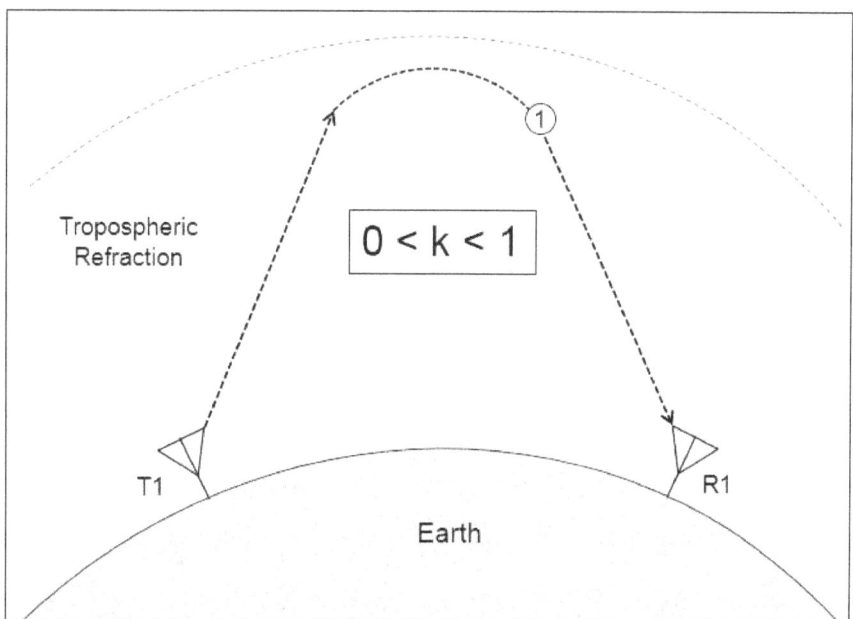

**Figure 5.4.c**

The formula in **Figure 5.4.d** can be used to estimate the LOS distance between transmitter and the receiver for tropospheric refraction.

$$d_{LOS} = d_t + d_r = \sqrt{(2.k.R_0.h_t)} + \sqrt{(2.k.R_0.h_r)}$$

where,

$$d_t = \sqrt{(2.k.R_0.h_t)}$$
$$d_r = \sqrt{(2.k.R_0.h_r)}$$

$d_{LOS}$ is the LOS distance between transmitter and receiver
$k$     is k-factor
$h_t$    is the height of transmitting antenna
$h_r$    is the height of receiving antenna
$R_0$    is real earth radius

**Figure 5.4.d**

## 5.4.1     Microwave Link Example

Let us now calculate the LOS distance for a microwave link. As shown in **Figure 5.4.1.a**, a value of k = 1.33 is used for standard atmospheric conditions. Substituting the values for antenna heights and earth radius, the LOS distance is calculated as 43 km.

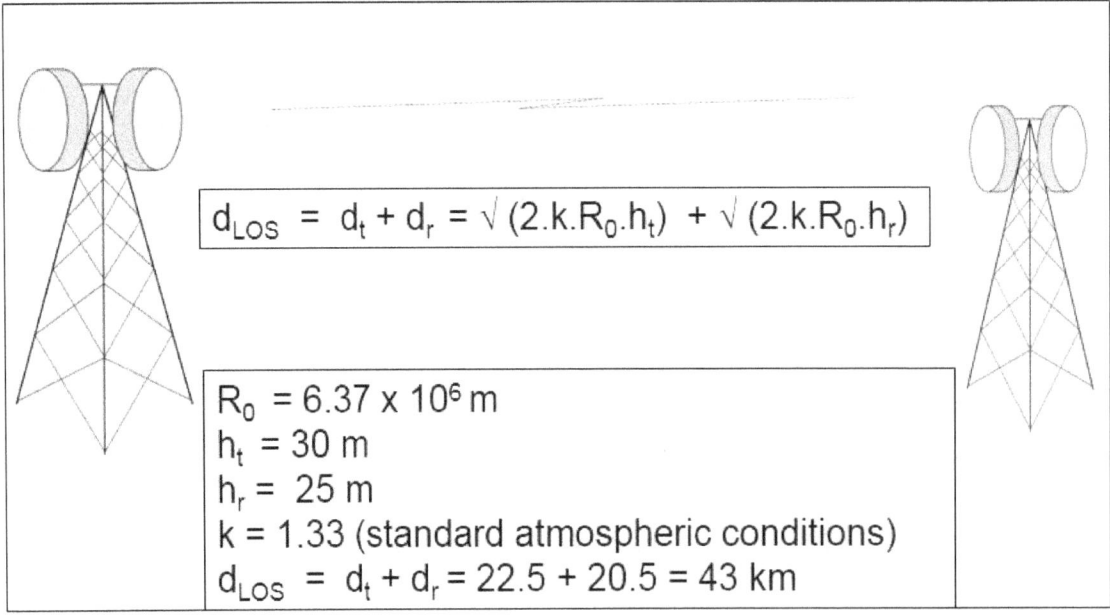

$$d_{LOS} = d_t + d_r = \sqrt{(2.k.R_0.h_t)} + \sqrt{(2.k.R_0.h_r)}$$

$R_0 = 6.37 \times 10^6$ m
$h_t = 30$ m
$h_r = 25$ m
$k = 1.33$ (standard atmospheric conditions)
$d_{LOS} = d_t + d_r = 22.5 + 20.5 = 43$ km

**Figure 5.4.1.a**

## 5.5  Scattering

RF waves are *scattered* by objects smaller than one wavelength, λ. The scattering objects can include utility poles, street signs, irregular terrain and trees. **Figure 5.5.a** shows how the rays are scattered in different directions by a rough surface.

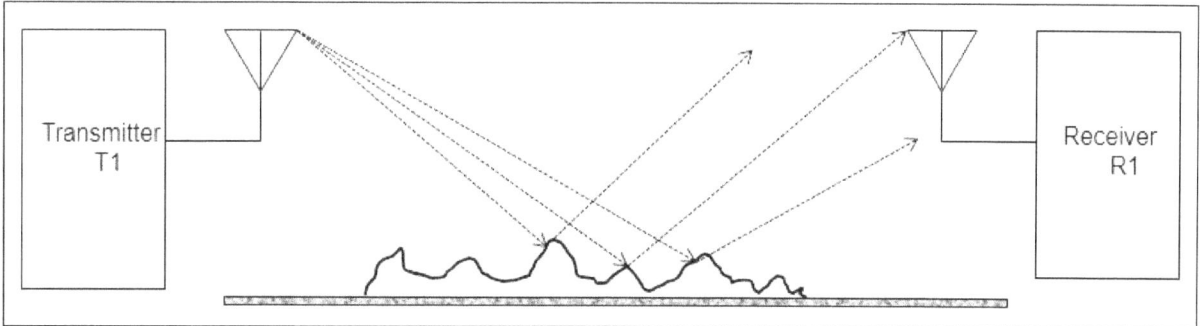

**Figure 5.5.a**

If a surface is smooth, then it is expected to reflect a ray of radio frequency. If there are multiple rays incident on a smooth surface, then the path difference is 0 between any two rays that are incident at the same angle. A difference in path of 0 implies a phase difference of 0. In other words, the 2 rays will arrive at the receiver with no phase difference.

However, if the surface is rough, then the path difference between the 2 reflected rays will be anywhere between 0 and half wavelength (λ/2). In other words, the phase difference between the 2 rays will be anywhere between 0 and 180°. If the phase difference is exactly 180 deg, the 2 rays will completely cancel out each other, and if it is less than 180 deg, they will cancel out to a lesser extent. This cancelling out of RF energy by a rough surface causes *scattering* loss. The RF energy is therefore scattered in all directions by a rough surface, as shown in **Figure 5.5.a**.

In order to measure the electrical roughness of a surface, *Rayleigh Criterion* is used. This criterion states that for a surface to be electrically smooth, the path difference should be less than a quarter wavelength, and for a surface to be electrically rough, the path difference should be greater than a quarter wavelength. *Rayleigh Criterion* is illustrated in **Figure 5.5.b**, where Ray 1 is incident at an angle Θ on an obstacle of height H, and Ray 2 is incident at the same angle on the ground surface. The obstacle of height H introduces a path difference between Rays 1 and 2.

Also shown in **Figure 5.5.b** is a formula that relates the obstacle height H with angle of incidence or grazing angle, Θ and the RF wavelength, λ. This relationship helps us arrive at some interesting results.

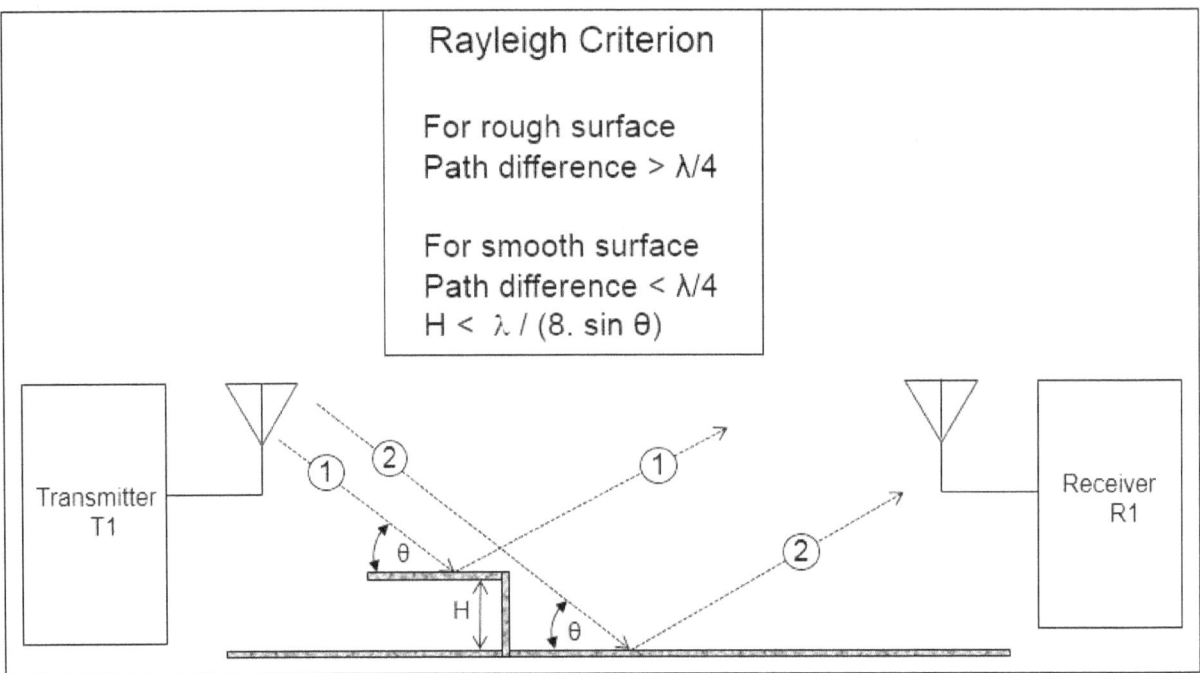

**Figure 5.5.b**

**Table 5.5.c** shows calculated limiting values of H for an electrically smooth surface. It shows the maximum height, H that an obstacle can have for a given frequency (or wavelength) and for a given grazing angle, before it begins to scatter the RF waves.

| For smooth surface, $H < \lambda / (8 . \sin \theta)$ | | |
|---|---|---|
| $\theta$ | 900 MHz, $\lambda = 0.33$ m | 1800 MHz, $\lambda = 0.16$ m |
| 1° | H < 2.36 m | H < 1.18 m |
| 30° | H < 0.08 m | H < 0.04 m |
| 60° | H < 0.05 m | H < 0.02 m |

**Table 5.5.c**

From **Table 5.5.c**, we observe that if the grazing angle is high enough at 60 deg for a frequency of 900 MHz, even a small discontinuity of 5 cm makes the surface electrically rough and causes scattering loss. Also, at the same grazing angle of 60 deg, if the frequency is doubled to 1800 MHz, then even a smaller discontinuity of 2 cm makes a surface electrically rough.

## 5.6 Atmospheric Losses

The presence of oxygen, water vapor, fog, rain, mist, clouds etc. in the atmosphere causes absorption of radio waves. This leads to loss of RF power during propagation in the atmosphere. The longer the path of propagation through the atmosphere more will be the atmospheric losses. A *loss scaling factor*, measured in dB/km, is defined for losses due to oxygen and water in the atmosphere. **Figure 5.6.a** shows the formula to calculate the total atmospheric losses, if the loss scaling factors are known for oxygen and water, in addition to the distance between the transmitter and receiver.

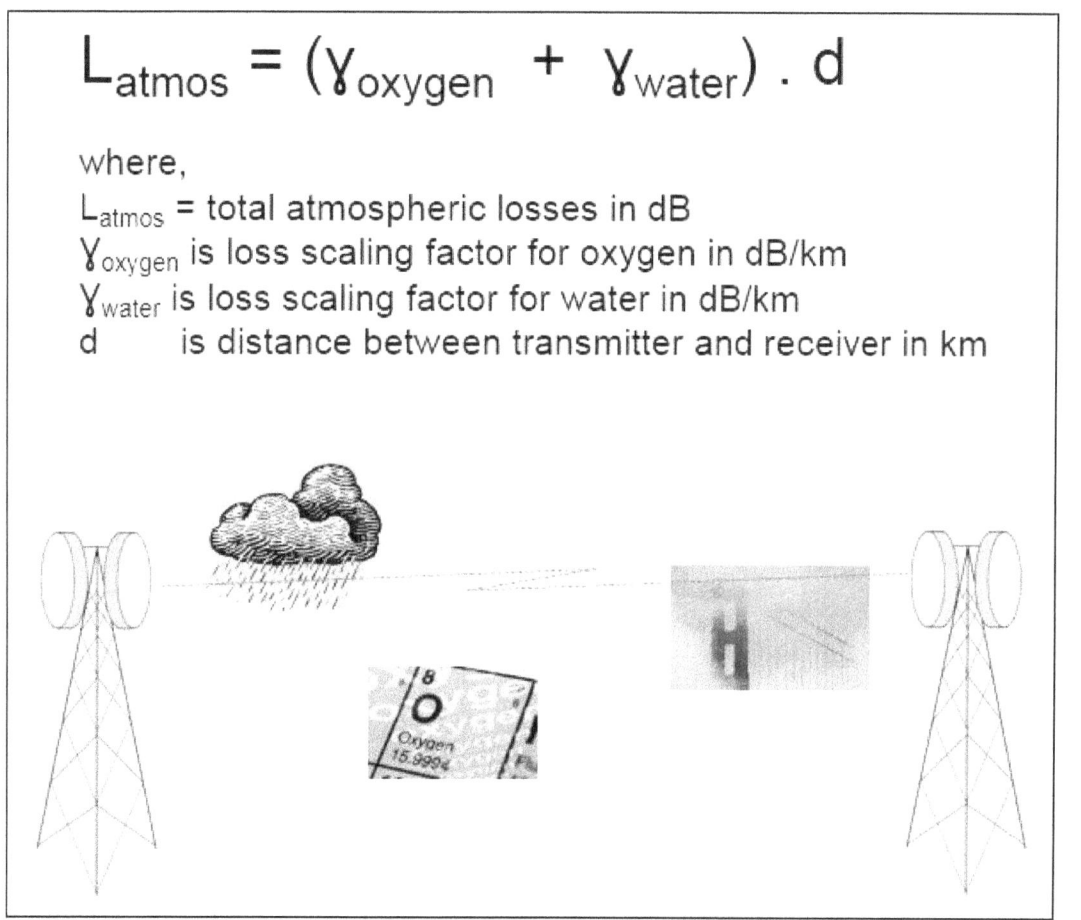

$$L_{atmos} = (\gamma_{oxygen} + \gamma_{water}) \cdot d$$

where,

$L_{atmos}$ = total atmospheric losses in dB
$\gamma_{oxygen}$ is loss scaling factor for oxygen in dB/km
$\gamma_{water}$ is loss scaling factor for water in dB/km
d is distance between transmitter and receiver in km

**Figure 5.6.a**

The graph in **Figure 5.6.b** can be used to figure out the *loss scaling factor*, which is also called *specific attenuation*, in dB/km for a particular radio frequency. It can be observed from the graph, that *loss scaling factor* is much more pronounced for frequencies higher than 10 GHz.

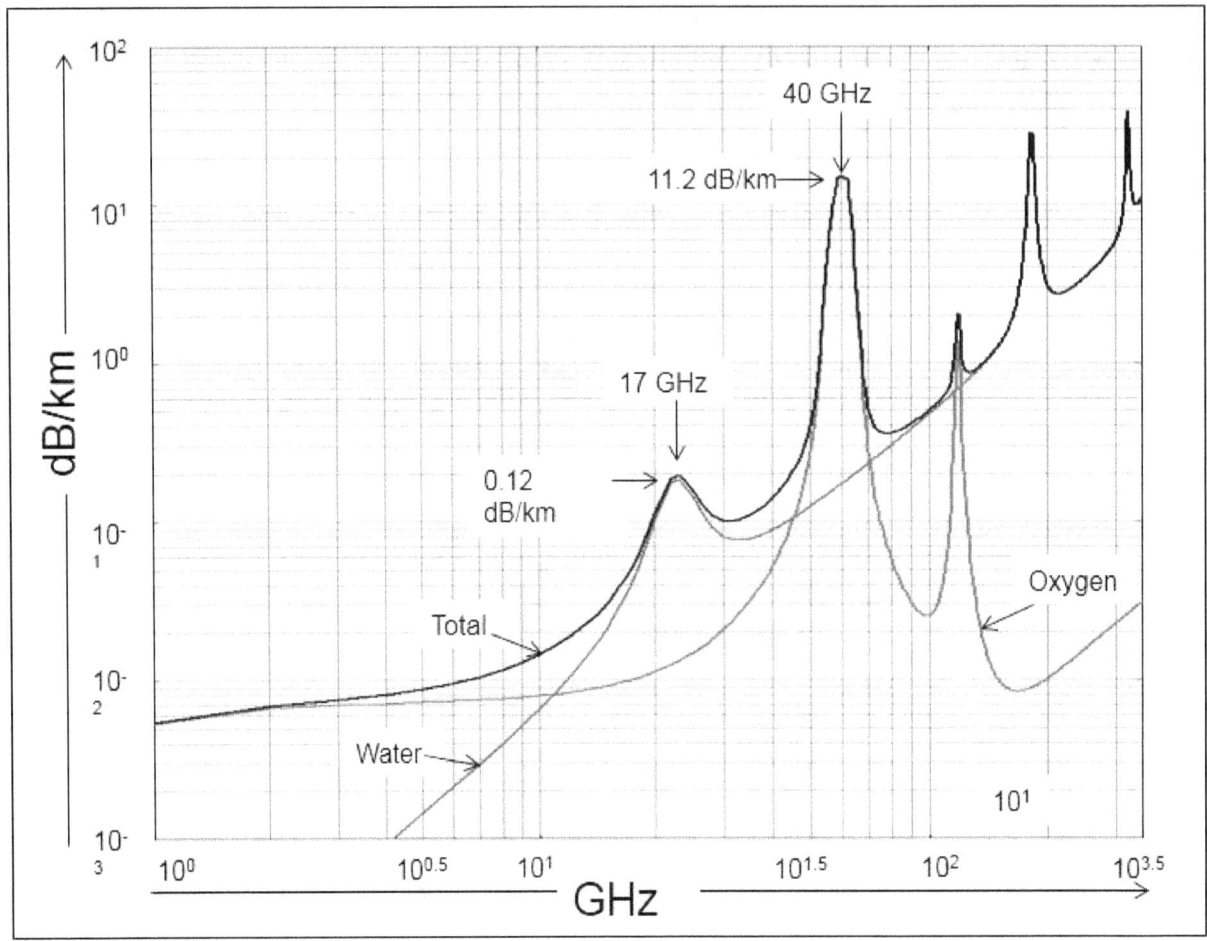

**Figure 5.6.b**

# 6.   Antenna Basics

In this chapter, we will study about the basics of RF antennas. The design of antenna is of utmost importance for proper functioning of the radio link between base station and the cell phone, between access point and the lap top, or between a satellite receiver and the communications satellite. We will explore various antenna design parameters, such as *input impedance, half power beam width, gain, effective aperture* and *EIRP*. We will also explore *size, weight* and *power* (SWAP) considerations for different antenna applications.

## 6.1   What is an Antenna?

A simple answer is, that antenna is a device used to propagate radio waves into space, from the transmitter to the receiver. Antenna is an essential component of any wireless communications system. **Figure 6.1.a** shows a wireless communications system with a transmitter and a receiver, and both have antennas connected to them represented by inverted triangles. The antenna on the transmitter transmits the radio signals that are received by the antenna on the receiver. The figure also shows real dish like antennas mounted on towers.

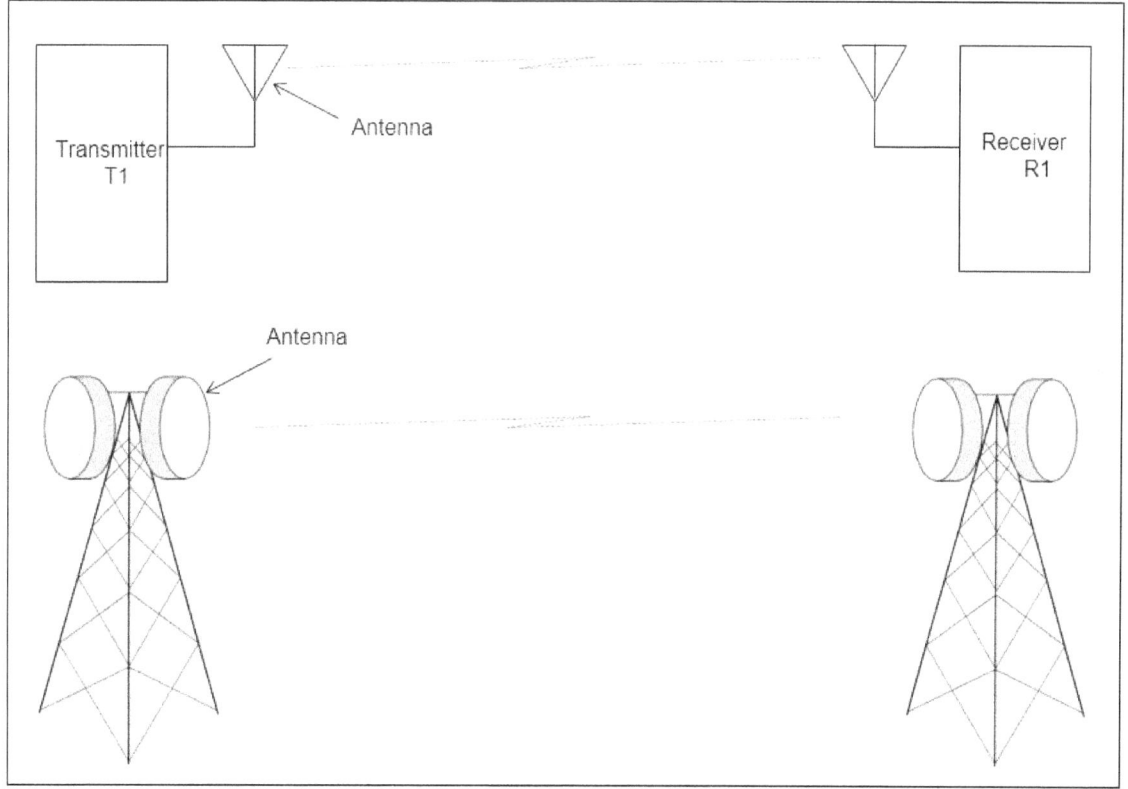

**Figure 6.1.a**

Antennas are connected to the transmitter or receiver equipment with a transmission line or a waveguide. The transmission line is usually in the form of a coaxial cable of appropriate impedance. A transmitter antenna can also be considered as a *transducer* that converts guided *electromagnetic* (EM) wave travelling over a transmission line or waveguide, into a propagating EM wave in free space. The reverse is true for a receiver antenna that is a *transducer* that converts EM wave propagating in free space into a guided EM wave over a transmission line.

**Figure 6.1.b** shows a base station antenna, a mesh antenna, a dish antenna, a horn antenna, a waveguide and a coaxial transmission line.

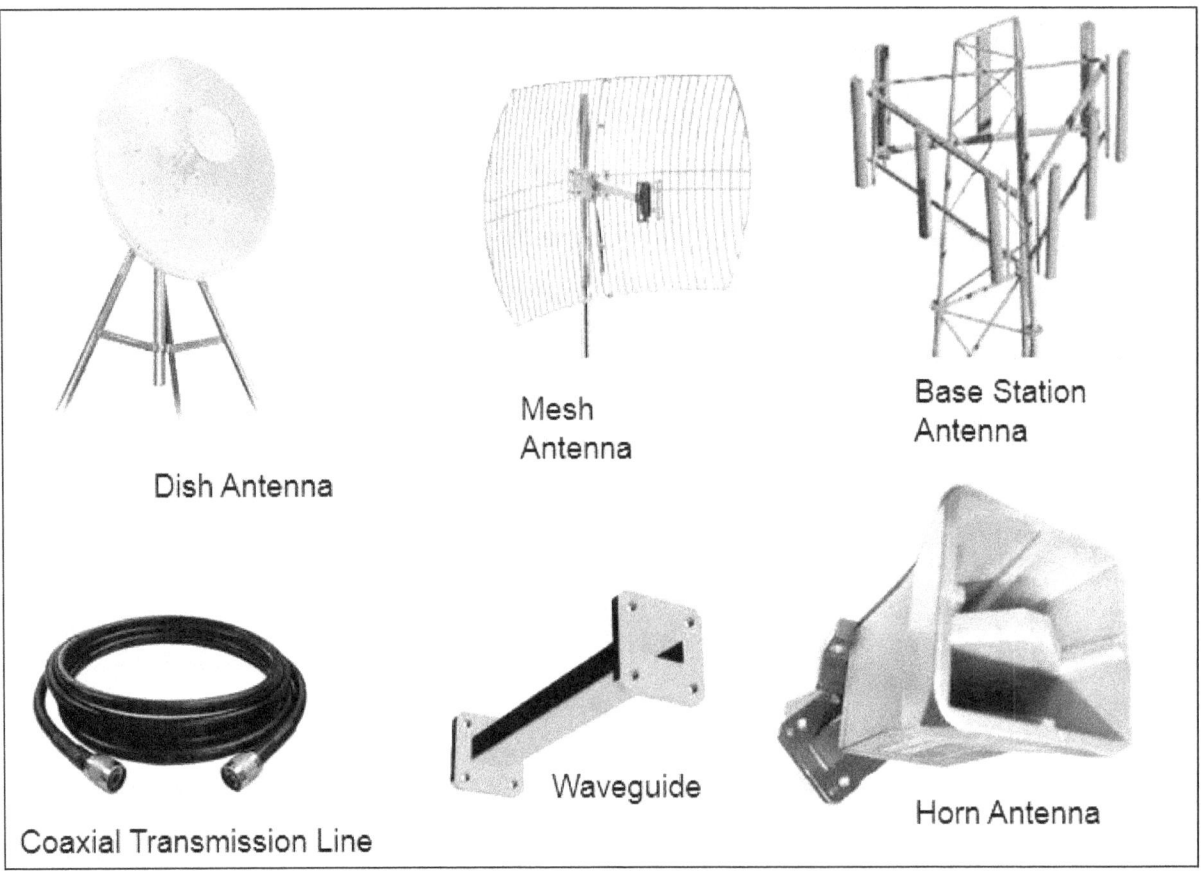

Dish Antenna

Mesh
Antenna

Base Station
Antenna

Coaxial Transmission Line

Waveguide

Horn Antenna

**Figure 6.1.b**

An antenna is *a reciprocal* device, which means the same antenna can transmit as well as receive radio signals. You must have noticed that all cell phones have a single antenna that is used to transmit radio signals to the base station as well as receive radio signals from the base station. However, transmit and receive signals must be separated out from the antenna so that they can be connected to the transmitter and receiver electronics separately. A *duplexer* is a device used to perform this function.

**Figure 6.1.c** shows how a quarter wave ($\lambda/4$) transformer is used to isolate transmit (Tx) and receive (Rx) signals. A real *duplexer* has a common coaxial connector for the antenna, and separate coaxial connectors, one for Tx, and another for Rx.

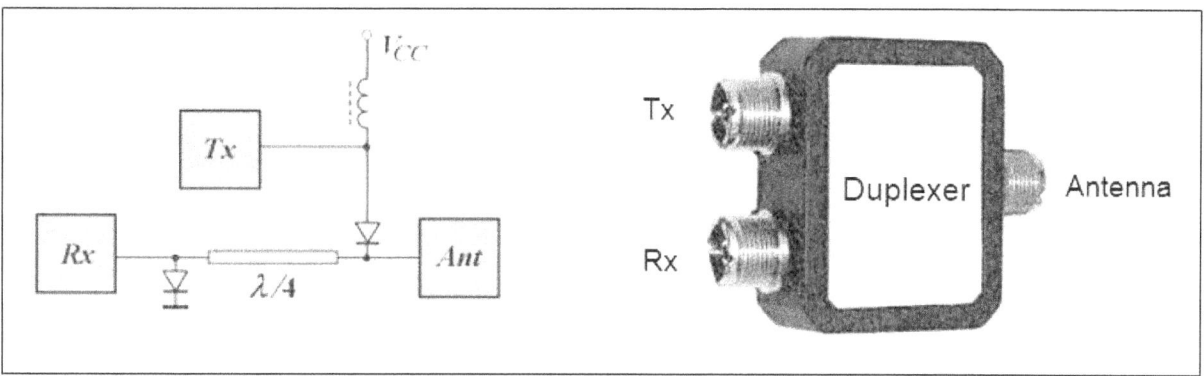

**Figure 6.1.c**

An antenna is also an *impedance transformer*. The free space impedance is 377 ohms, whereas the typical input or output impedance of transmitter or receiver equipment is 50 ohms. The antenna connects the input/output ports of *transceiver* (transmitter + receiver) equipment with free space for transmission and reception of radio waves. So, the antenna transforms the input/output impedance of radio equipment (50 ohms) to the free space impedance of 377 ohms and hence can be called an *impedance transformer*, as shown in **Figure 6.1.d**.

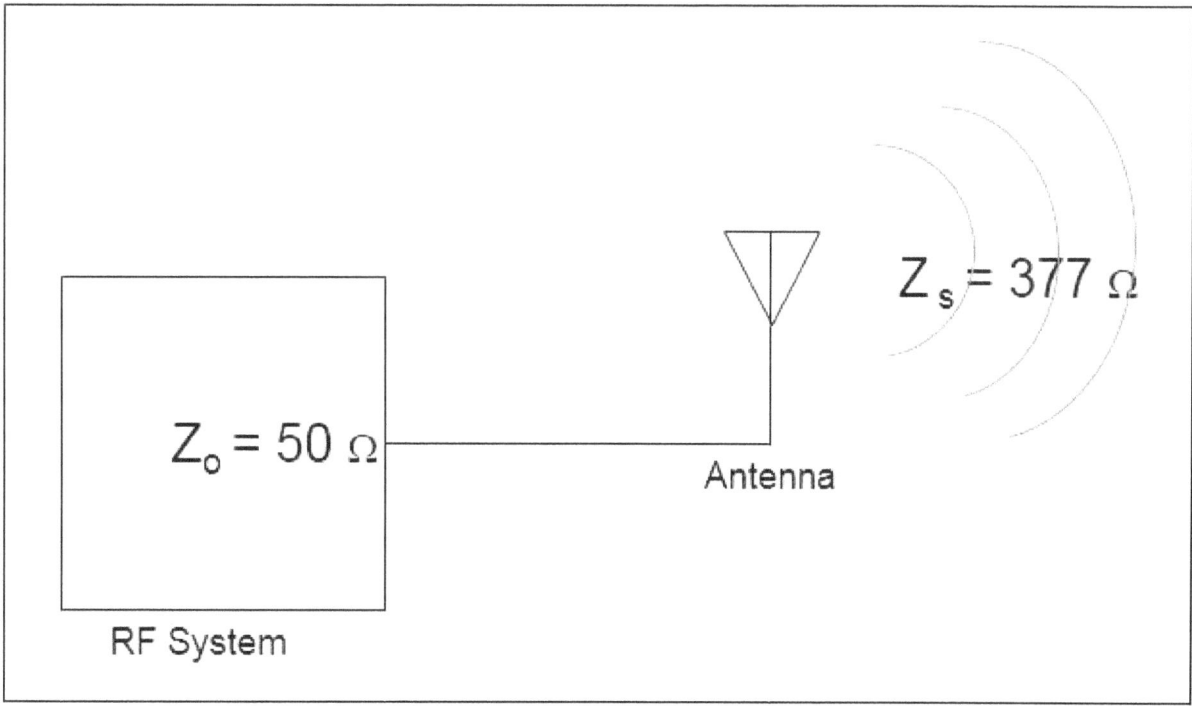

**Figure 6.1.d**

An antenna can also be considered as a *band pass filter*. Antennas are designed for specific frequency range, such as from 2.3 GHz to 2.5 GHz. This is called the *pass band* for the antenna, which means that the antenna is designed to transmit and receive radio frequencies within this specified frequency range. It will pass only the frequencies within the narrow frequency band of 2.3 to 2.5 GHz and will reject all frequencies outside this range, acting just like a *band pass filter* with frequency response peaking at the *center frequency* of 2.4 GHz, as shown in **Figure 6.1.e**.

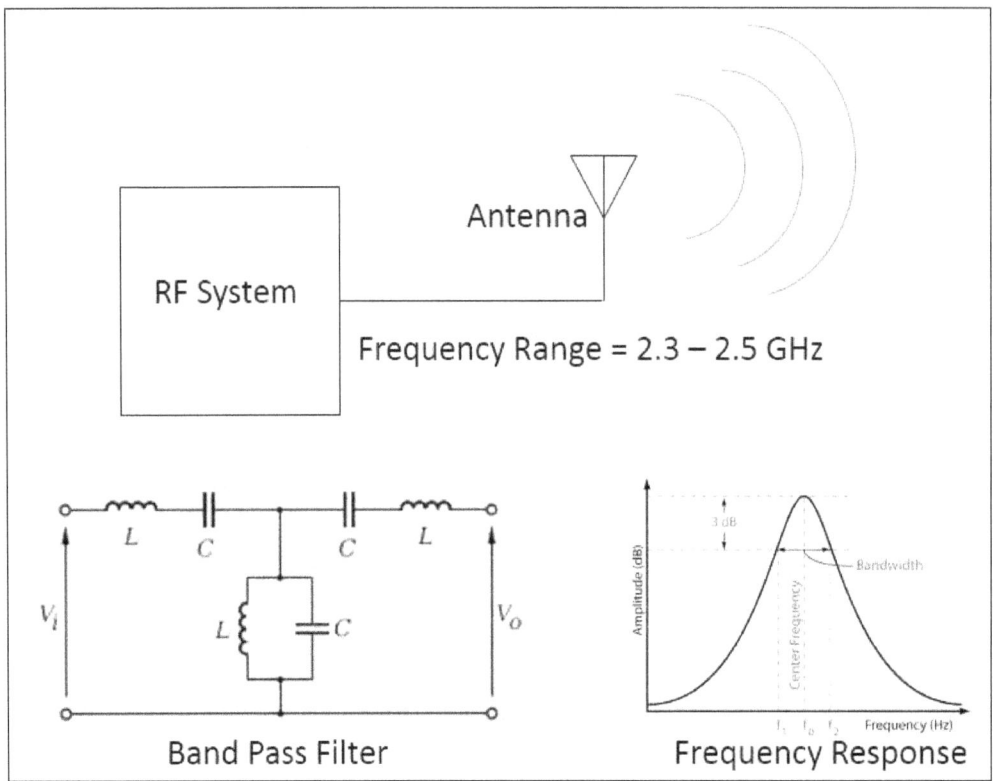

**Figure 6.1.e**

## 6.2   Fields around Antenna

We will now study the *electromagnetic* (EM) fields that exist around an antenna. As shown in **Figure 6.2.a,** the EM fields around an antenna are of three types; the *reactive near field*, the *radiating near field* and the *radiating far field*.

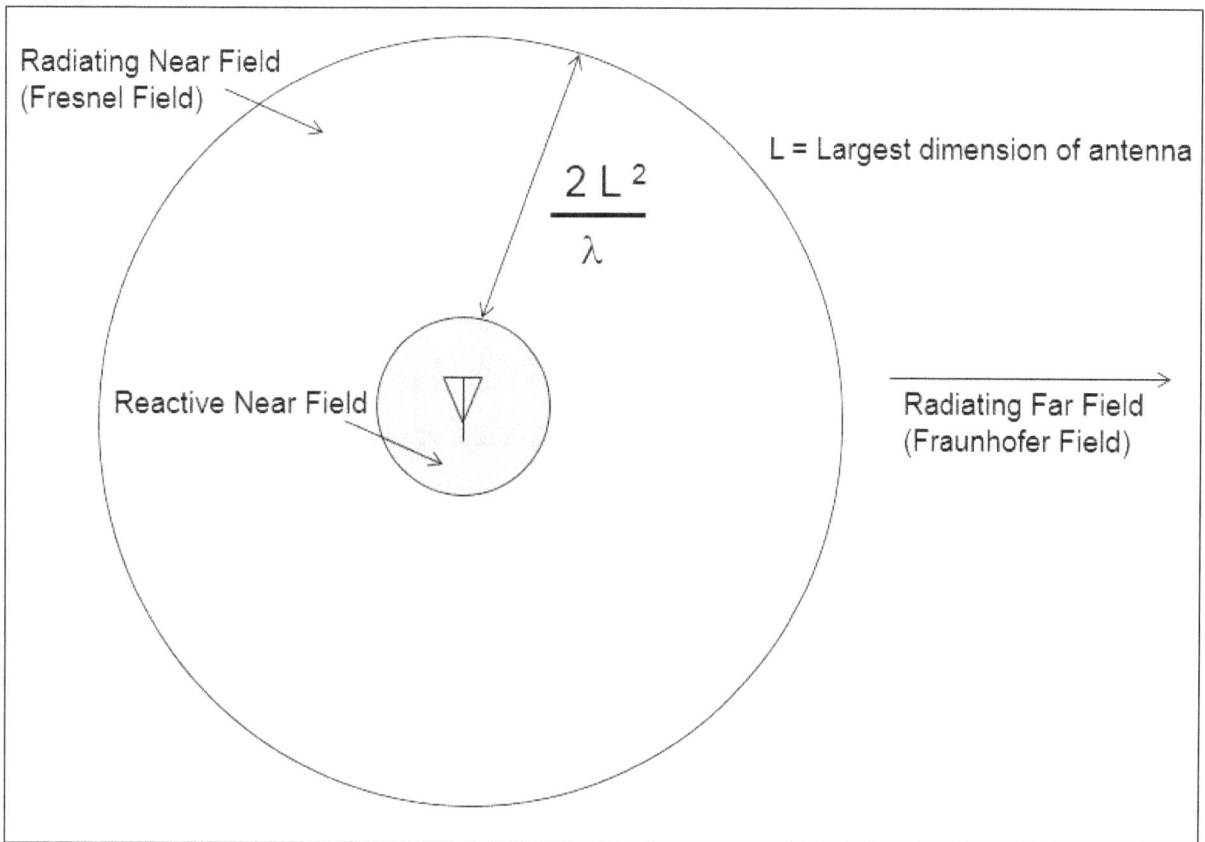

Radiating Near Field
(Fresnel Field)

L = Largest dimension of antenna

$$\frac{2\,L^2}{\lambda}$$

Reactive Near Field

Radiating Far Field
(Fraunhofer Field)

**Figure 6.2.a**

The *reactive near field* is also sometimes called the *induction field*, because the magnetic component of the EM field is usually stronger than the electric component. The antenna, in this field can be considered equivalent to a lumped inductor or capacitor that stores energy in the *reactive near field* instead of propagating into free space. For a simple wire antenna, the reactive near field is up to a distance of half wavelength ($\lambda/2$) from the center of antenna. When making antenna measurements, the probe must never be placed in the reactive near field, because it will establish a mutual coupling with the antenna, causing a distortion of the EM field.

Beyond the *reactive near field, the radiating near field* or the *Fresnel Field* exists up to a radius defined by the formula in **Figure 6.2.a**, that is twice the square of largest dimension of the antenna, divided by the wavelength.

The *radiating far field* or the *Fraunhofer Field* exists beyond the *radiating near field*. The *radiating far field* consists of only transverse EM waves, for which the RF power density is inversely proportional to the square of distance. (See **Section 5.1**).

It should be clear by now that the antenna measurement probe must be placed several wavelengths away from the antenna to avoid any mutual coupling.

## 6.3   Input or Feed Point Impedance

The impedance at a given point of an antenna is equal to the ratio of voltage to current at that point. For example, if there is 100 V of RF voltage and 1.5 A of RF current at a point on an antenna and both voltage and current are in phase, then the impedance at that point is said to be 67 ohms.

The *input impedance* or *feed point impedance* of an antenna is measured at the point where the transmission line is attached to the antenna as shown in **Figure 6.3.a** for a half wave dipole wire antenna

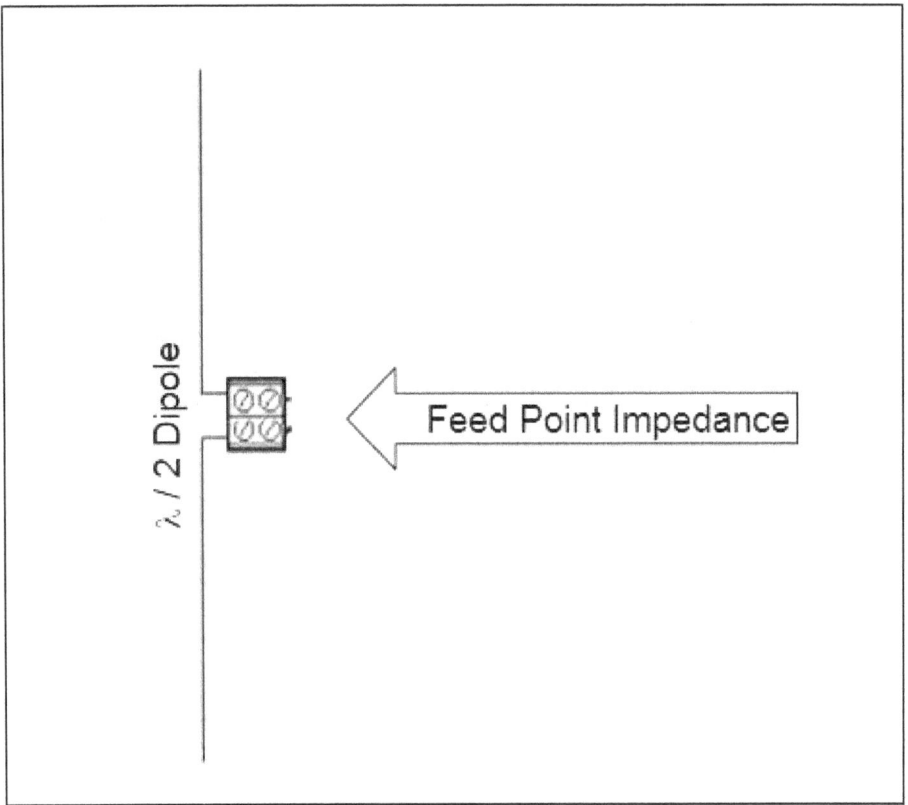

**Figure 6.3.a**

The input impedance of an antenna is not constant. It changes with the physical construction of the antenna, wavelength, as well as height of the antenna from ground. The earth is a big electrical conductor and reflector of radio waves, so it does have a bearing on the input impedance of an antenna. The table in **Figure 6.3.b** shows how the input impedance changes with height from ground for vertical and horizontal half wave dipole antennas.

| Height Above Ground | Input Impedance ($\Omega$) for Vertical Half Wave Dipole Antenna | Input Impedance ($\Omega$) for Horizontal Half Wave Dipole Antenna |
|---|---|---|
| 1.00 $\lambda$ | 75 | 75 |
| 0.85 $\lambda$ | 73 | 83 |
| 0.75 $\lambda$ | 73 | 76 |
| 0.60 $\lambda$ | 70 | 58 |
| 0.50 $\lambda$ | 68 | 68 |
| 0.40 $\lambda$ | 69 | 90 |
| 0.30 $\lambda$ | 71 | 96 |
| 0.20 $\lambda$ | 78 | 65 |
| 0.10 $\lambda$ | 87 | 20 |
| 0.00 $\lambda$ | 98 | 0 |

$\lambda$ / 2 Dipole

**Figure 6.3.b**

You will notice from this table that the input impedance for both vertical and horizontal half wave dipole is the same at 75 ohms at a height of one wavelength above the ground. However, at ground level, the vertical antenna has a high impedance of 98 ohms, whereas horizontal antenna has an impedance of zero.

The input impedance of an antenna must match the impedance of transmission line or coaxial cable that connects the feed point of antenna with the RF equipment. The impedance matching is required to avoid power losses due to reflected EM waves. **Figure 6.3.c** shows a half wave dipole antenna that has a *balanced* impedance of 75 ohms. The symmetrical physical structure of the two elements of dipole antenna makes its impedance *balanced*, because each element has same impedance with respect to ground. However, the coaxial cable that connects the RF equipment to the antenna, has a central conductor and an outer conducting shield, and has an *unbalanced* impedance of 50 ohms. The outer shield of the coaxial cable usually connects to the ground on the side of RF equipment.

The outer and inner surfaces of the shield on coaxial cable act like two different conductors of RF current due to *skin effect*, which causes the alternating RF currents to travel on the inner and outer surfaces of the shield. The outer surface is therefore a third conductor and it can cause the RF current to flow into the enclosure of RF equipment, causing impairment of the EM field pattern.

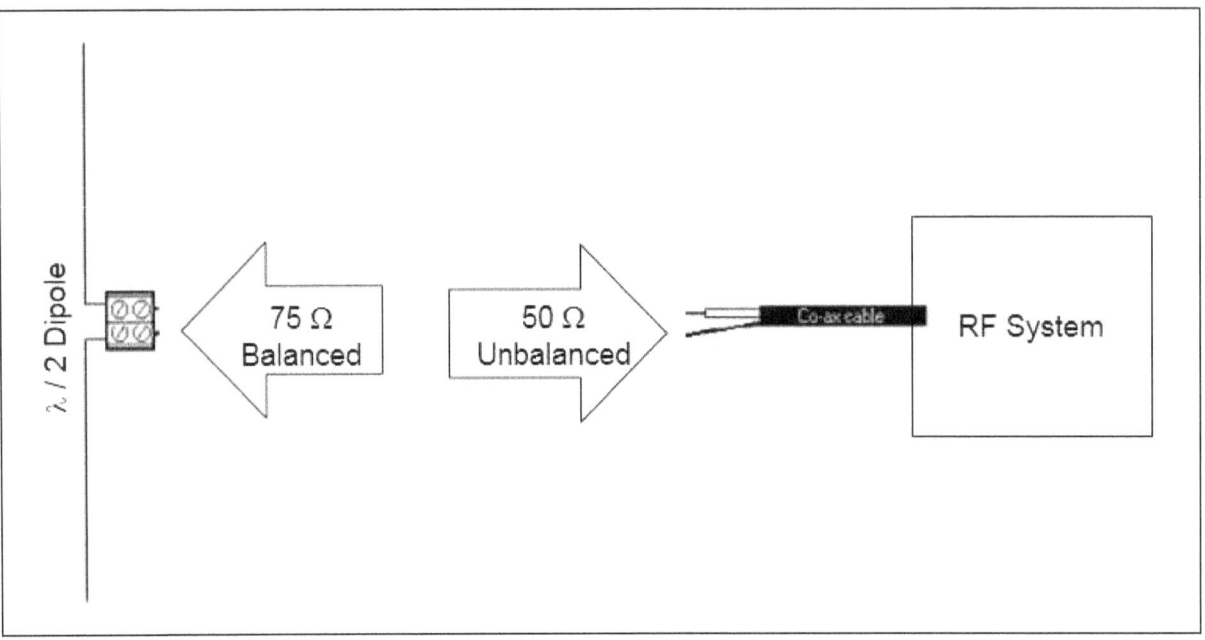

**Figure 6.3.c**

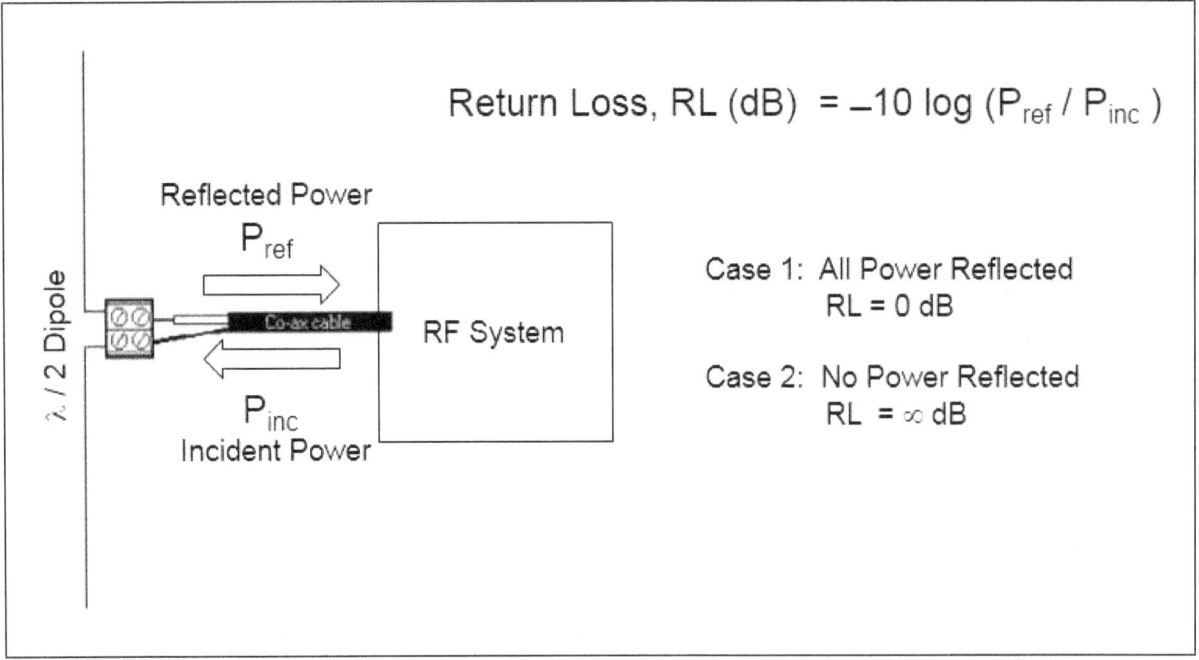

**Figure 6.3.d**

Let us see what happens when a coaxial cable with unbalanced impedance of 50 ohms is connected to an antenna with balanced impedance of 75 ohms. We have *impedance mismatch*, as shown in **Figure 6.3.d**. This impedance mismatch causes part of RF power to reflect from the feed point. So, all of RF power does not get transferred from the RF System to antenna, and hence the radiated power from antenna is less than expected.

*Return Loss (RL)* is defined as the negative 10 log of the ratio of Reflected power to Incident Power and its units are in dB. (See **Figure 6.3.d**). It provides a measure of the power reflected due to impedance mismatch. Reflected power is always less than the incident power, so RL is always positive.

For RL, two extreme cases can be considered. If all RF power is reflected, then RL = 0 dB. However, if no power is reflected, then RL is calculated to be infinite.

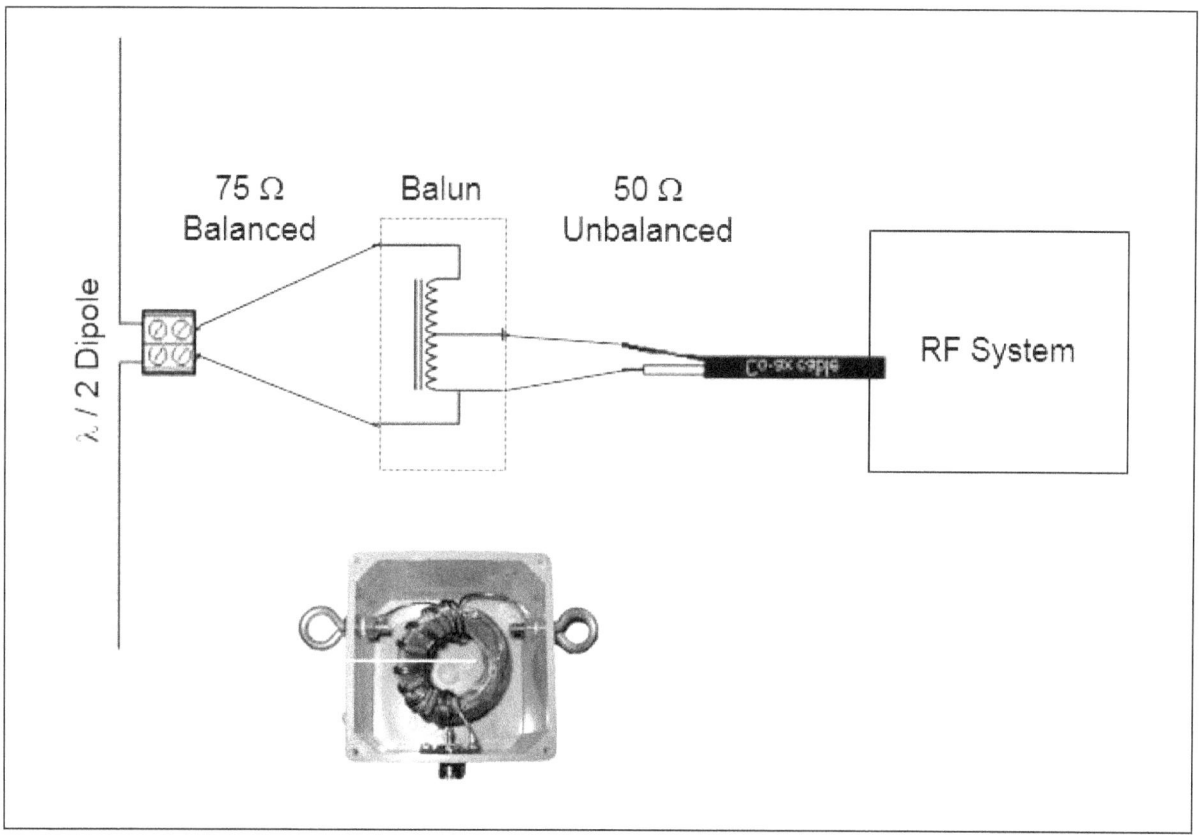

**Figure 6.3.e**

*Baluns* (short for balanced to unbalanced transformer) can be used to match balanced impedance of antenna to unbalanced impedance of coaxial cable, as shown in **Figure 6.3.e**. The picture shows a balun built from few turns of copper wire wound around a ferrite core. The baluns connect between the feed point of antenna and the coaxial cable, and they allow RF power to be transferred between balanced and unbalanced portions of an antenna system in either direction.

## 6.4  Size, Weight and Power

An important consideration when designing an antenna for a specific application is SWAP, that is, *size* of antenna, *weight* of antenna and radiated p*ower* from the antenna. As a matter of fact, if the frequency is kept same, then antennas with higher gain are bigger in size. Also, if size is kept the same, then higher frequencies yield antennas with higher gain.

A base station covers the area of a cell that may be a few miles across, whereas for a WLAN access point, the area covered may be less than a few hundred feet. For base station antenna, as well as access point antenna, SWAP is a minor concern, because they are usually fixed at a physical location (**Figure 6.4.a**).

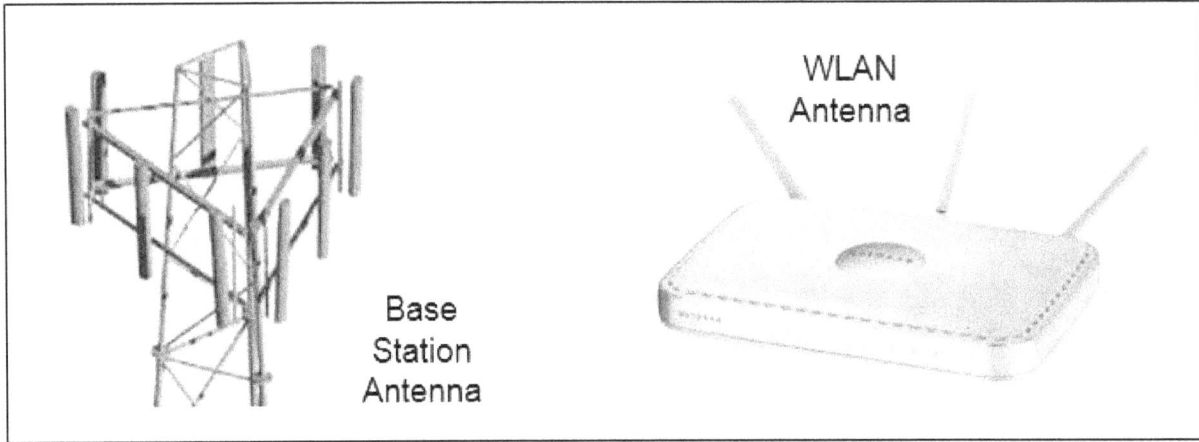

**Figure 6.4.a**

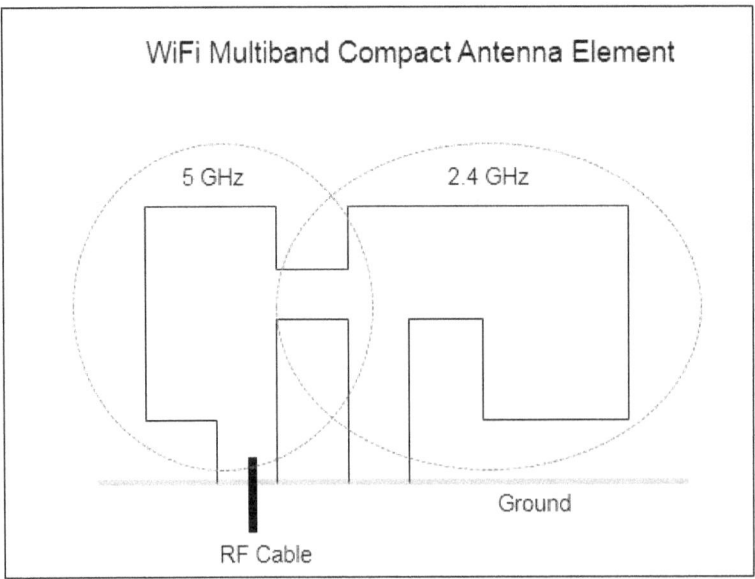

**Figure 6.4.b**

For portable WiFi WLAN devices, such as laptops, smartphones, tablets and ebook readers, SWAP is a critical concern. Antennas are packed with several other peripherals which are conductive and lossy, and also reduce the efficiency of antenna. The performance parameters, such as *operating bandwidth*, *impedance matching*, *peak* and *average gain*, *radiation patterns*, *efficiency* and *specific absorption rate* are important. In addition to that, for all these hand held devices, the effects of human body on the antenna also need to be considered. Therefore, SWAP is a critical concern when designing antennas for portable devices and trade-offs are made between *design, performance* and *placement* of the antenna. **Figure 6.4.b** shows a WiFi multiband compact antenna, which is fabricated on a plane surface. It has actually two antenna elements, one for 5 GHz WiFi frequency and the second for 2.4 GHz frequency. Notice that the antenna elements are designed to physically fit into the form factor of a portable device enclosure.

SWAP is also a critical concern for antennas installed on aircraft and satellites (**Figure 6.4.c**). In this case, *size*, *weight* as well as *power* comes at a premium, and if not designed carefully, the functioning of wireless communications link may be impaired.

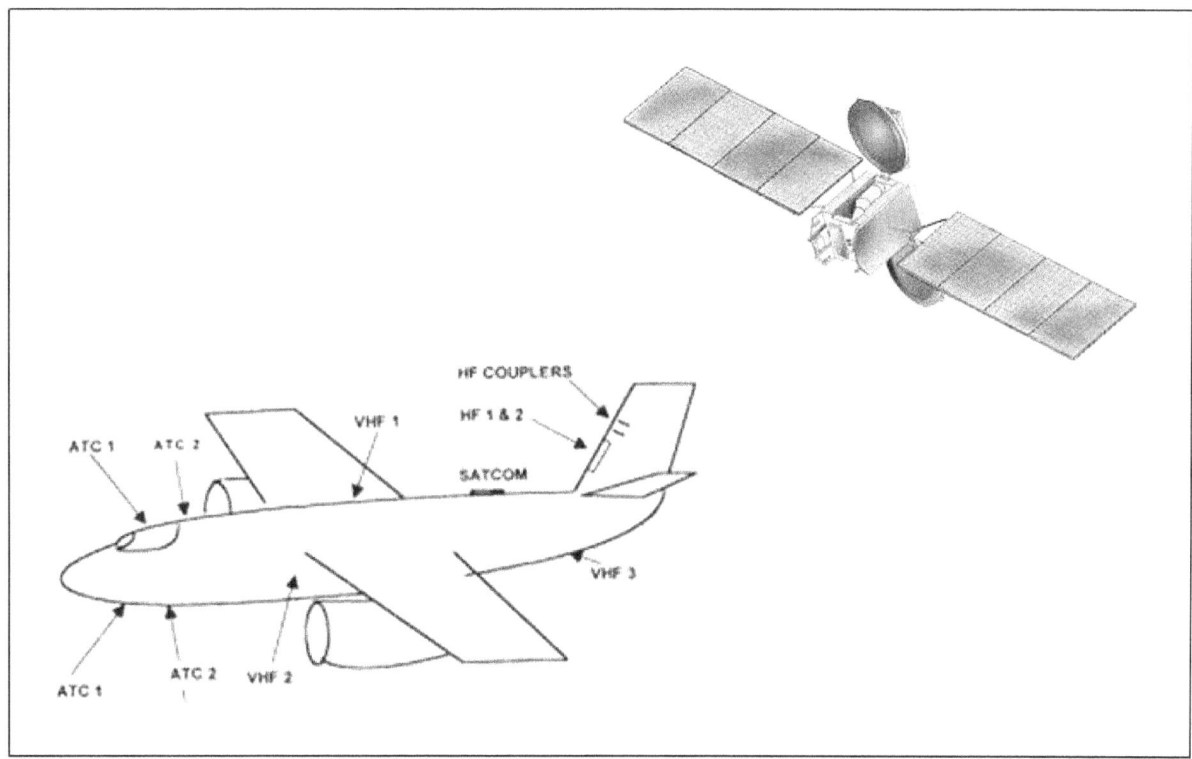

**Figure 6.4.c**

## 6.5   Field and Power Patterns

**Figure 6.5.a** shows field and power patterns of an antenna, which are a graphical representation of the direction of propagation of RF energy. The pattern consists of *nulls* and *lobes*. *Nulls* are the angles at which antenna gain is minimum. *Lobes* are the angles at which antenna gain is maximum. There are two types of lobes, *minor side lobes* and the *main lobe*. The lobe with highest gain is called the *main lobe* and is in the direction of propagation of transverse electromagnetic (EM) waves. The electric field, E and the magnetic field, H are transverse (or perpendicular) to each other.  The *minor side lobes* have lower gain and represent propagation of RF energy on the sides of the *main lobe*.

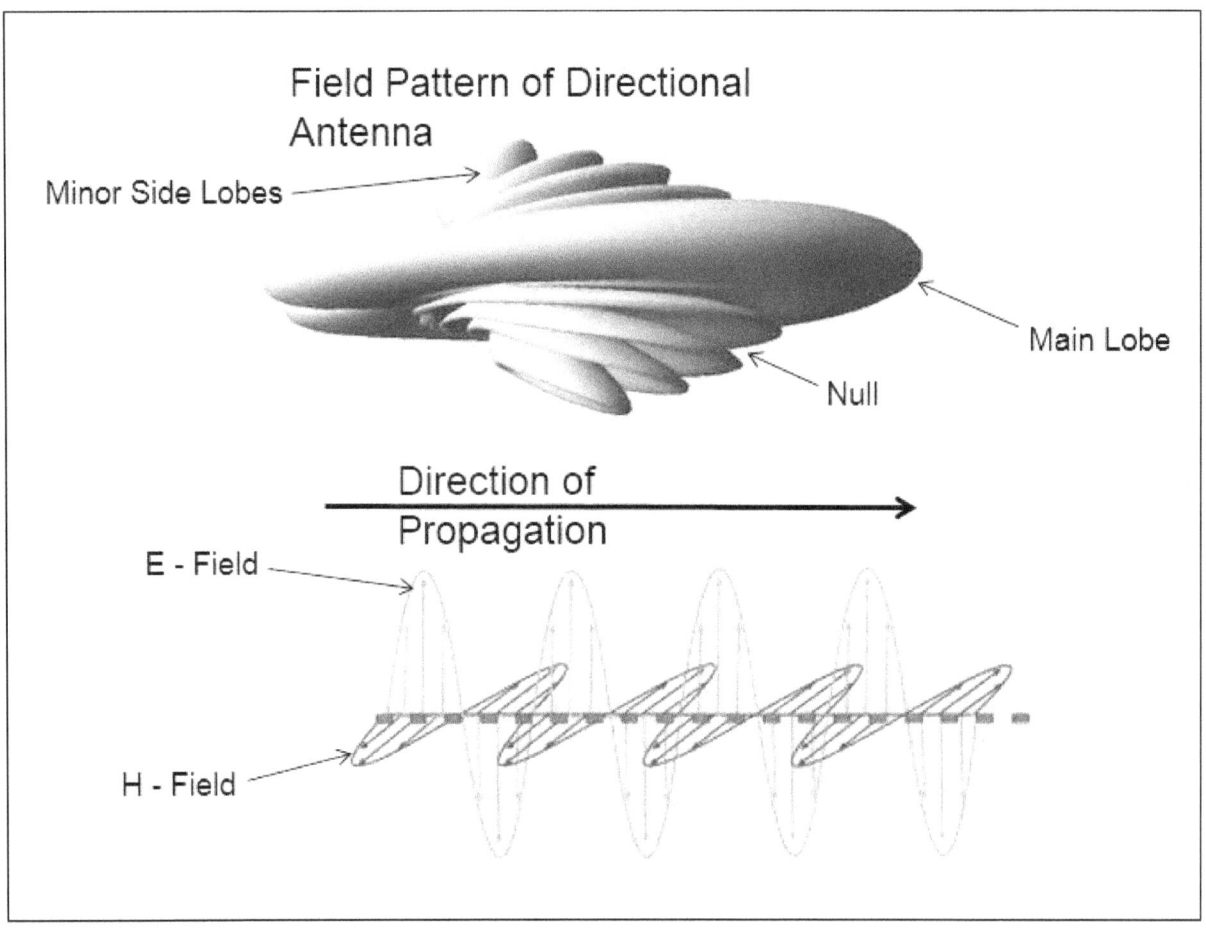

**Figure 6.5.a**

**Figure 6.5.b** shows the polar coordinates that are used to represent a point in the field and power pattern of an antenna. The various angles associated with this coordinate system are shown with reference to x, y and z axes.

**Table 6.5.c** shows the formulas used to represent the pattern in the far field of antenna. These formulas represent the normalized values of electric field and power at a specific polar coordinate, with respect to their maximum values in the direction of propagation. Notice that in the far field of antenna, there is no radial component of the electric field.

**Table 6.5.d** shows the units used to represent *electric field*, *magnetic field* and the instantaneous power density, also called the *Poynting Vector*.

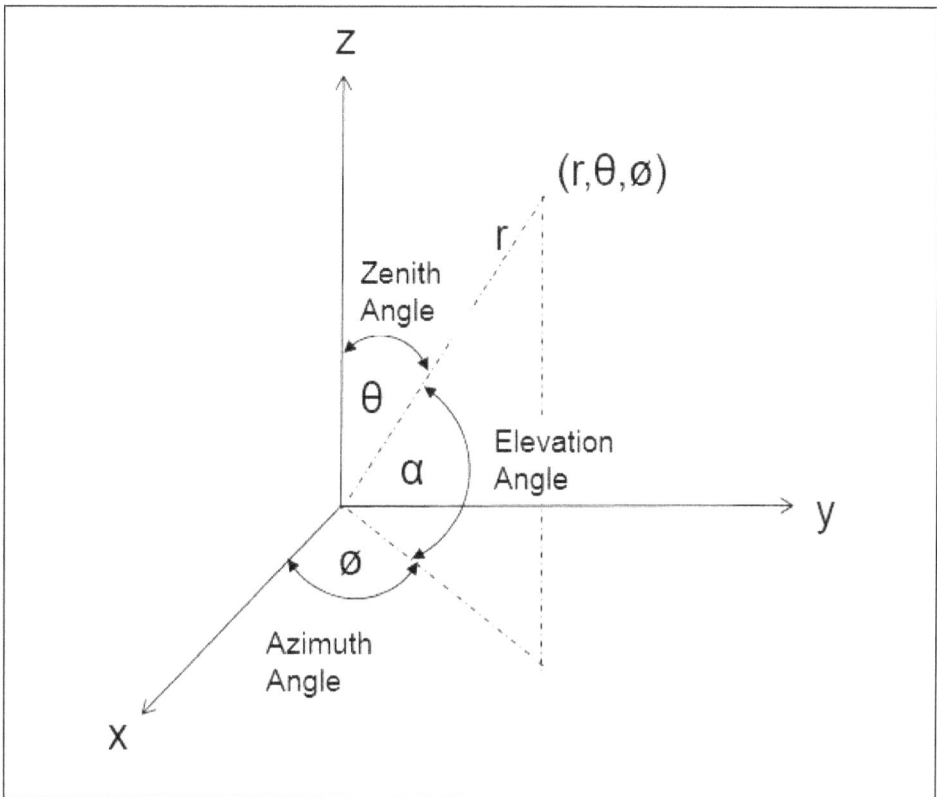

**Figure 6.5.b**

| Normalized Zenith ($\theta$) Pattern of E field | $E_\theta (\theta,\phi)_n = E_\theta (\theta,\phi) / E_\theta (\theta,\phi)_{max}$ |
|---|---|
| Normalized Azimuth ($\phi$) Pattern of E field | $E_\phi (\theta,\phi)_n = E_\phi (\theta,\phi) / E_\phi (\theta,\phi)_{max}$ |
| Normalized Power Pattern | $P_n (\theta,\phi)_n = S (\theta,\phi) / S (\theta,\phi)_{max}$ |
| Poynting Vector (Instantaneous Power Density) | $S (\theta,\phi) = [E_\theta^2 (\theta,\phi) + E_\phi^2 (\theta,\phi)] / Z_0$ <br><br> where $Z_0$ is free space impedance (376.7 $\Omega$) |

**Table 6.5.c**

| Electric Field Strength | V/m (Volts per meter) |
|---|---|
| Magnetic Field Strength | A/m (Amperes per meter) |
| Poynting Vector (Instantaneous Power Density) | W/m² (Watts per meter square) |

**Table 6.5.d**

## 6.6 Half Power Beam Width (HPBW)

*Half Power Beam Width or HPBW* is an important antenna parameter which represents the width of radio beam for an antenna, whether transmitting or receiving. It is defined as the angle between the directions where the received or transmitted power is half of the maximum power. **Figure 6.6.a** shows a radio beam for a directional antenna with main lobe, side lobes and rear lobe. The main lobe is in the direction of maximum power. The HPBW is the angle measured between the half power points on the main lobe as shown in the figure. A reduction in power by half represents 3 dB reduction. It may be noted that at half power points the electric field is reduced to 0.707 of its maximum value on the main lobe axis.

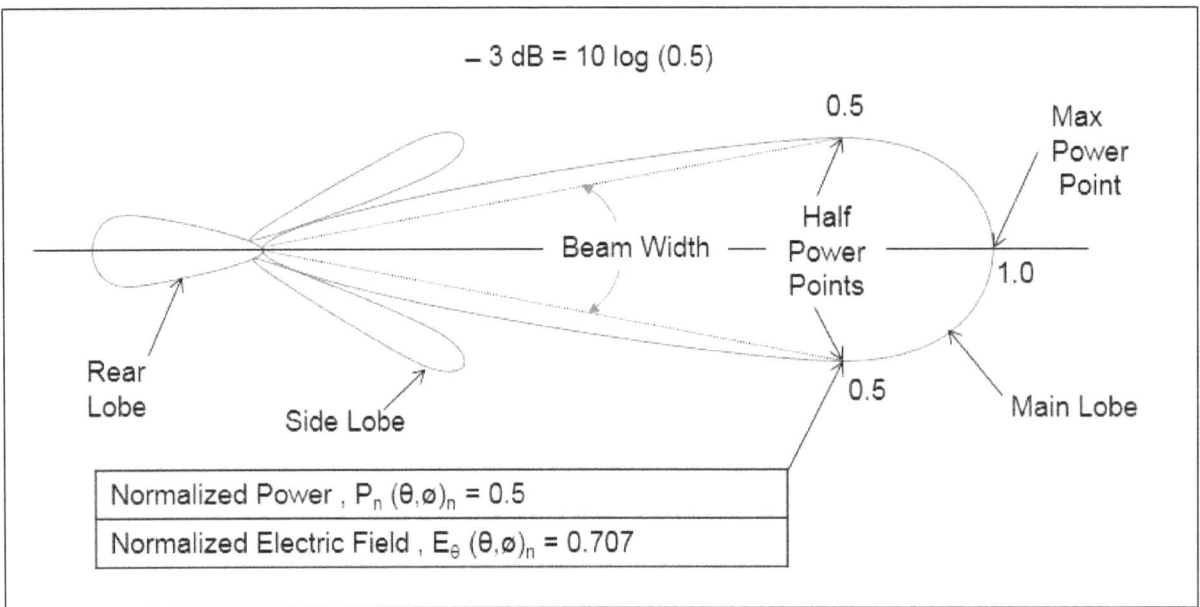

**Figure 6.6.a**

## 6.6.1    Beam Width and Gain

There is a relationship between the *beam width* and *gain* of an antenna. The antennas with high gain have narrower beam widths. A half wave dipole antenna typically has 78 deg beam width, as shown in **Figure 6.6.1.a**.

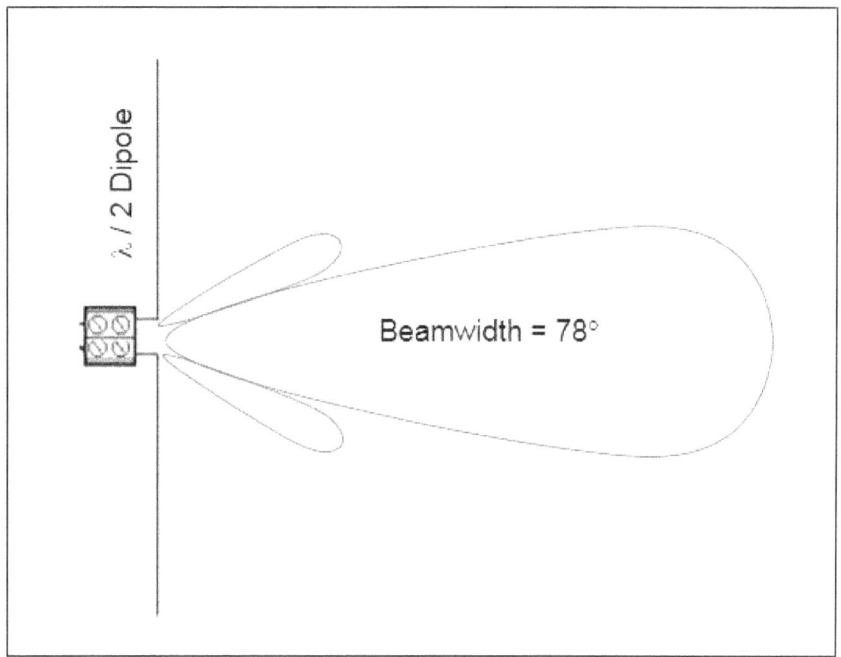

**Figure 6.6.1.a**

A parabolic dish antenna may typically have a beam width of only 1 deg, but much higher gain, because the beam is very narrow and highly directional. **Figure 6.6.1.b** shows the beam for a dish antenna.

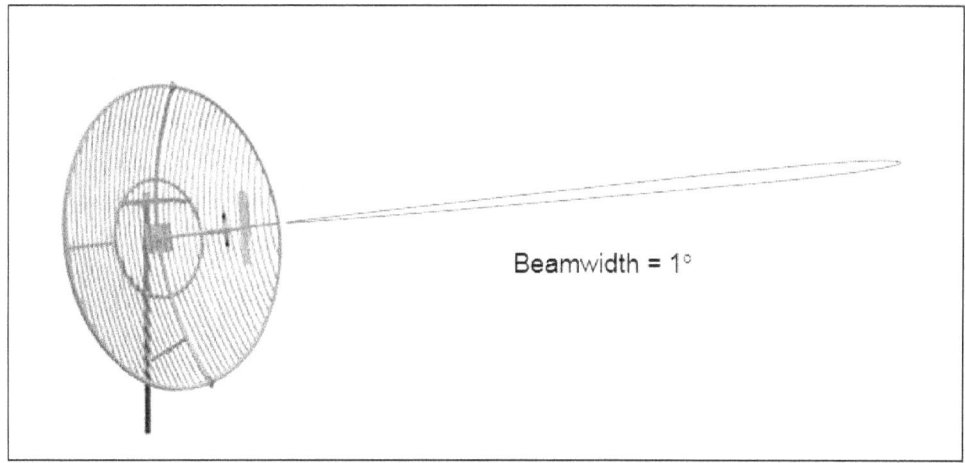

**Figure 6.6.1.b**

## 6.7 Isotropic Radiator

An *isotropic radiator* is a theoretical antenna which uniformly radiates in all directions. It is a point source of *electromagnetic* (EM) waves at the center of a sphere. The field and power pattern of an isotropic radiator has uniform field strength and power density over a spherical surface with isotropic antenna at the center, as shown in **Figure 6.7.a**. The concept of *isotropic radiator* is used as a reference for real antennas.

The gain of an isotropic antenna is 0 dBi. The gain of a real antenna is expressed in comparison with that of an isotropic antenna.

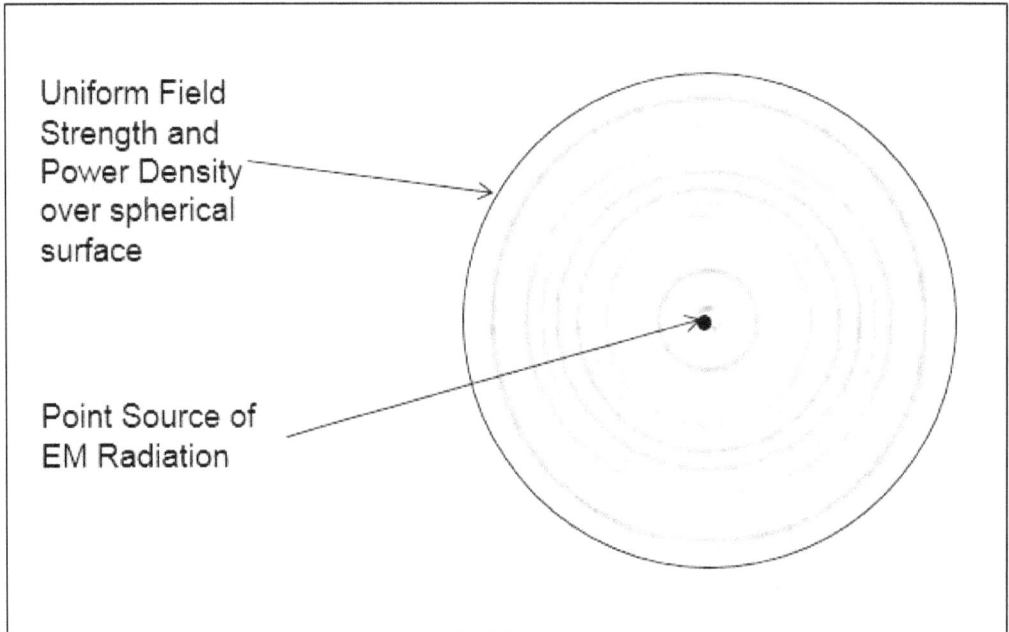

**Figure 6.7.a**

## 6.8 Directivity

*Directivity, D* of an antenna measures its ability to radiate more strongly in one direction compared to others. It is defined as the ratio of maximum power density of the radiation to its power density when averaged over the spherical surface of an equivalent isotropic radiator, as shown in **Figure 6.8.a**.

The value of D = 1 for an isotropic radiator, and D = 1.64 for a half wave dipole antenna.

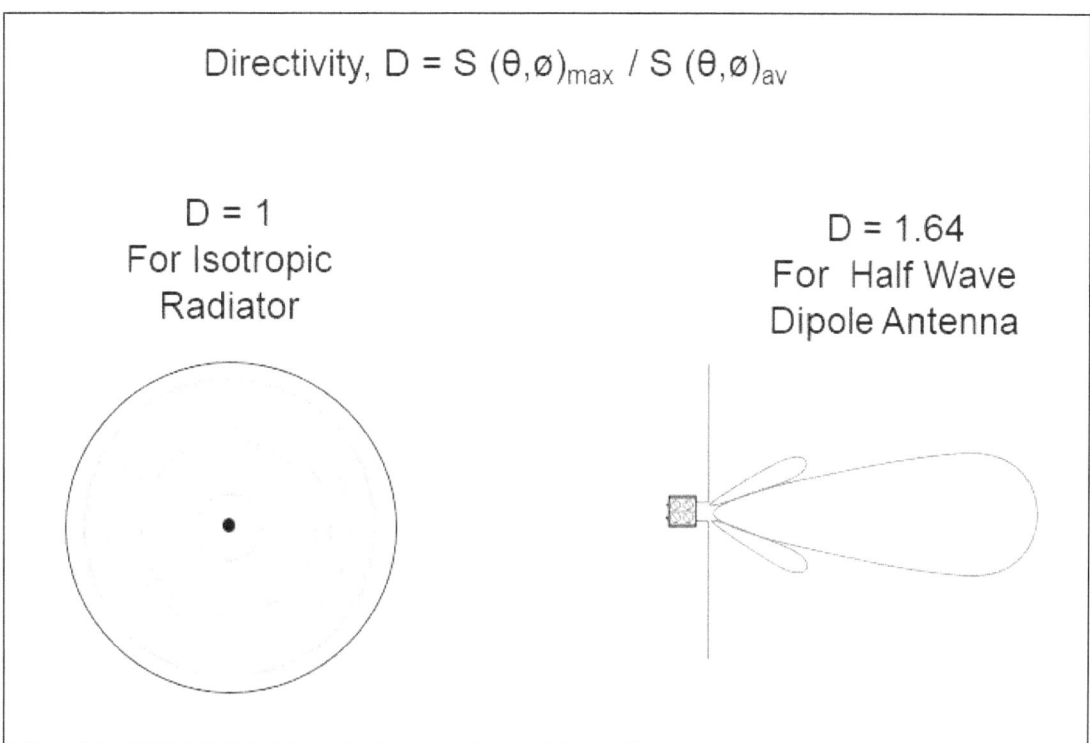

Directivity, D = S $(\theta, \emptyset)_{max}$ / S $(\theta, \emptyset)_{av}$

D = 1
For Isotropic
Radiator

D = 1.64
For Half Wave
Dipole Antenna

**Figure 6.8.a**

# 6.9 Gain

The radiated power of an antenna is less than its input power. This is because of various losses in the antenna feed and the elements of antenna. So, an *antenna efficiency factor*, k is defined, which is the ratio of radiated power to input power. The value of k can be anywhere between 0 and 1.

*Gain, G* of an antenna is defined as k times Directivity, D, or G = k . D. It is obvious that gain takes into account the antenna efficiency, along with directivity, and therefore has more practical value. It is usually expressed as dBi, which is gain in decibels over an isotropic radiator.

**Figure 6.9.a** shows that for isotropic radiator, G = 0 dBi, and for half wave dipole antenna, G = 2.16 dBi. Also shown is an approximate formula to calculate gain from the measured field and power pattern of an antenna. If the HPBW is known for E plane and H plane, the formula shown in the figure can be used to calculate gain with reasonable accuracy.

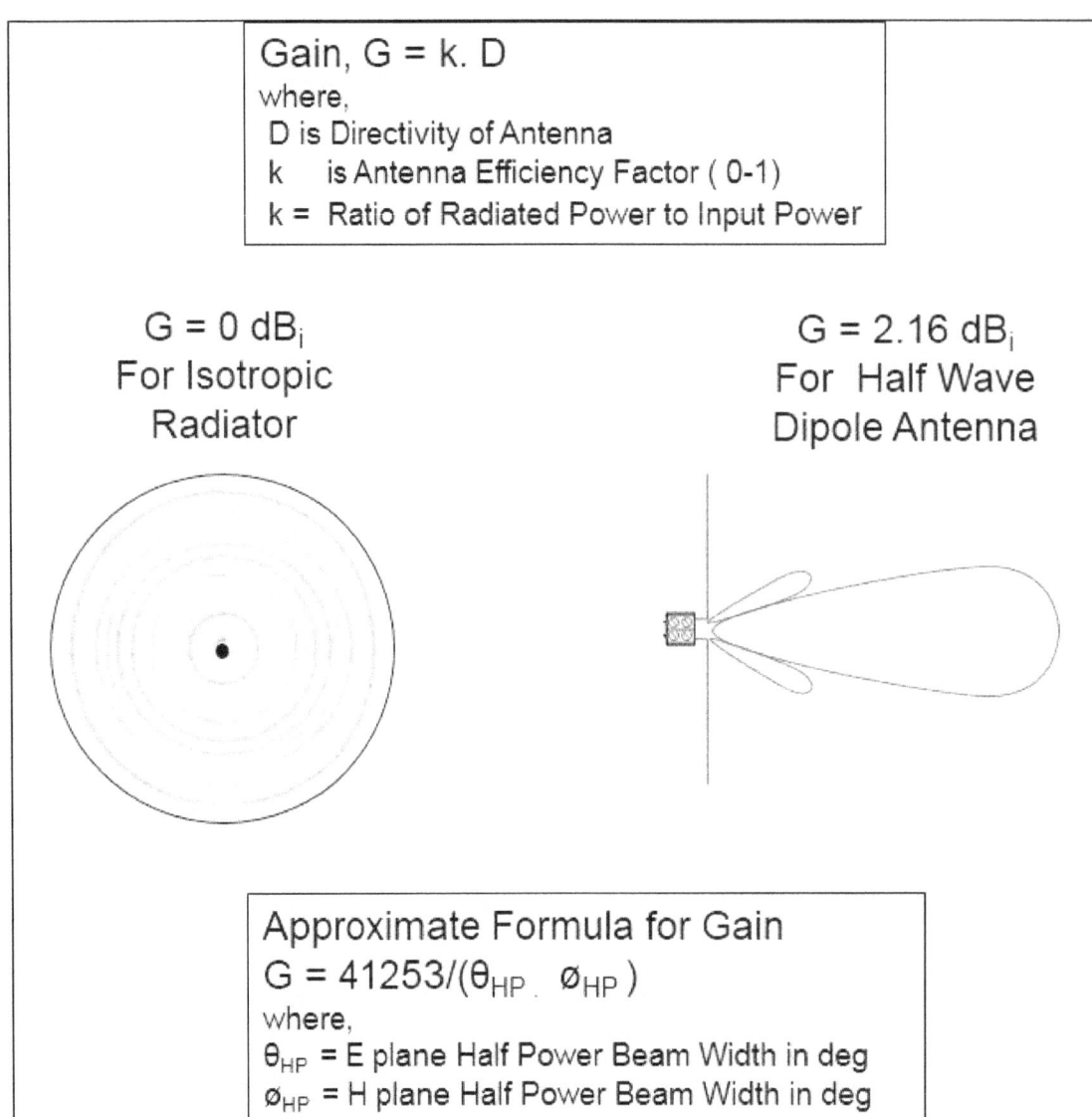

Gain, G = k. D
where,
D is Directivity of Antenna
k    is Antenna Efficiency Factor ( 0-1)
k = Ratio of Radiated Power to Input Power

G = 0 dB$_i$
For Isotropic
Radiator

G = 2.16 dB$_i$
For  Half Wave
Dipole Antenna

Approximate Formula for Gain
G = 41253/($\theta_{HP}$ . $\emptyset_{HP}$ )
where,
$\theta_{HP}$ = E plane Half Power Beam Width in deg
$\emptyset_{HP}$ = H plane Half Power Beam Width in deg

**Figure 6.9.a**

## 6.10 Effective Aperture

The electrical capture area of an antenna is different from its physical area. **Figure 6.10.a** shows radio waves incident on the physical dimensions of a half wave dipole antenna with power density, $P_d$ measured in Watts per square meter, and output power $P_o$ measured in Watts that is received at the output of the antenna.

*Effective Aperture*, $A_e$ is defined as the ratio of received output power from antenna to the incident power density.

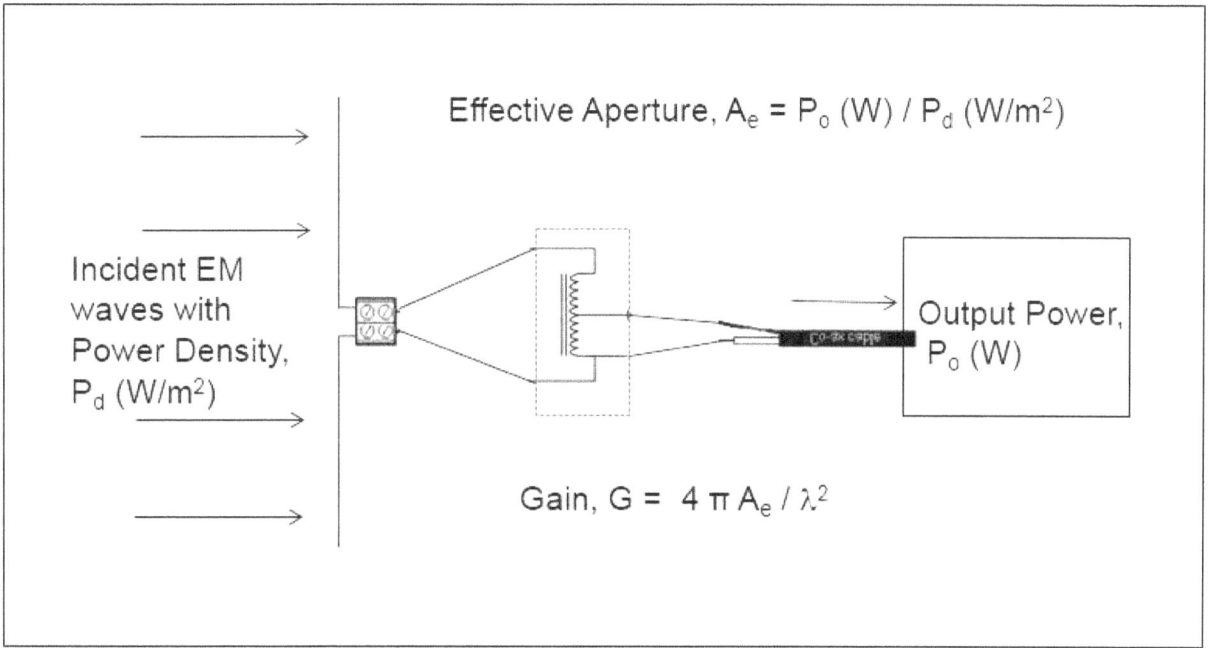

**Figure 6.10.a**

Figure **6.10.a** also shows a relationship between Gain, G of the antenna and its effective aperture. It is obvious that for higher gain of antenna, the effective aperture must be large. Also, the gain, G is inversely proportional to the square of wavelength, λ. This implies that gain of an antenna is higher for smaller wavelengths, or higher frequencies.

## 6.11 Effective Isotropic Radiated Power (EIRP)

*EIRP* is defined as the product of *gain* of antenna and the net power input to the antenna after deducting the transmission losses, as shown in **Figure 6.11.a**. When *EIRP* is expressed in the units of dB, the gain in dBi is added to the net input power in dB. It may be noted that dBi is the gain of antenna with reference to an isotropic radiator.

**Figure 6.11.b** shows an example for the calculation of *EIRP*. The transmitter power output is 1W or 0 dB. It is assumed that the power output from the transmitter suffers a transmission loss of 0.5 W or -3 dB. The power input to the antenna is 0.5 W or -3dB and the antenna gain is given as 34 dBi. Therefore, *EIRP* is calculated as 31 dB.

$$\text{EIRP} = G_T \cdot P_T$$

$$\text{EIRP}_{dB} = G_{T_{dBi}} + P_{T_{dB}}$$

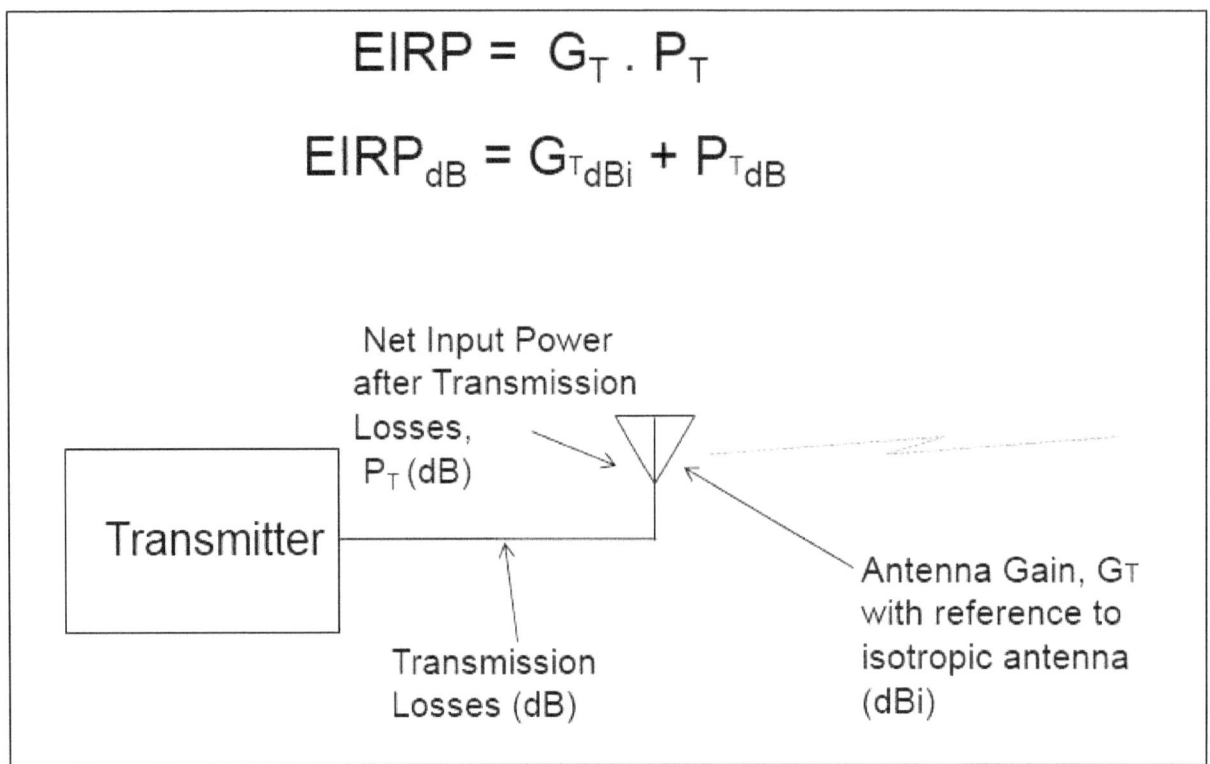

**Figure 6.11.a**

$$\text{EIRP}_{dB} = G_{T_{dBi}} + P_{T_{dB}}$$

$$\text{EIRP}_{dB} = 34 \text{ dBi} - 3 \text{ dB} = +31$$

Net Input Power after
Transmission Losses,
$P_T$ (dB) = 0.5 W ( – 3 dB)

Antenna Gain = 34 dBi

1 W (0 dB)
Transmitter

Transmission Losses = 0.5 W ( – 3 dB )

**Figure 6.11.b**

## 6.12 Effective Radiated Power (ERP)

*ERP* is similar to EIRP, except for one difference. For ERP, the gain of antenna is expressed in dBd, that is with reference to the gain of a dipole antenna, instead of an isotropic radiator. See **Figure 6.12.a** and a calculation example in **Figure 6.12.b**.

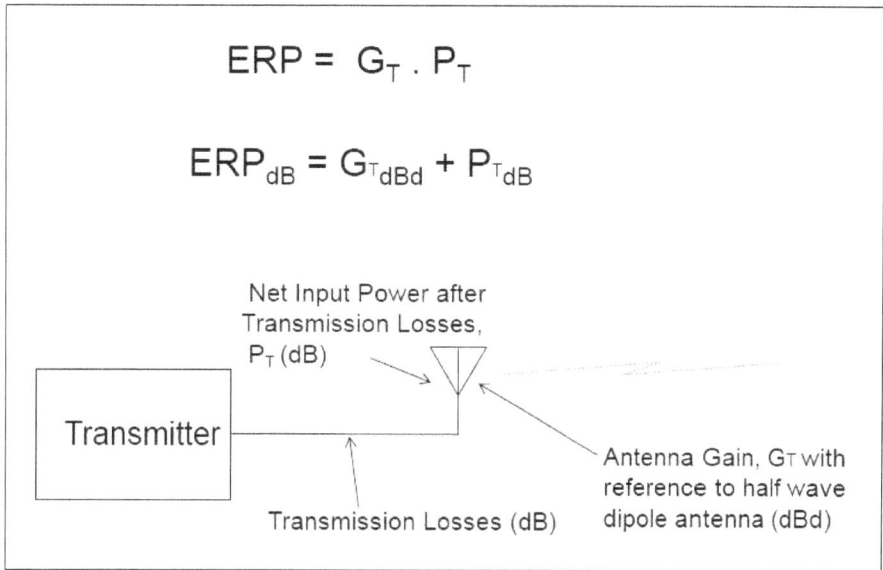

**Figure 6.12.a**

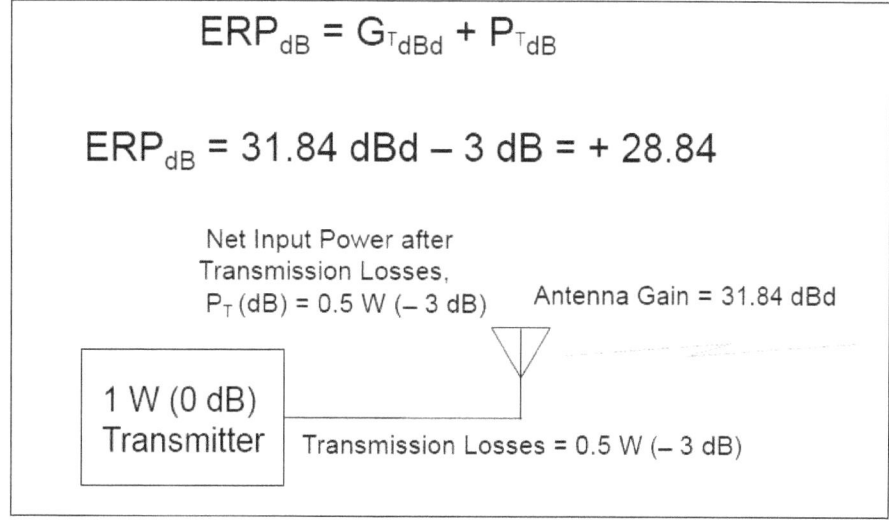

**Figure 6.12.b**

## 6.13 EIRP vs ERP

For *EIRP*, the antenna gain is expressed with reference to an isotropic radiator, and for *ERP*, the antenna gain is expressed with reference to a half wave dipole antenna. However, a half wave dipole antenna has a gain of 2.16 dBi with reference to that of an isotropic radiator. So, *ERP* is always less than *EIRP* by 2.16 dBi, as shown in **Figure 6.13.a**. Many antenna design professionals use the term ERP, because it compares with a real half wave dipole antenna instead of a theoretical isotropic radiator. It is important to understand the difference between *ERP* and *EIRP* for correct antenna gain calculations.

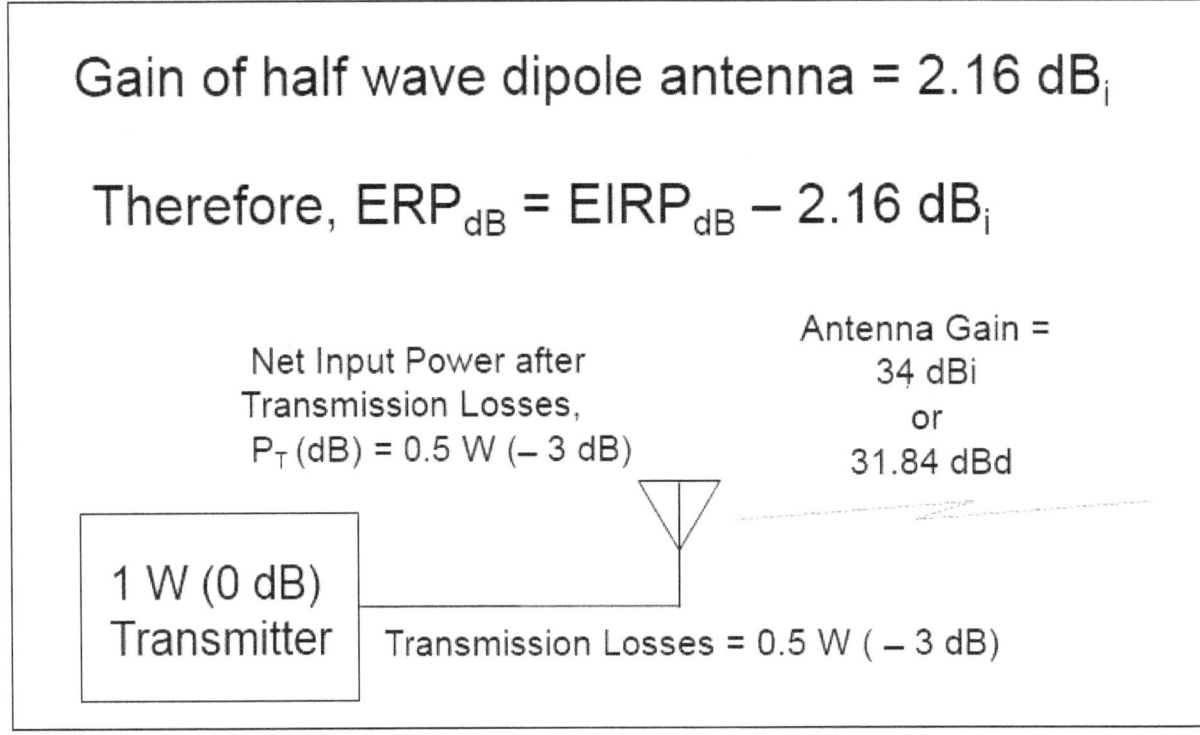

**Figure 6.13.a**

## 6.14 Polarization

*Polarization* of an antenna refers to the orientation of electric field, E in the far field region of the antenna. It is primarily of three types, *linear*, *circular* and *elliptical*.

**Figure 6.14.a** shows *linear polarization (LP) vertical* and Figure **6.14.b** shows *linear polarization (LP) horizontal*. Notice that the orientation of E-field is parallel to the antenna element, through which the RF electric current flows. *LP vertical* and *LP horizontal* are said to be *cross pol* of each other, as shown in **Figure 6.14.c**.

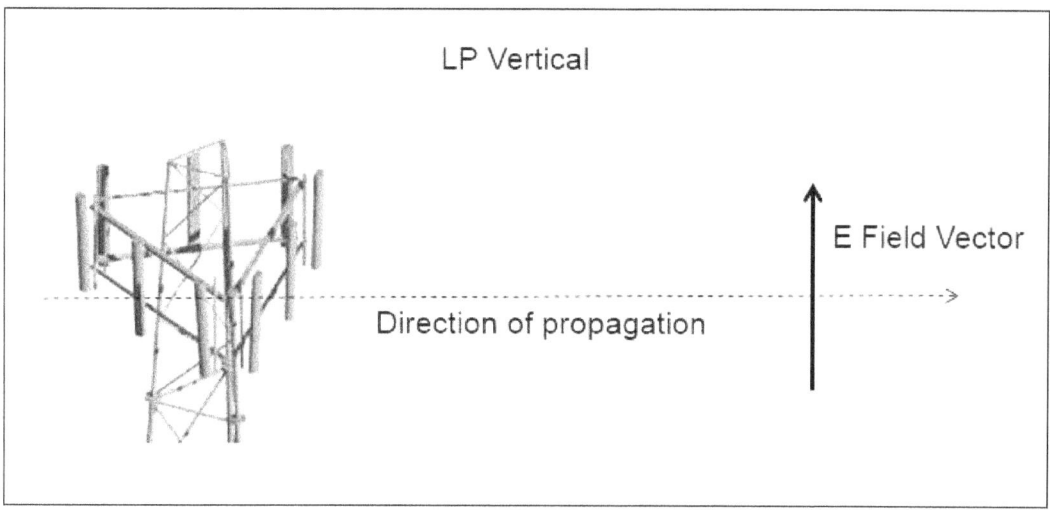

**Figure 6.14.a**

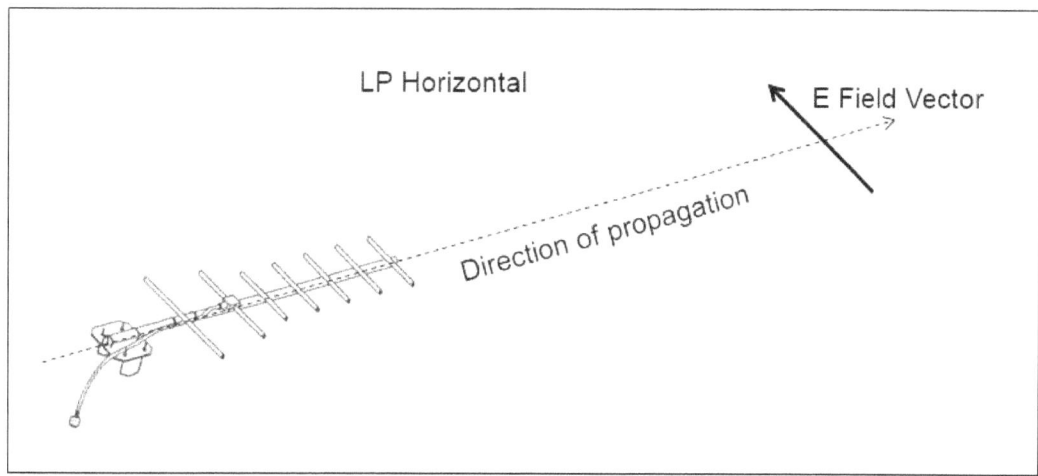

**Figure 6.14.b**

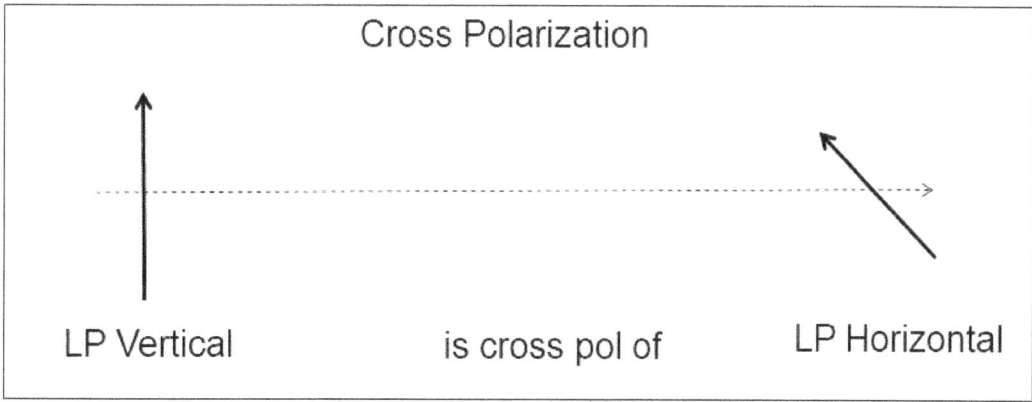

**Figure 6.14.c**

**Figure 6.14.d** shows *right hand circular polarization (RHCP),* in which the electric field vector rotates clockwise in a circle in the direction of propagation. The antenna consists of vertical and horizontal elements that transmit a rotating electric field. **Figure 6.14.e** shows *left hand circular polarization (LHCP)*, in which the electric field rotates counterclockwise in a circle in the direction of propagation.

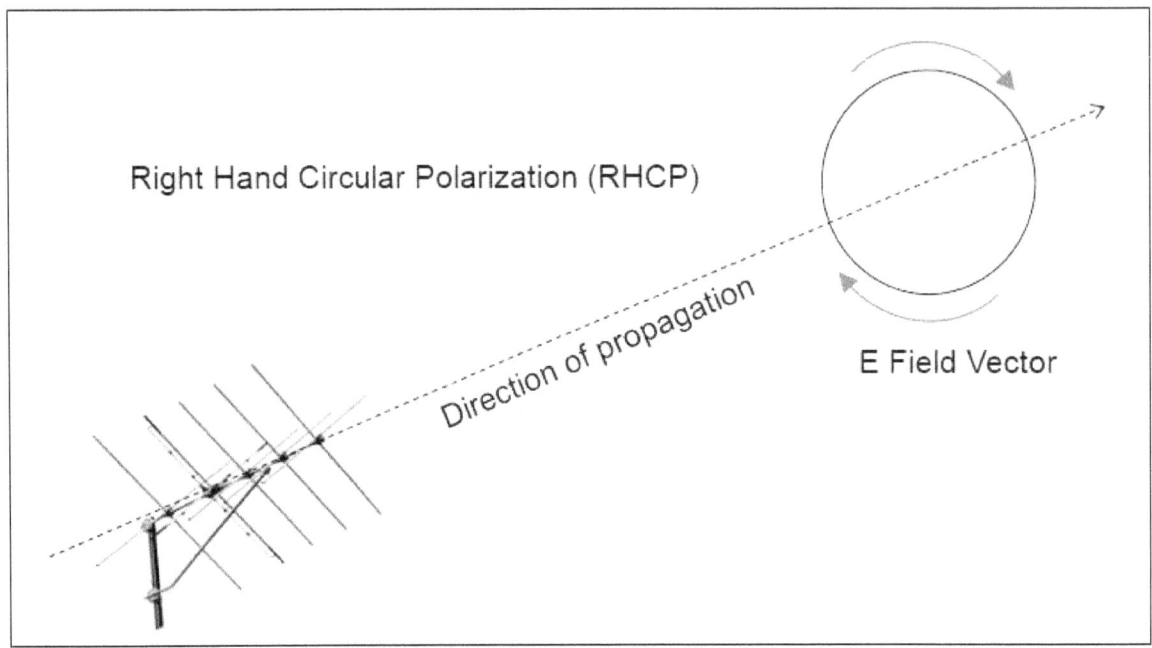

**Figure 6.14.d**

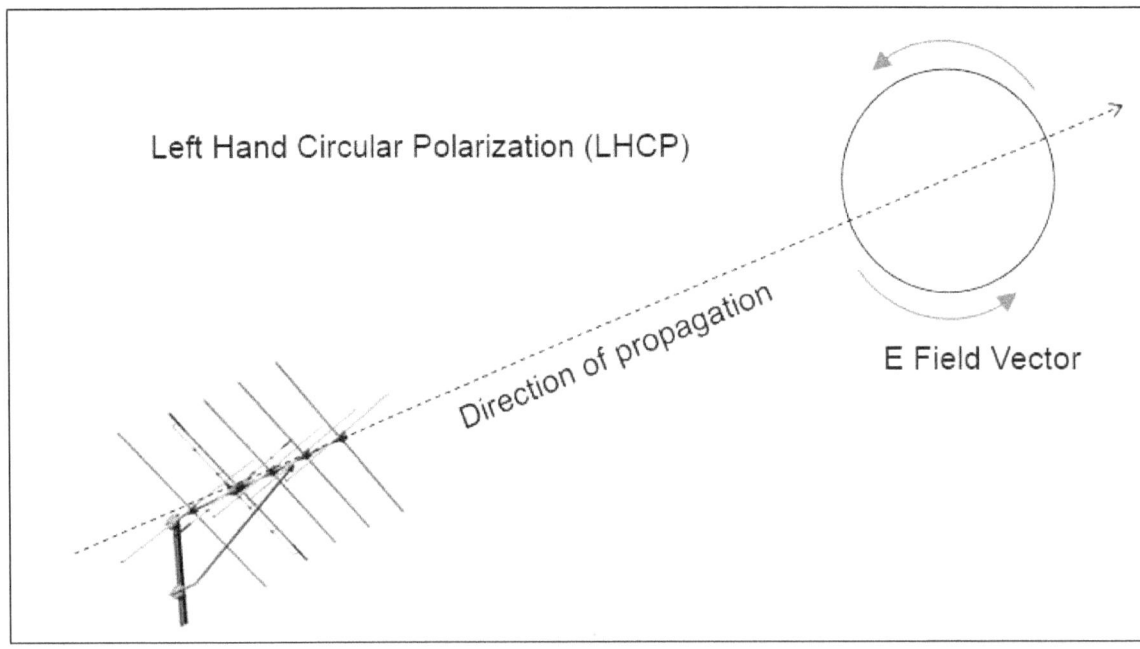

**Figure 6.14.e**

**Figure 6.14.f** shows *right hand elliptical polarization (RHEP)*, in which the electric field vector rotates clockwise in an ellipse in the direction of propagation. **Figure 6.14.g** shows *left* *hand elliptical polarization (LHEP)*, in which the electric field vector rotates counterclockwise in an ellipse in the direction of propagation.

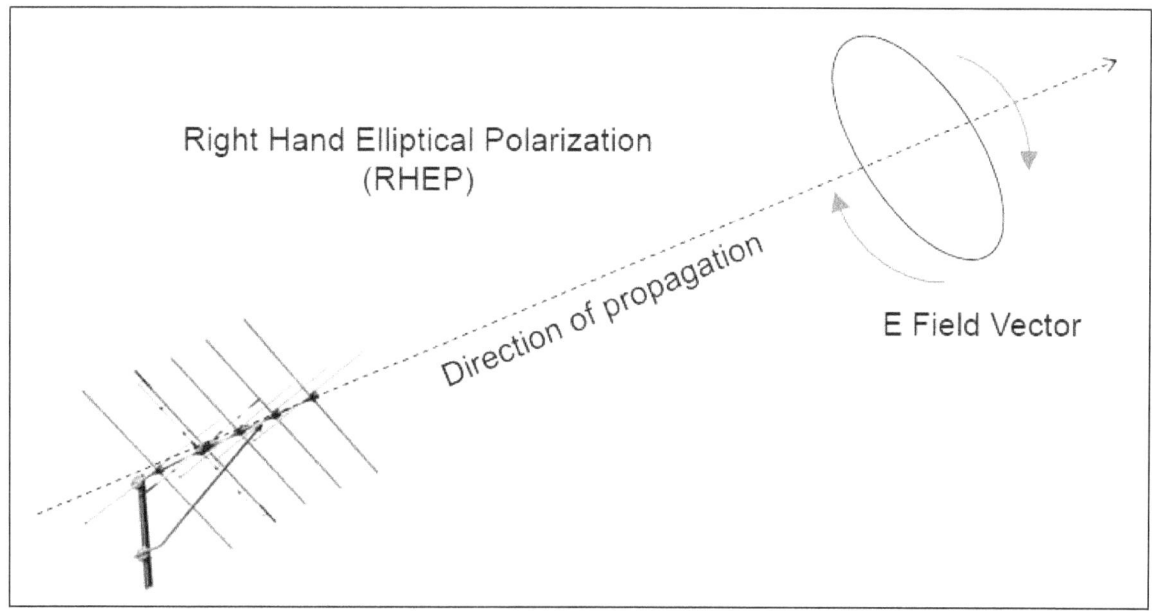

**Figure 6.14.f**

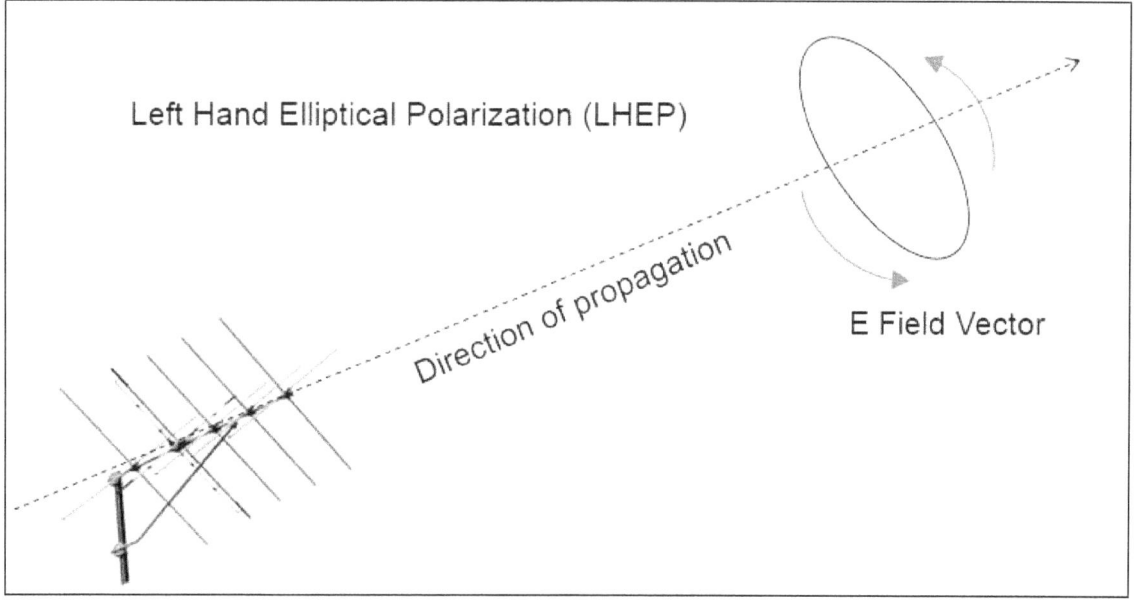

**Figure 6.14.g**

**Figure 6.14.h** shows the *cross pol* relationship between RHCP and LHCP, and between RHEP and LHEP.

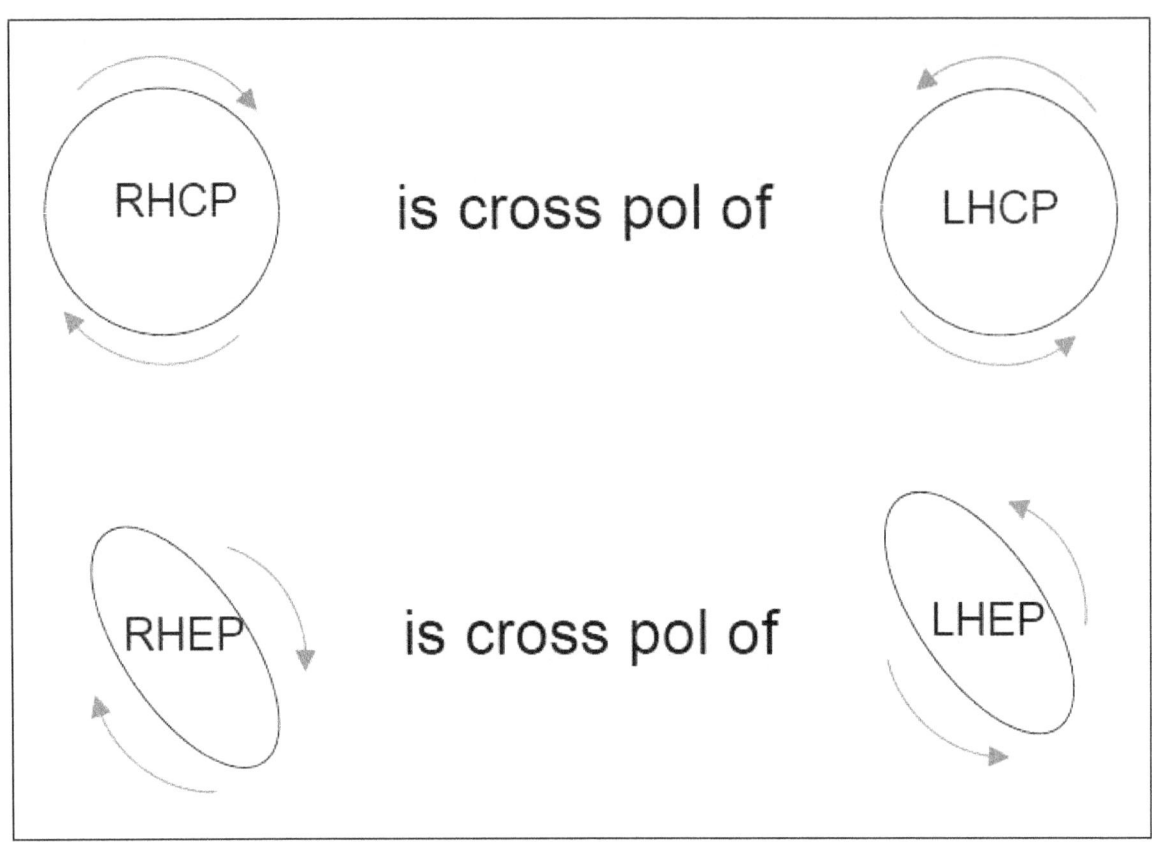

**Figure 6.14.h**

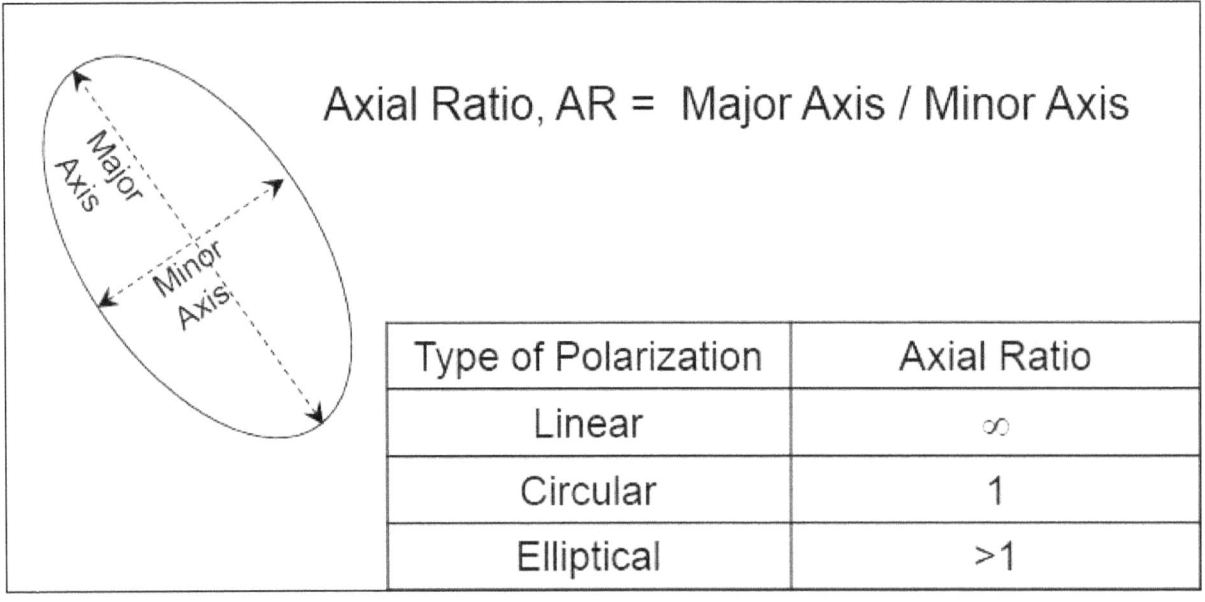

**Figure 6.14.i**

The *axial ratio (AR)*, is a measure of polarization and is defined as the ratio of major axis to minor axis of the ellipse, as shown in **Figure 6.14.i**. For *linear polarization*, the minor axis is

0, so AR is infinity. For *circular polarization*, the major and minor axes are of equal length, so AR is 1. For *elliptical polarization*, AR is greater than 1.

A *polarization mismatch* occurs when the electrical elements of transmitting and receiving antennas are not aligned to the same types of polarization. We will now consider three cases here.

**Figure 6.14.j** shows Case 1 for *polarization mismatch*. The transmitting antenna is LP horizontal whereas receiving antenna is LP vertical. Due to this incompatible alignment, no signal is received.

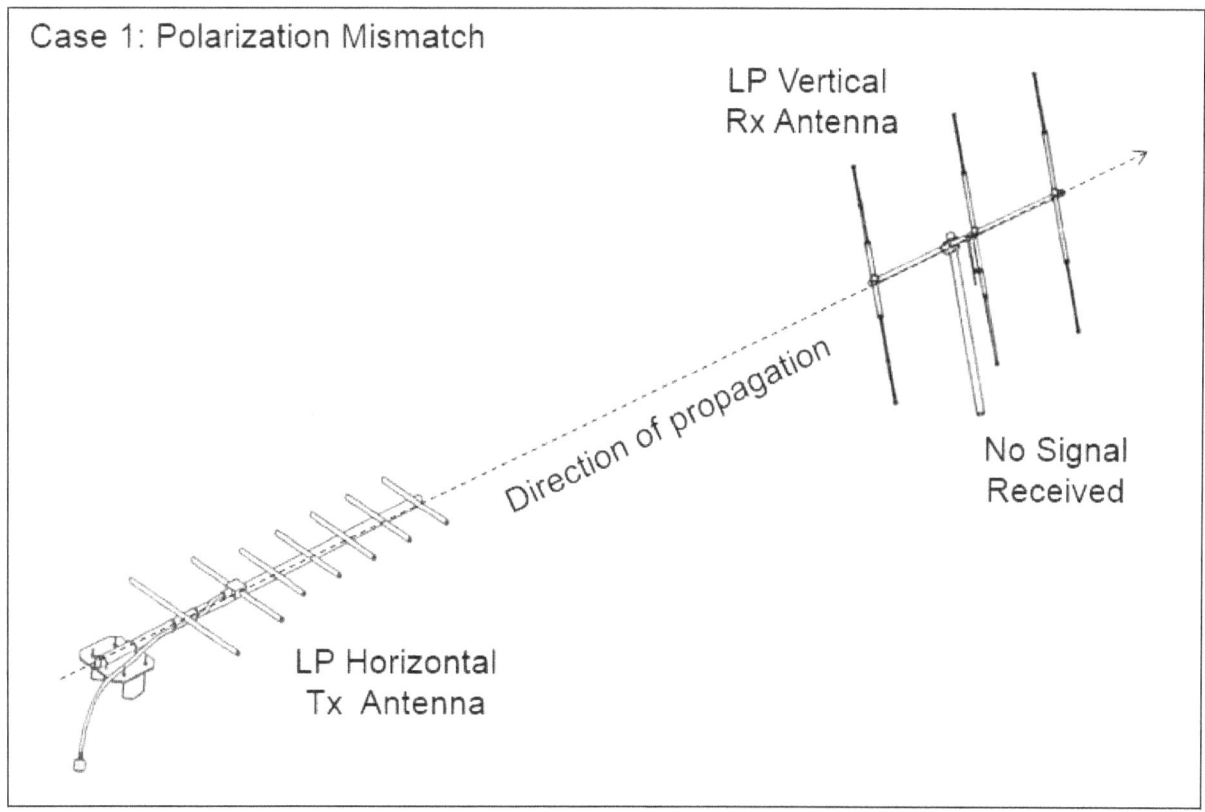

**Figure 6.14.j**

**Figure 6.14.k** shows Case 2 for *polarization match*, where the transmitting as well as receiving antenna are LP horizontal. Therefore, strongest signal is received in this alignment.

**Figure 6.14.l** shows Case 3 for *partial polarization match*. In this case the transmitting antenna uses *circular polarization* (CP), and it transmits vertical as well as horizontal components of E field vector. However, the receiving antenna is LP vertical and can only receive the vertical component of the E field vector transmitted by the CP antenna. Therefore, in this case, the received power is reduced by half or 3 dB.

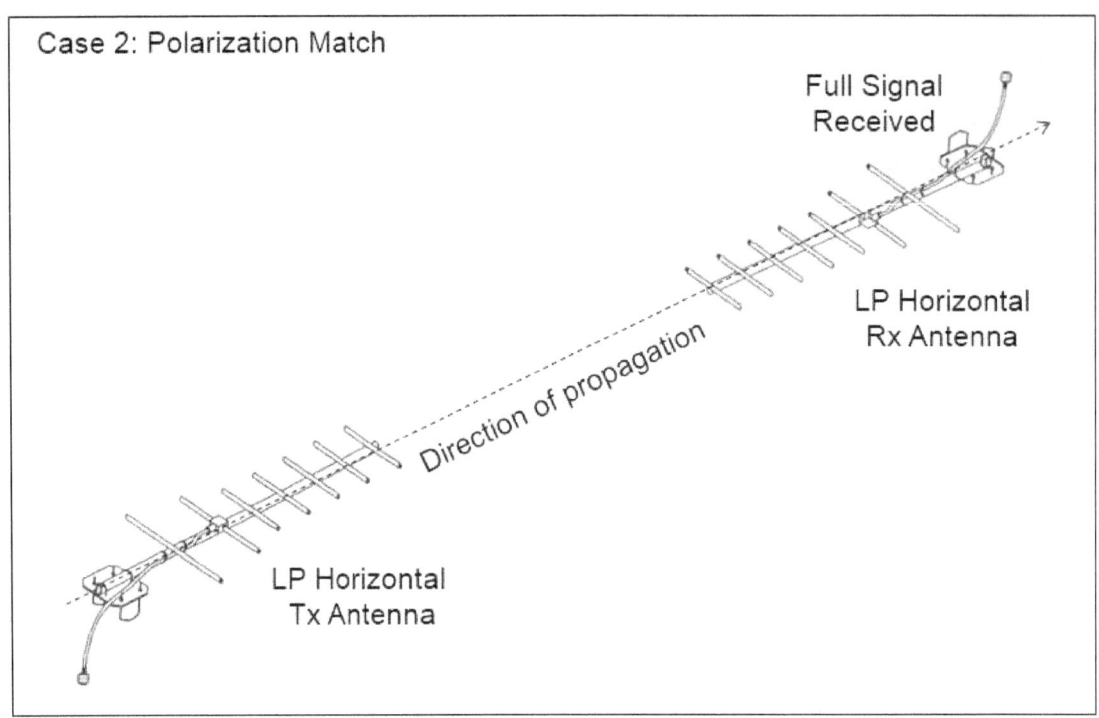

**Figure 6.14.k**

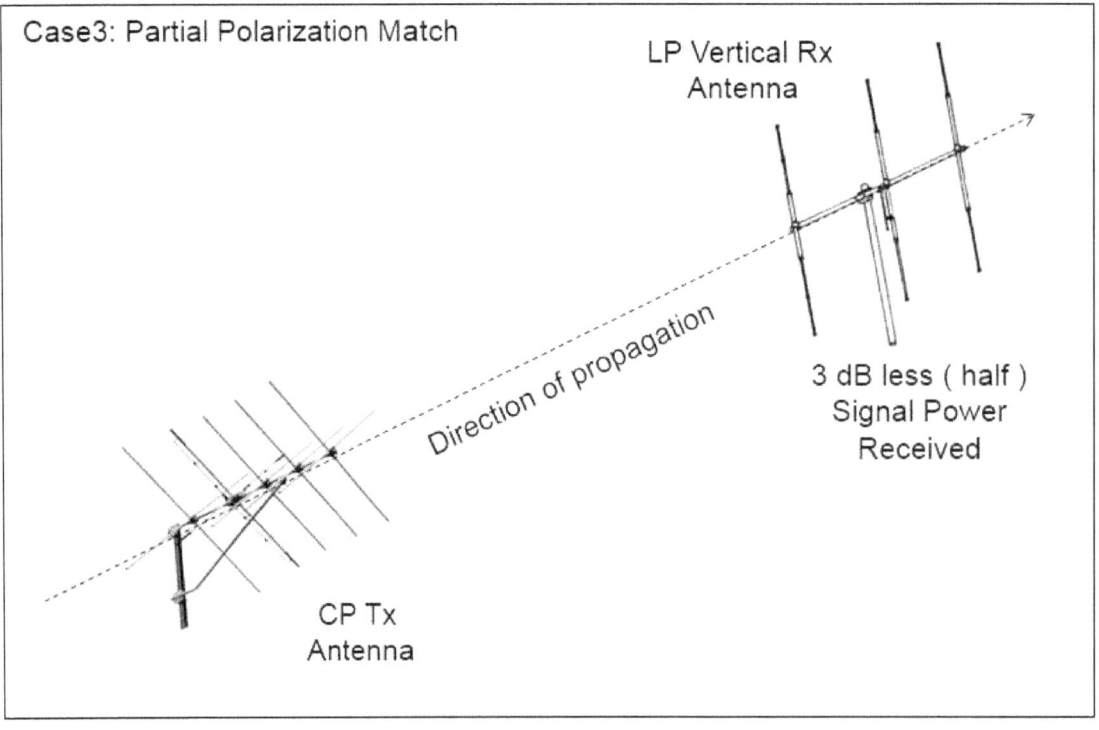

**Figure 6.14.l**

# 7.  Antenna Types

Antennas can be classified in several different ways. We can classify them based on bandwidth, such as narrowband and wideband, or by their electrical size, such as large aperture or small aperture, or by their physical size, or even by their mode of operation, such as travelling wave or guided wave. The most common way to classify antennas is by their physical design. In this chapter we will get familiar with various antenna types based on their physical design, especially, *wire antennas, aperture antennas and fractal antennas*

## 7.1  Wire Antennas

*Wire antennas* come in different shapes and sizes. These are in general suitable for lower gain applications, from 2 – 3 dBi and are narrowband. **Figure 7.1.a** shows a *square loop* and a *circular loop*. This type of wire antenna fits very well inside the radio equipment.

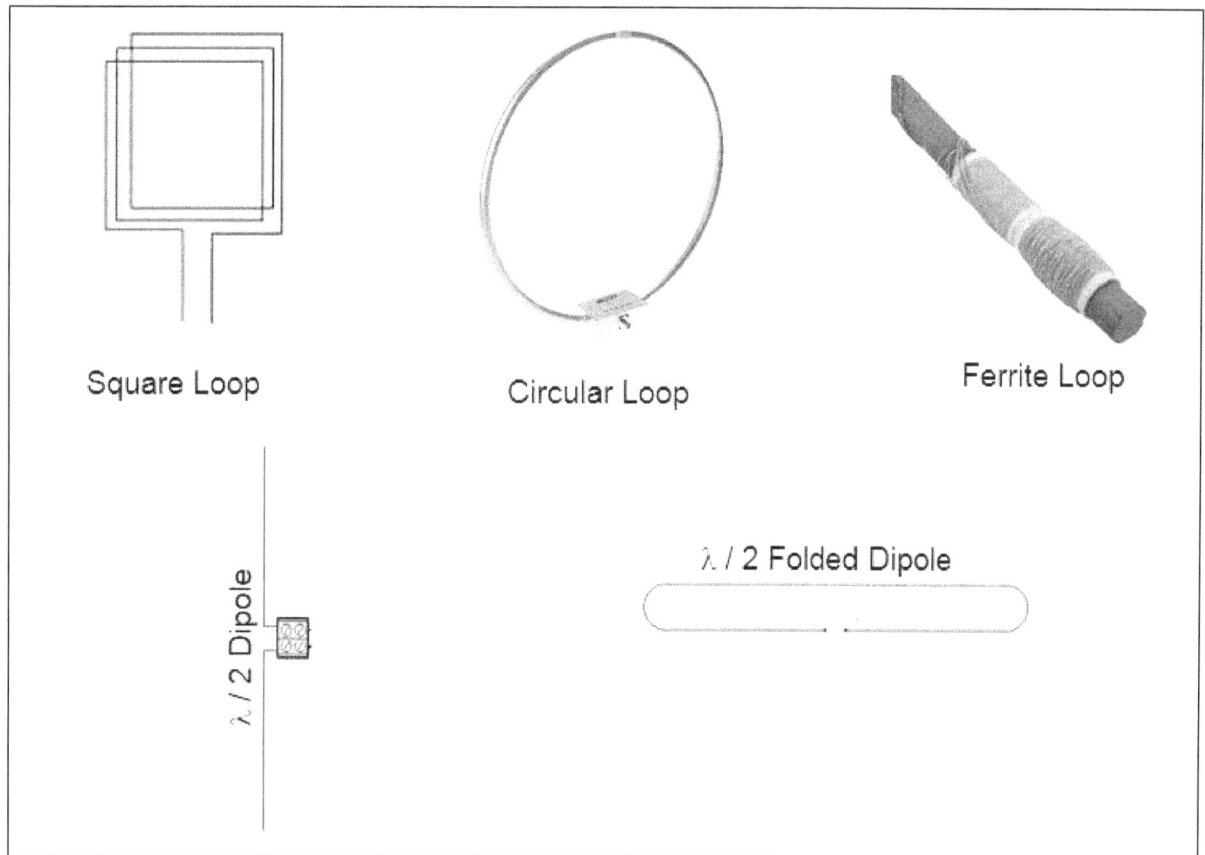

**Figure 7.1.a**

A *ferrite loop* consists of a coil of wire wound around an adjustable core of ferrite material. This type of antenna is common in conventional AM/FM radio receivers. *Half wave dipole* and *folded dipole* antennas are used in VHF/UHF radio receivers. They can be seen in residential neighborhoods, and are used for broadcast TV reception. The utility companies also use them for wireless communication with power meters and other equipment.

**Figure 7.1.b** shows *long wire, twin lead and rhombic antennas*, which are used for MF (Band 6) and HF (Band 7) frequency bands that have longer wavelength. Also shown is a *helical antenna* in the shape of a spiral and a reflector, which has moderate gain from 8 – 15 dBi.

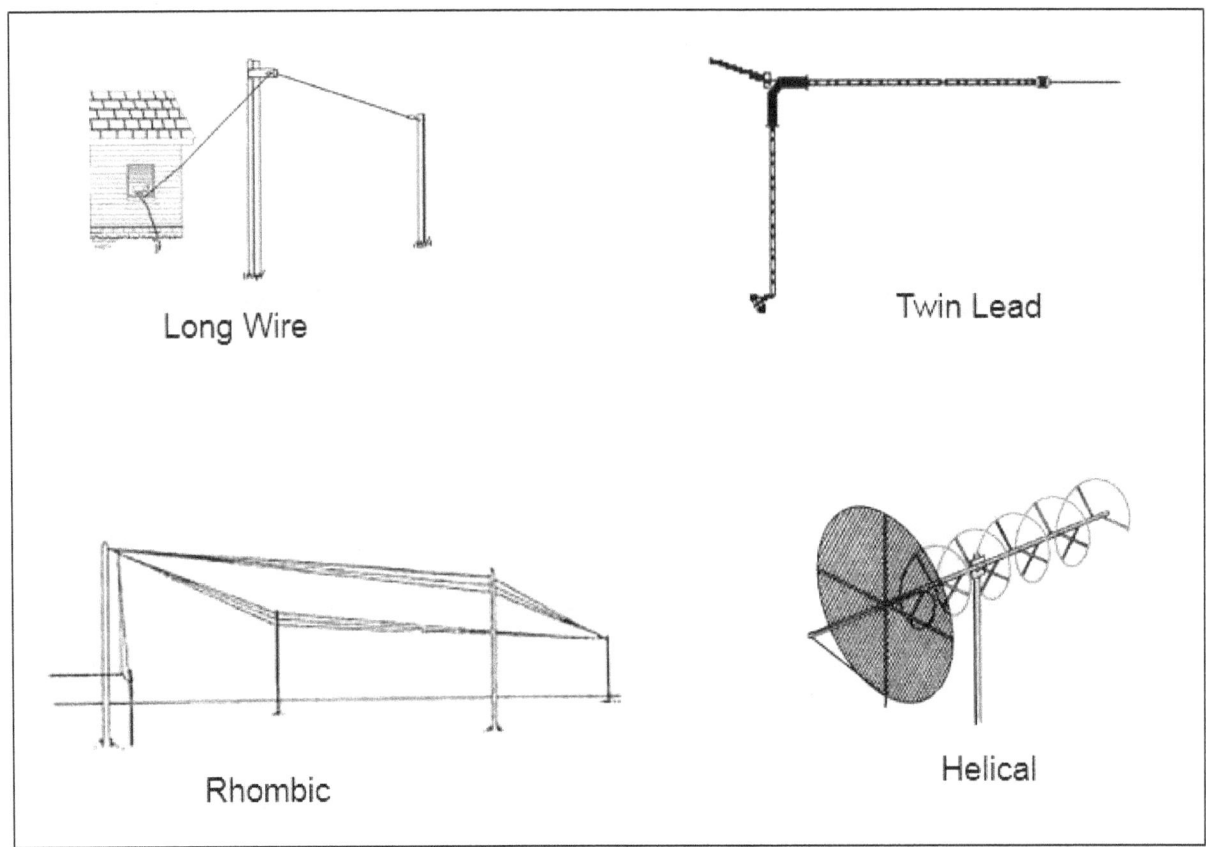

**Figure 7.1.b**

**Figure 7.1.c** shows wire antennas used in automobiles and portable devices, telescopic antennas and a Yagi Uda antenna. Yagi Uda antenna is a multi element antenna with a half wave dipole, a reflector element and several director elements. Because of their broad bandwidth, Yagi antennas are widely used for broadcast TV signal reception in VHF (Band 8) and UHF (Band 9) bands.

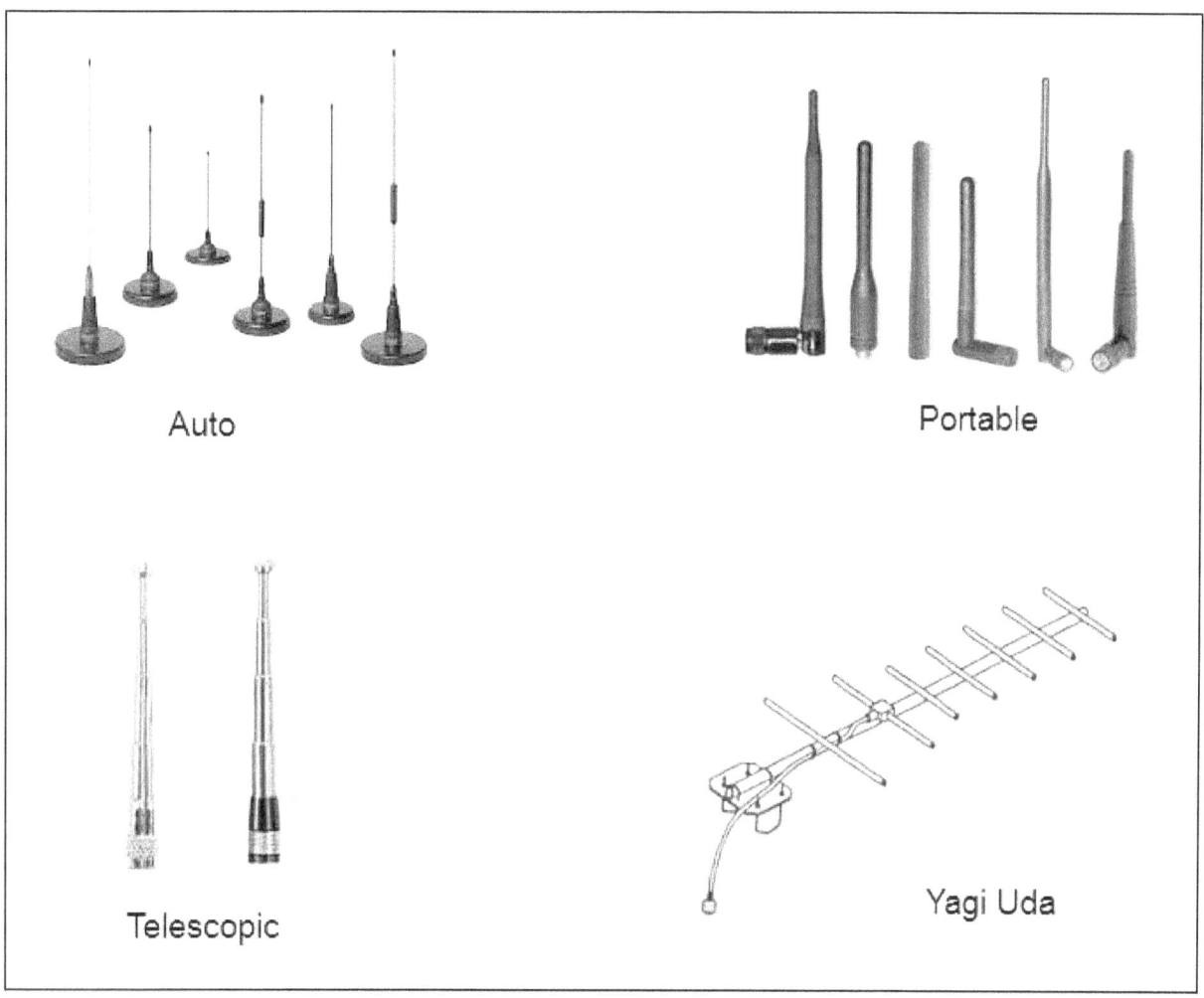

**Figure 7.1.c**

**Figure 7.1.d** shows a *log periodic dipole array (LPDA)*, which is one of a family of frequency independent antennas. These antennas also have broad bandwidth. The structure of LPDA consists of several linear elements. The longest element is half wavelength long at the lowest design frequency, and the shortest element is half wavelength long at a frequency much higher than the maximum operating frequency. LPDA antennas are used for frequency range of 14 – 30 MHz, which lies in the HF (Band 7) band.

**Figure 7.1.e** shows wire antennas made in the shape of a *rectangular spiral* and a *circular spiral*.

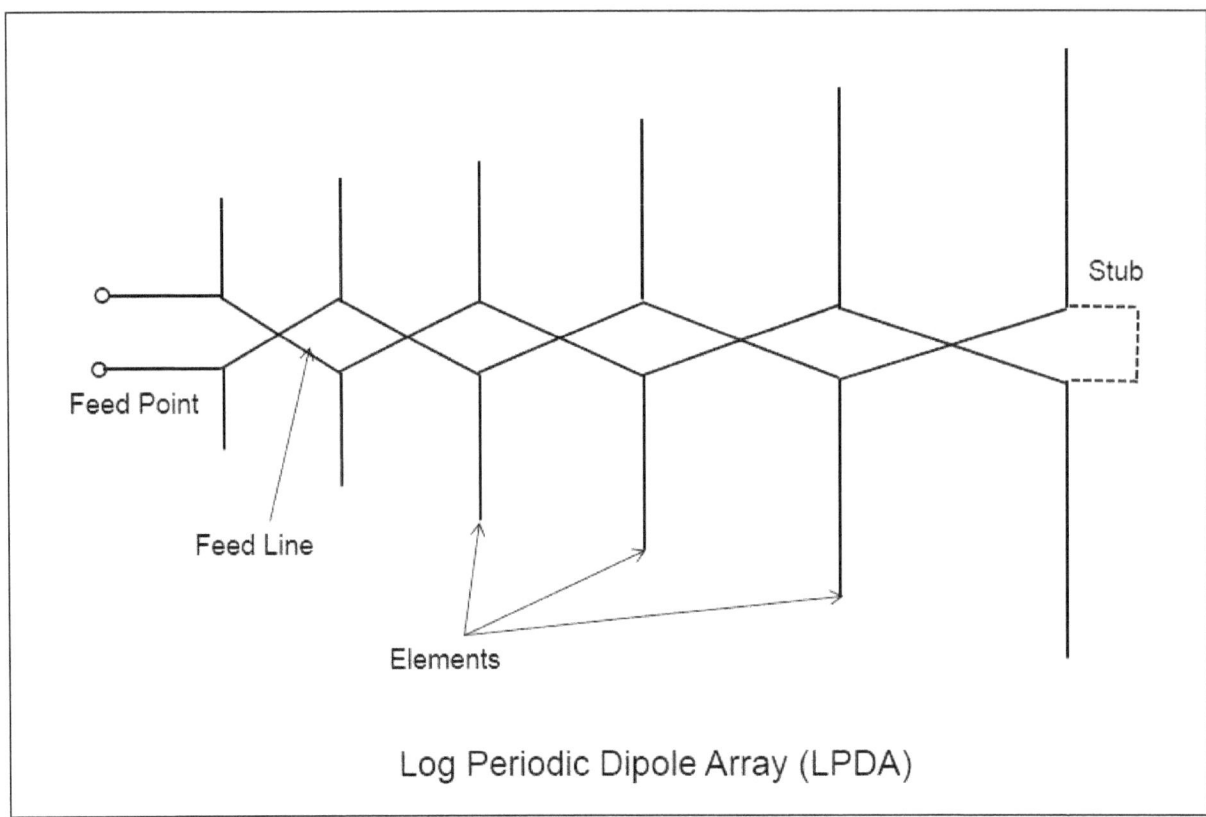

Log Periodic Dipole Array (LPDA)

**Figure 7.1.d**

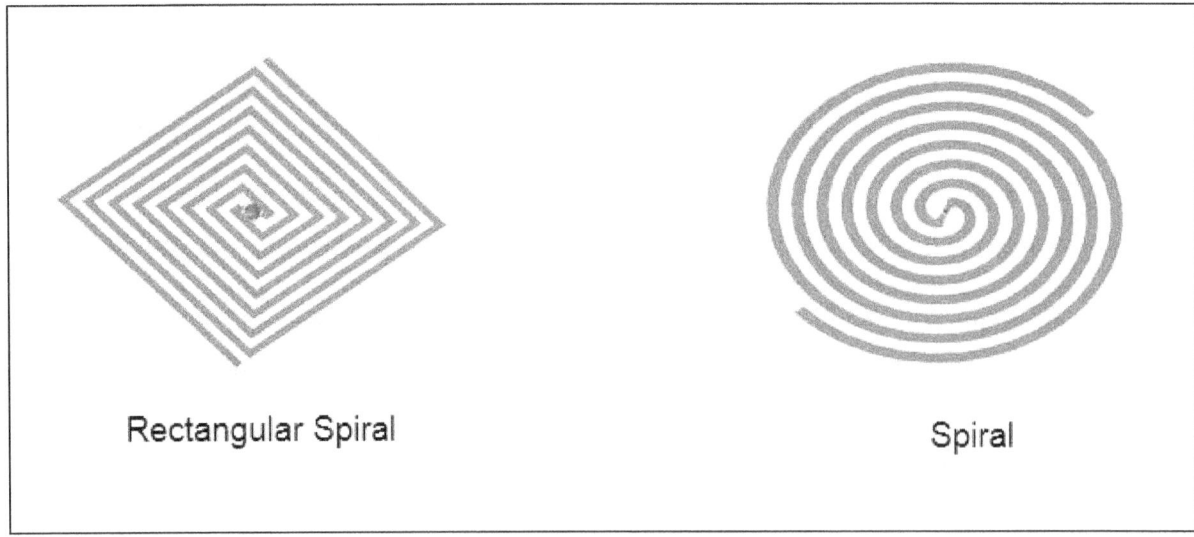

Rectangular Spiral                    Spiral

**Figure 7.1.e**

## 7.2  Aperture Antennas

*Aperture antennas* are highly directional antennas with very narrow beam widths, typically few degrees. They have moderate to high gain, typically from 3 dBi – 20 dBi, and find applications in RADAR, satellite and space communications, microwave relays and astronomy.

*Aperture antennas* come in several different shapes and sizes. **Figure 7.2.a** shows aperture antennas in the shape of a parabolic dish, a grid reflector, a horn, or a reflector with multiple feed horns that are fed by microwave waveguides.

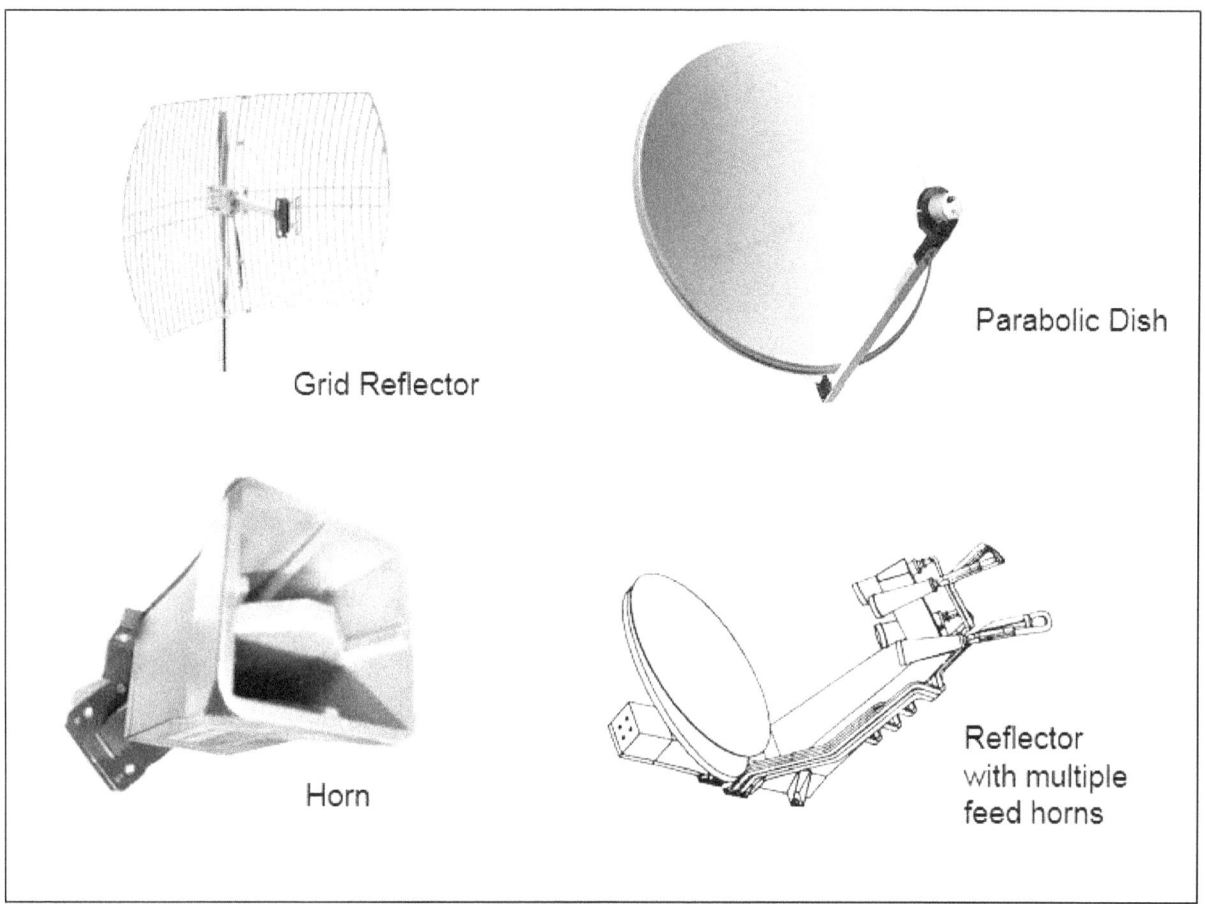

**Figure 7.2.a**

**Figure 7.2.b** shows another aperture antenna with hyperboloid reflector in the front and a parabolic reflector at the back. The arrows show the direction of propagation of reflected EM waves.

**Figure 7.2.c** shows a conical log spiral antenna and a flat panel antenna. The flat panel antennas are widely used for broadcast TV reception and WLAN applications. These can be used indoors as well as outdoors.

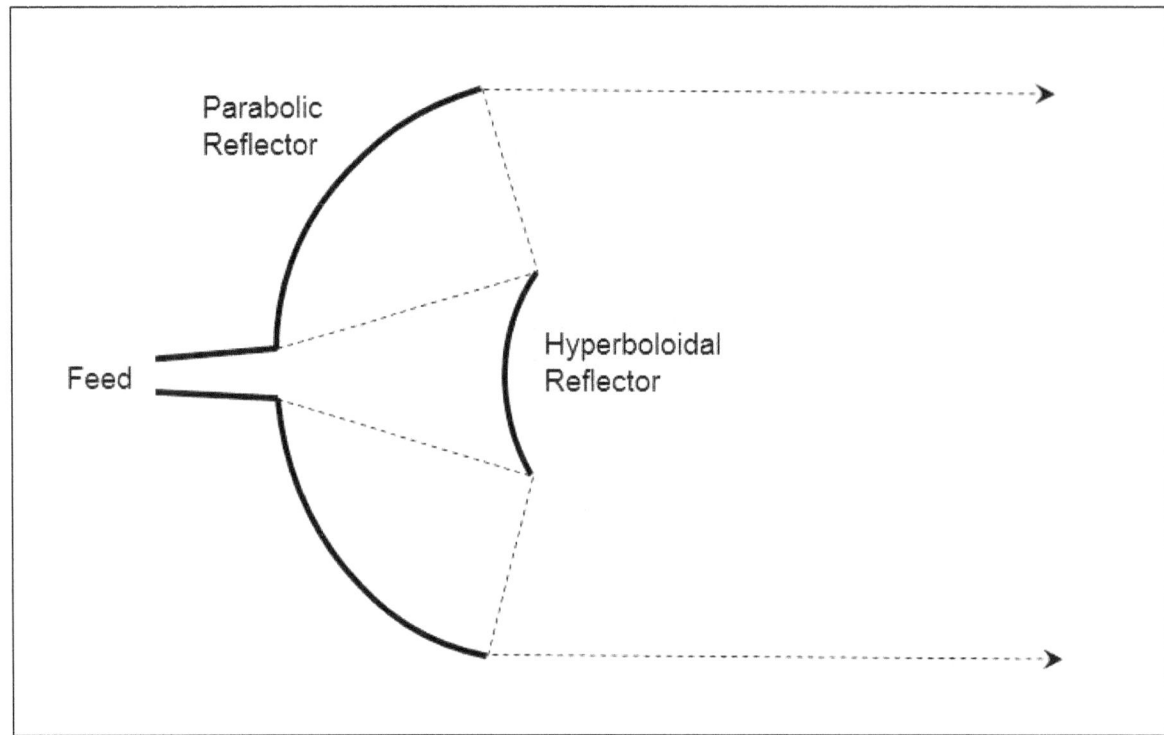

**Figure 7.2.b**

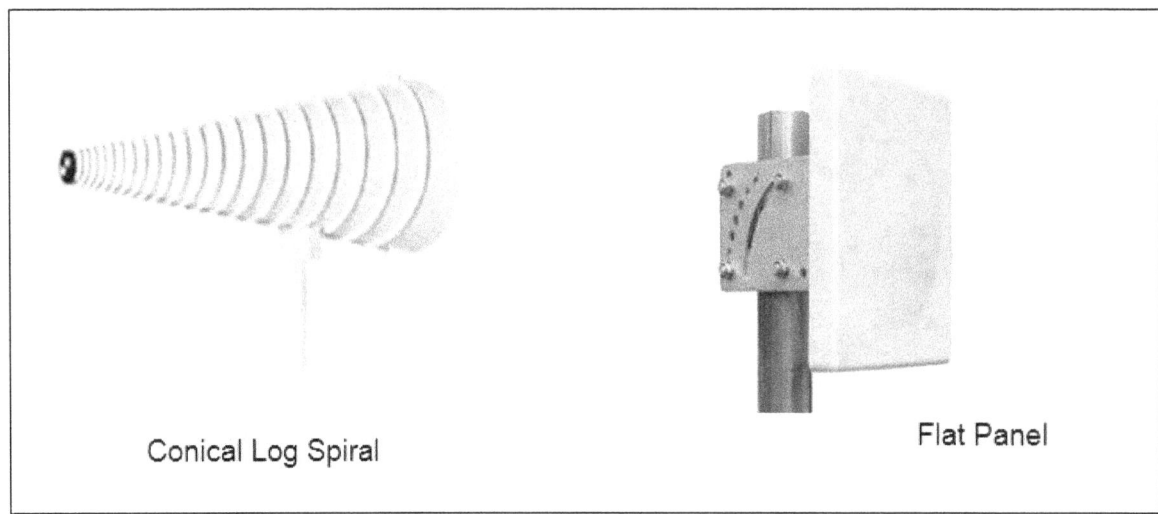

**Figure 7.2.c**

**Figure 7.2.d** shows a *waveguide microwave lens*, which follows the principle of optical lens. Dielectric lenses or metal plate lenses are typically used to refract the EM waves. A point at the focus of the lens produces a plane EM wave on the other side of the lens. *Polyethylene*, *polystyrene*, *plexiglass* and *teflon* are suitable materials for making small microwave lenses, which have low losses. A converging lens is thicker in the middle than the outer edges, just like an optical lens.

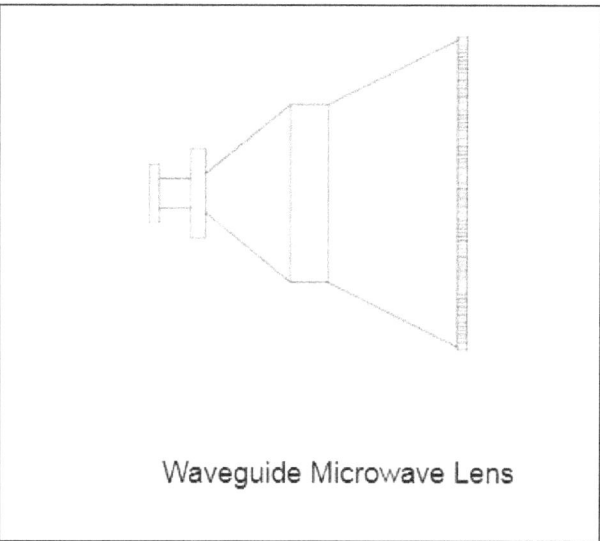

**Figure 7.2.d**

## 7.3  Fractal Antennas

**Figure 7.3.a** shows a different class of antennas called *fractal antennas*. These antennas are inspired by fractal geometry. The antennas are constructed from repetitive geometries. These antennas are very long, but physically they are much smaller, and can be fabricated on surfaces with small dimensions, such as portable electronic devices. Fractal antennas are excellent for multiband and broadband applications.

**Figure 7.3.a**

# 8. Phased Array Antennas

In this chapter, we will study *phased array antennas*, which are essentially arrays of antennas, either a linear array or a 2-dimensional array. The main benefit of a phased array antenna is that the radio beam can be steered electrically in the desired direction without physically moving the antenna. These antennas enable targeted wireless communication with users that are co-located in a geographical region. A phased array antenna can also be used for radio scan of the surrounding area, and therefore, it finds applications in weather and surveillance RADARS too. A 2 dimensional flat panel; antenna array is shown in **Figure 8.a**.

**Figure 8.b** shows a vertical linear phased array with a radio beam along the horizontal axis of the array. **Figure 8.c** shows the radio beam steered in the upward direction and **Figure 8.d** shows the radio beam steered in the downward direction.

16 Element Flat Panel

**Figure 8.a**

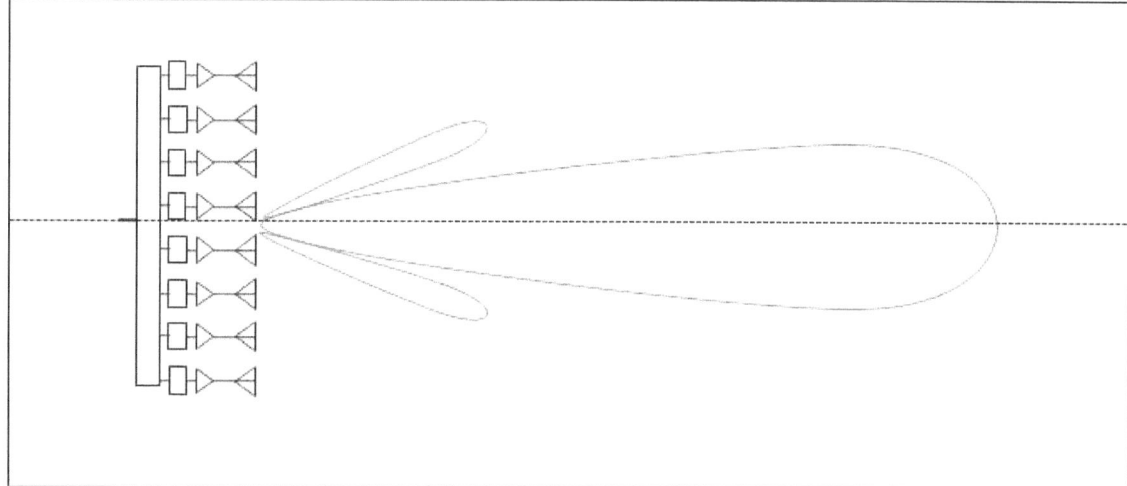

**Figure 8.b**

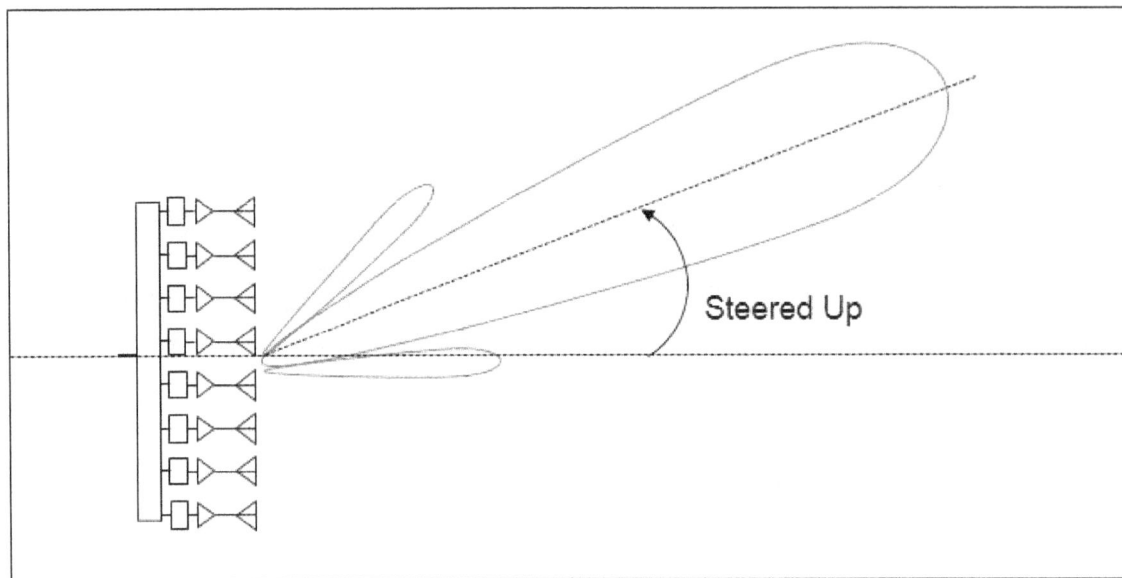

**Figure 8.c**

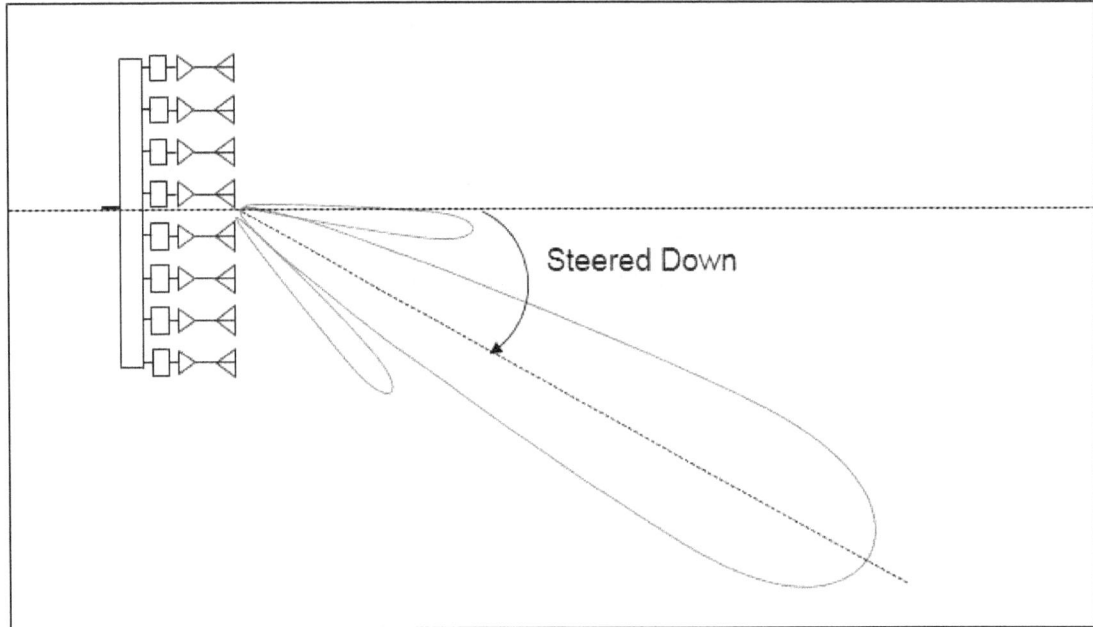

**Figure 8.d**

# 8.1  Construction

**Figure 8.1.a** shows the construction of a *linear phased array antenna.*

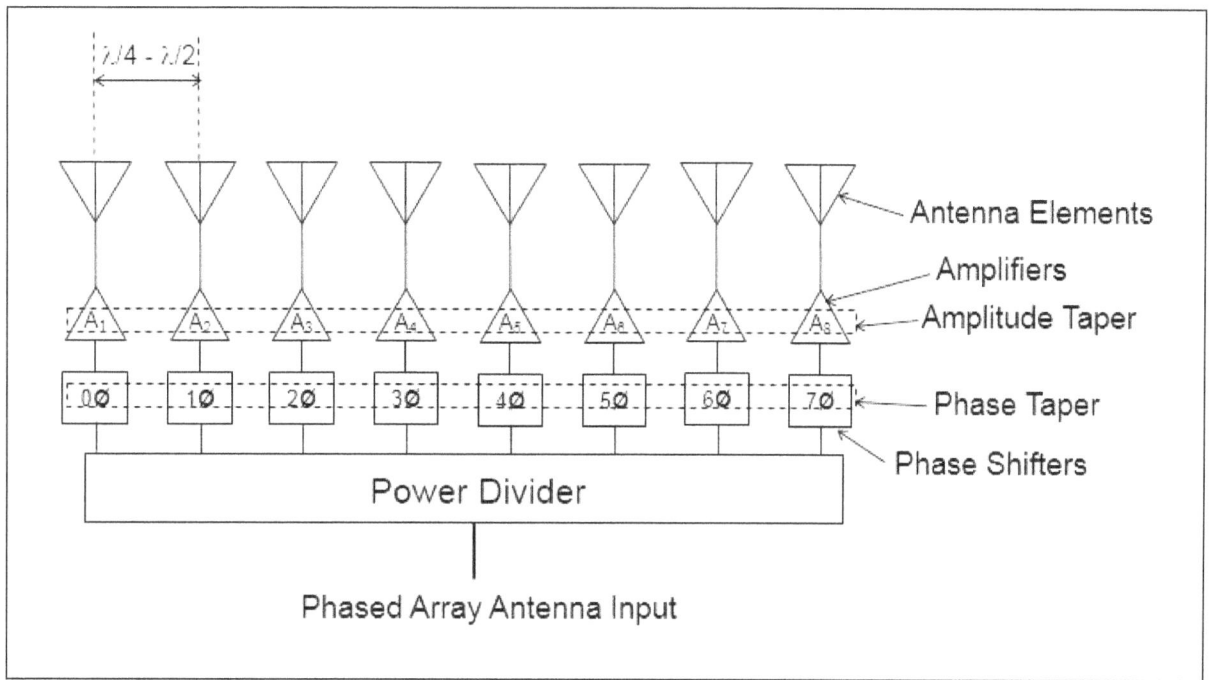

**Figure 8.1.a**

The elements of a phased array antenna are typically placed quarter to half wavelength apart. This is required to avoid mutual coupling between antenna elements in their *reactive near field.* A taper in amplitude and phase of the RF signal input to each element is introduced successively to achieve the desired antenna pattern in the far field of antenna and also to electrically steer the radio beam in the desired direction.

Notice that in **Figure 8.1.a**, each antenna element is fed through an amplifier and a phase shifter connected in tandem. A1, A2,...A8 represent the amplitude taper and φ1, φ2...φ8 represent the phase taper. The phased array antenna input is fed through a power divider that divides the power between the antenna elements.

## 8.2  Beam Steering

Let us now understand how the radio beam gets steered in a phased array antenna.

**Figure 8.2.a** shows the amplitude and phase taper between successive elements of a linear array. The taper causes the wave front radiated from the array to tilt in the direction of antenna element with maximum phase delay.

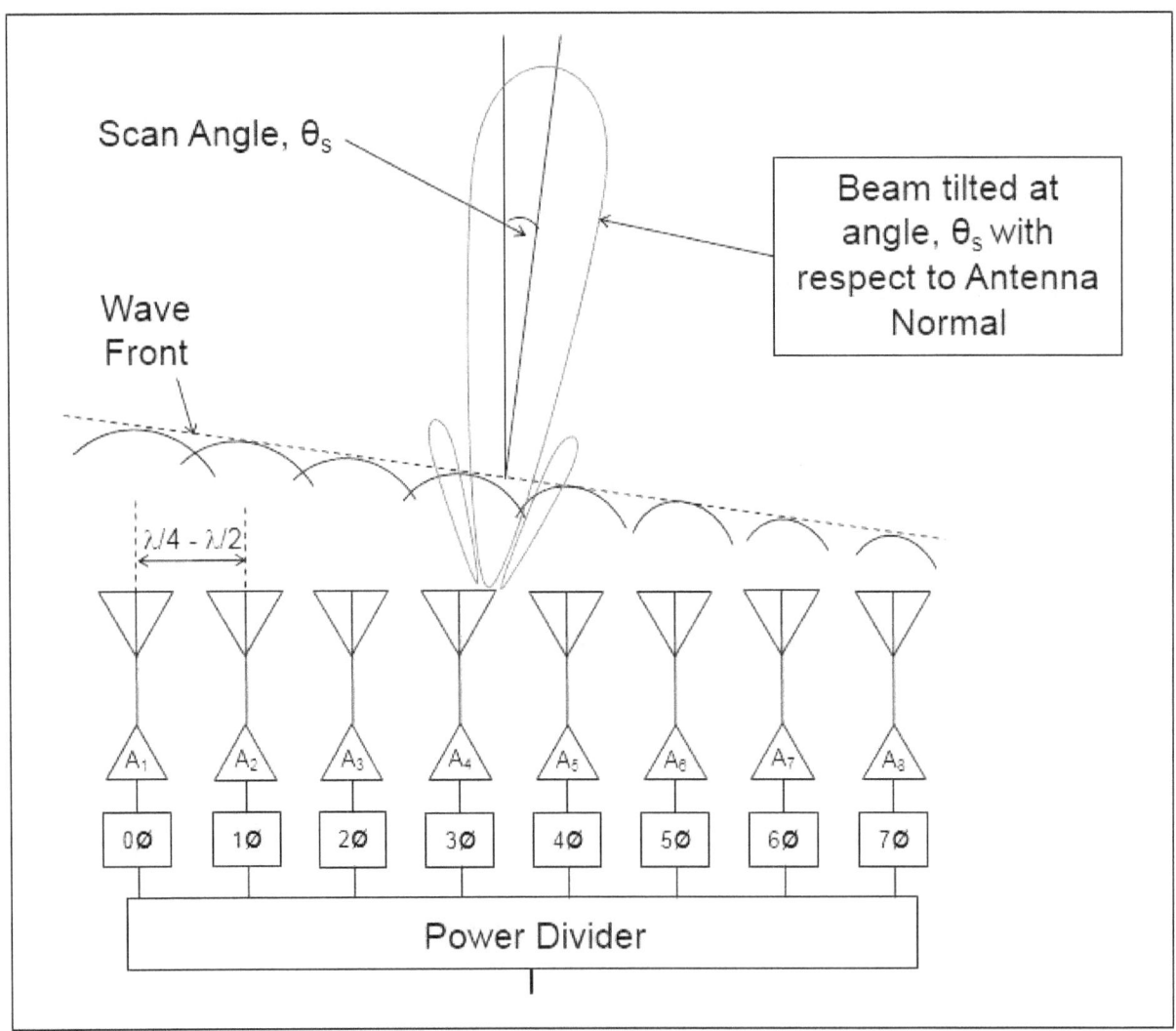

**Figure 8.2.a**

The angle by which the beam tilts with respect to the antenna normal is called the *scan angle*, as shown in **Figure 8.2.a**.

Now, let us see what happens when the spacing between successive antenna elements is increased to more than a wavelength.

**Figure 8.2.b** shows that *grating lobes* are formed on the two sides of the main lobe. These grating lobes are unwanted side lobes and they degrade antenna performance by transmitting and receiving signals from unwanted directions. Therefore, for acceptable performance, the spacing between antenna elements must be carefully controlled.

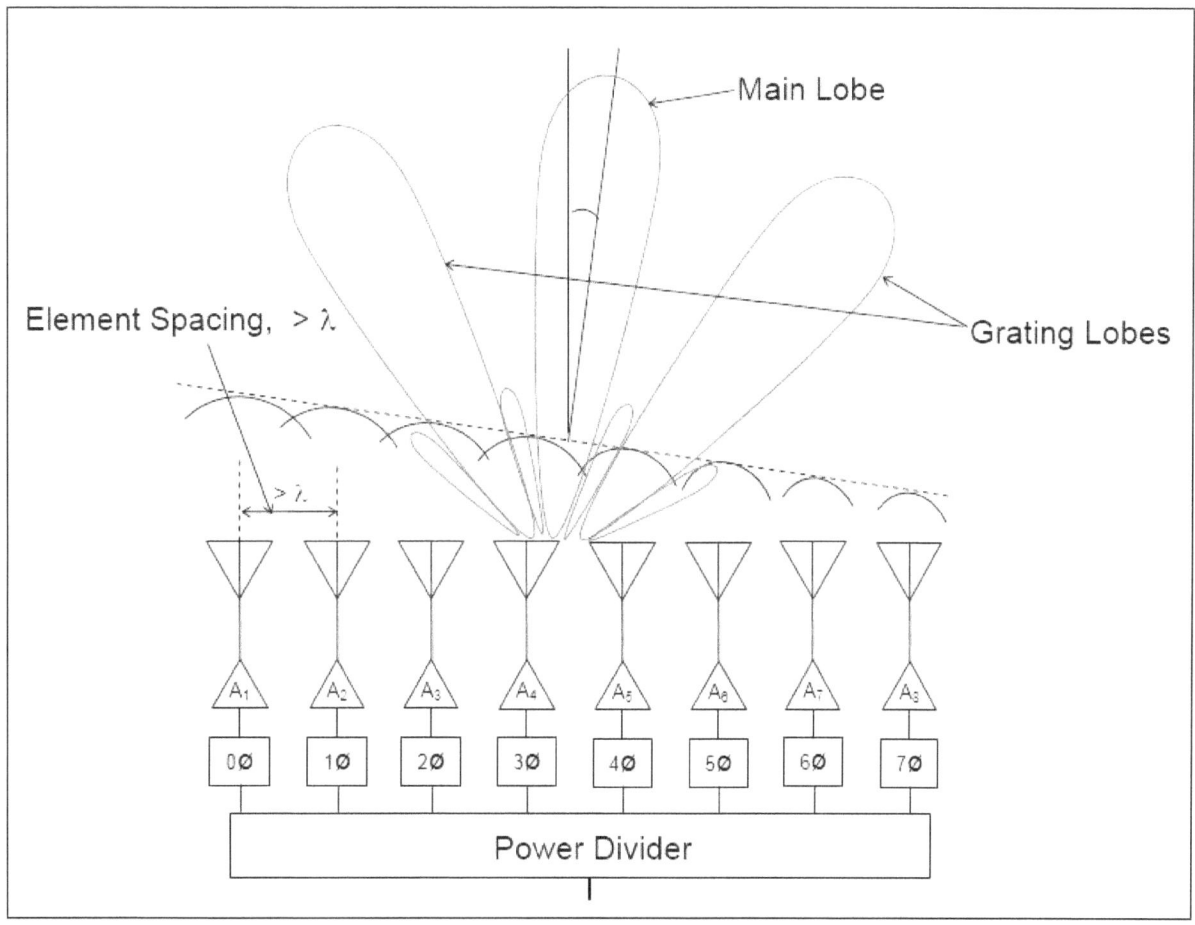

**Figure 8.2.b**

*Array factor* of an antenna array is defined as the antenna pattern for given amplitude and phase taper, assuming each antenna element is a point source. *Array factor* is used as a reference to achieve the target side lobe performance by adjusting the *element spacing, amplitude taper* and the *phase taper*.

For an antenna array with identical elements, the *array factor* is simply multiplied with the antenna element of one element, to achieve the composite array pattern.

For an antenna with non-identical elements, to build the composite array pattern, a vector sum must be performed of EM fields radiated from each antenna element in the far field region of the antenna array.

## 8.3  Multiple Independent Beamforming

Phased array antennas can be used to form multiple independent radio beams. Figure **8.3.a** shows multiple independent radio beams, Beam1, Beam 2 and Beam 3, which are formed by the same phased array antenna, with each beam electrically steered in a different direction. This can be achieved either by using sub-arrays or by using beam forming networks.

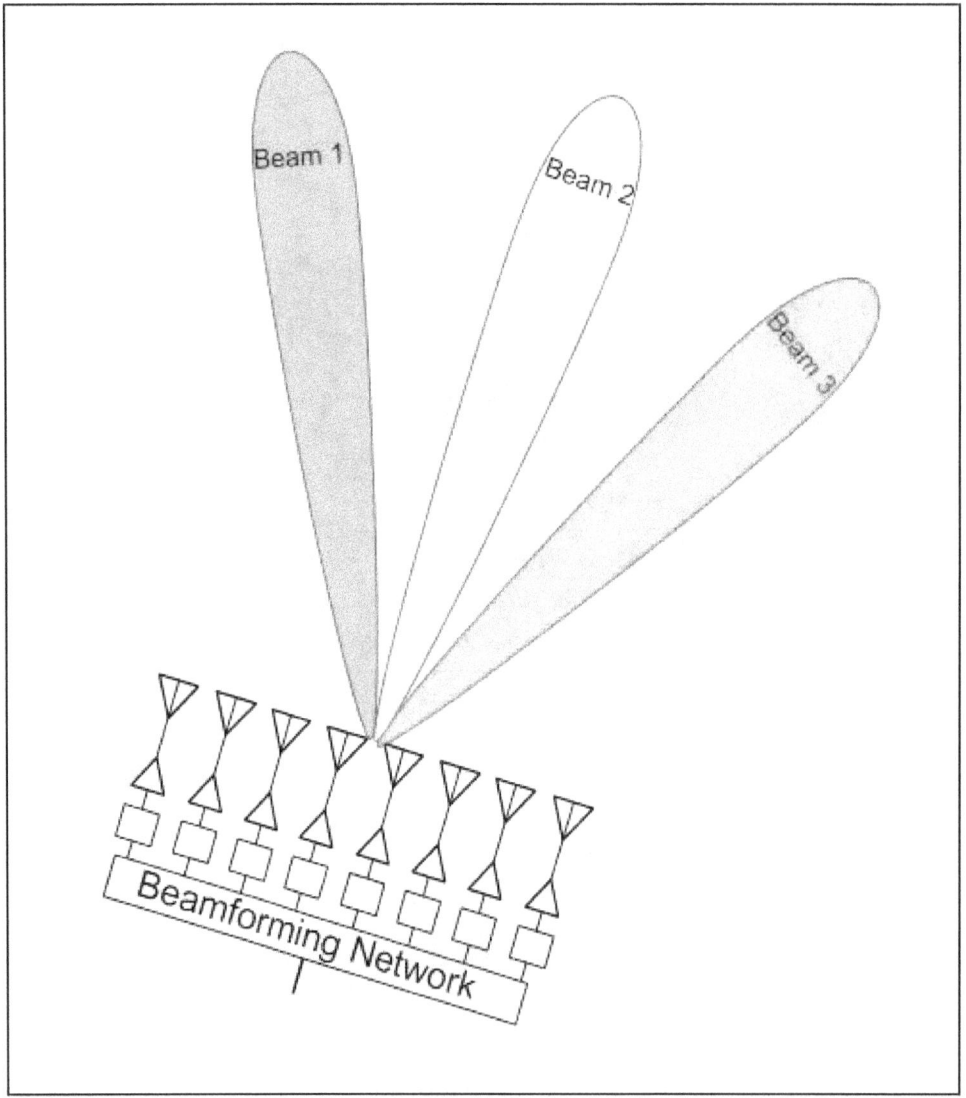

**Figure 8.3.a**

## 8.3.1    Sub Arrays

Sub arrays are formed by dividing an antenna array into independent sub arrays. **Figure 8.3.1.a**. shows four independent sub arrays formed by dividing antenna elements into four groups, Sub Array 1, Sub Array 2, Sub Array 3 and Sub Array 4. Each sub array has its own independent signal feed and its own independent radio beam that can be electrically steered.

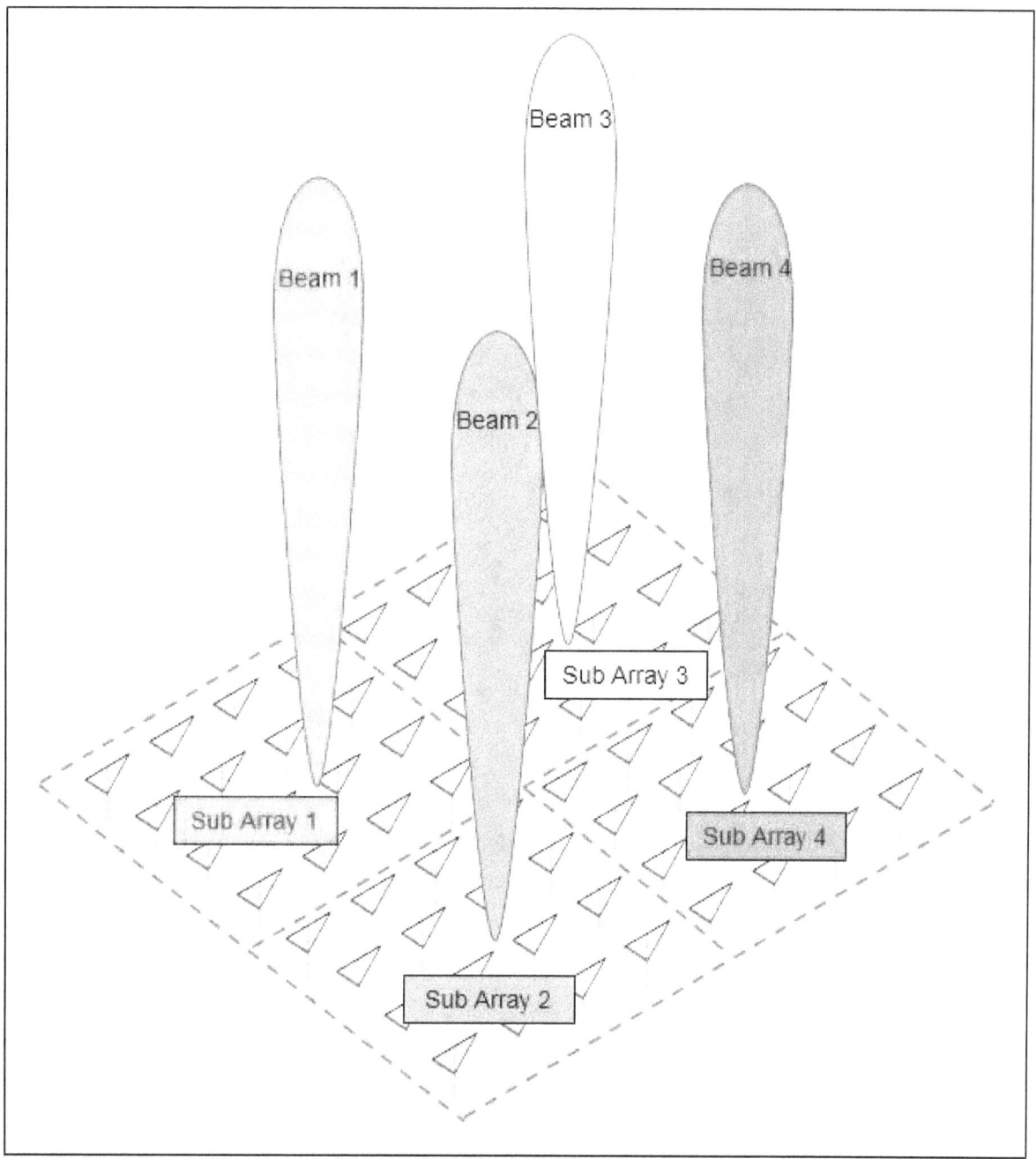

**Figure 8.3.1.a**

## 8.3.2    Beamforming Networks

A beamforming network sits between the antenna feed and the antenna elements and is used to provide the right amount of phase and amplitude taper to help form multiple independent beams. Some examples of beamforming networks are, Butler Matrix, Blass and Nolen Matrices, Wullenweber Array, McFarland 2D Matrix, Rotman Lens, Bootlace Lens, Dome lens. A detailed description of all these beamforming networks is beyond the scope of this book. However, **Figure 8.3.2.a** shows the construction of Butler Matrix, which is a typical beamforming network. It shows the circuit of an 8 element array that generates 8 independent beams. It uses 12 directional couplers, and 8 fixed phase shifters. The expressions in the figure show that for 8 elements, we require 12 directional couplers and 8 fixed phase shifters. It is obvious from these expressions that the complexity of Butler Matrix increases with the number of antenna elements. For example, if we have 64 elements, we require 192 directional couplers and 160 fixed phase shifters.

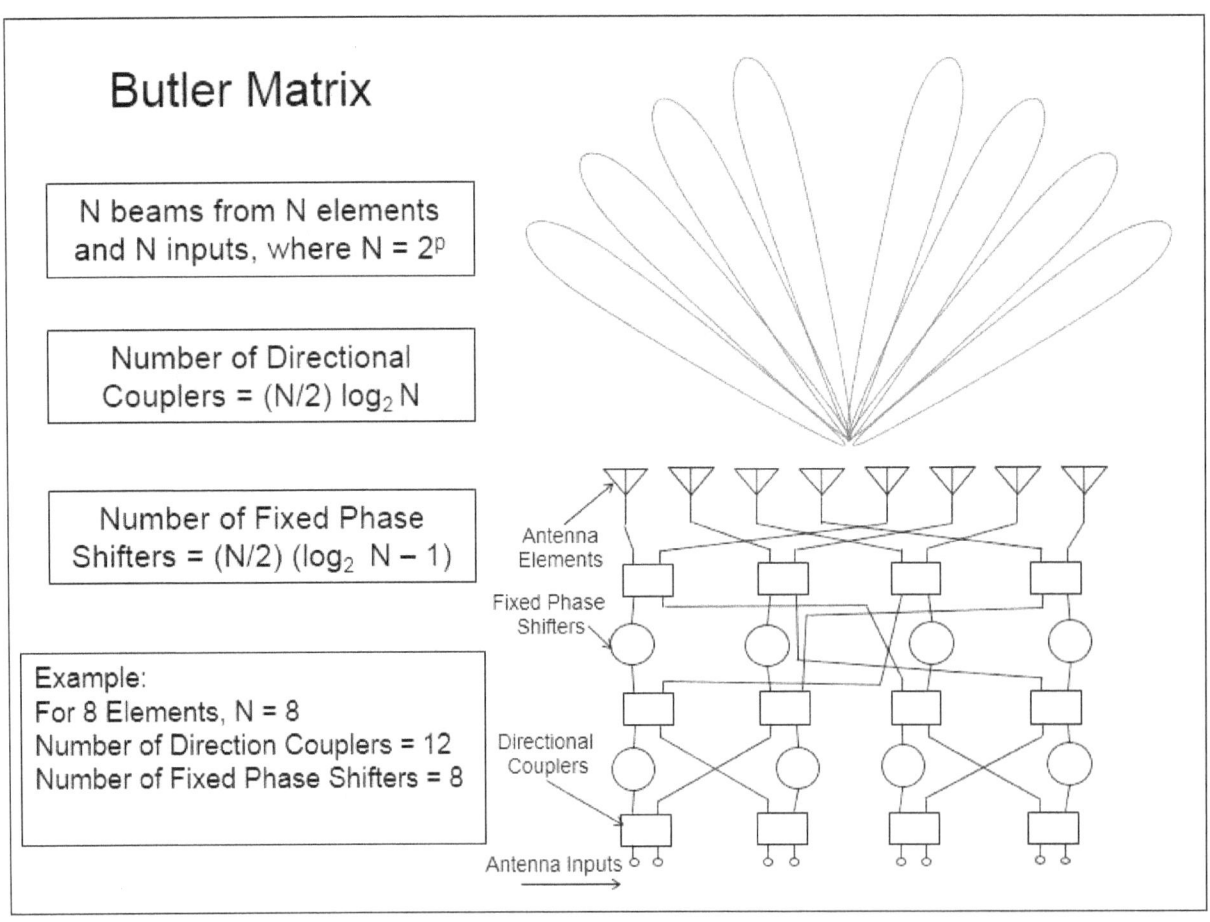

**Figure 8.3.2.a**

## 8.4 Smart Antennas

*Smart antennas* or *adaptive antenna arrays* process the received signals and shape the radio beam pattern dynamically to maximize *signal to noise ratio* (SNR), so that undesirable pickup from noise sources can be reduced. The adaptive antenna arrays can also compensate for failed antenna elements, or for other mechanical or electrical errors. They can also be used to compensate for blockage by any nearby physical structures, such as buildings, terrain and hills.

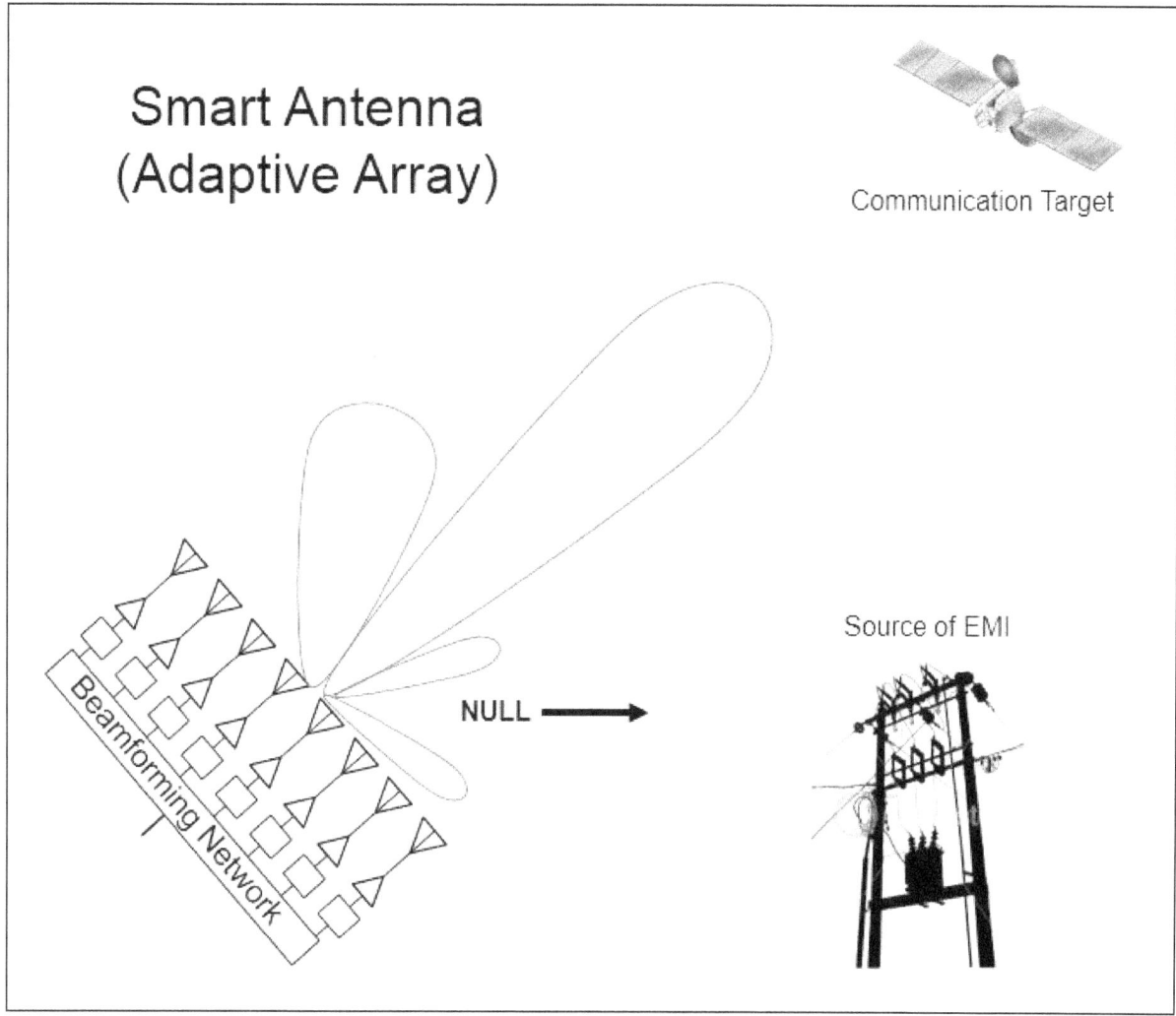

**Figure 8.4.a**

**Figure 8.4.a** shows an example of a smart antenna or an adaptive array. The antenna array in this figure has a satellite as the communication target. However, in the vicinity of antenna array, there is a strong source of *electromagnetic interference* (EMI), from a high voltage power line. In this situation, the smart antenna will process the received signals from the satellite as well as the EMI from power line. Next, it will determine the direction from which the EMI is received and then modify the radio beam pattern in such a way that a NULL is placed in the direction from which EMI is received. This will help improve the strength of radio signal received from the target communication satellite. Some of the signal processing algorithms that are used by smart antennas include, *least mean squares* (LMS), *sample matrix inversion* (SMI), *recursive least squares* (RLS), *conjugate gradient method* (CGM) and *constant modulus algorithm* (CMA). A detailed description of these algorithms is beyond the scope of this book.

# 9. Antenna Design and Measurements

In this chapter, we will study the process of antenna design and various measurements that are required to define the operational characteristics of an antenna.

## 9.1 Antenna Design Process

The best way to understand the process involved in the design of an antenna is to follow the flow chart shown in **Figure 9.1.a**.

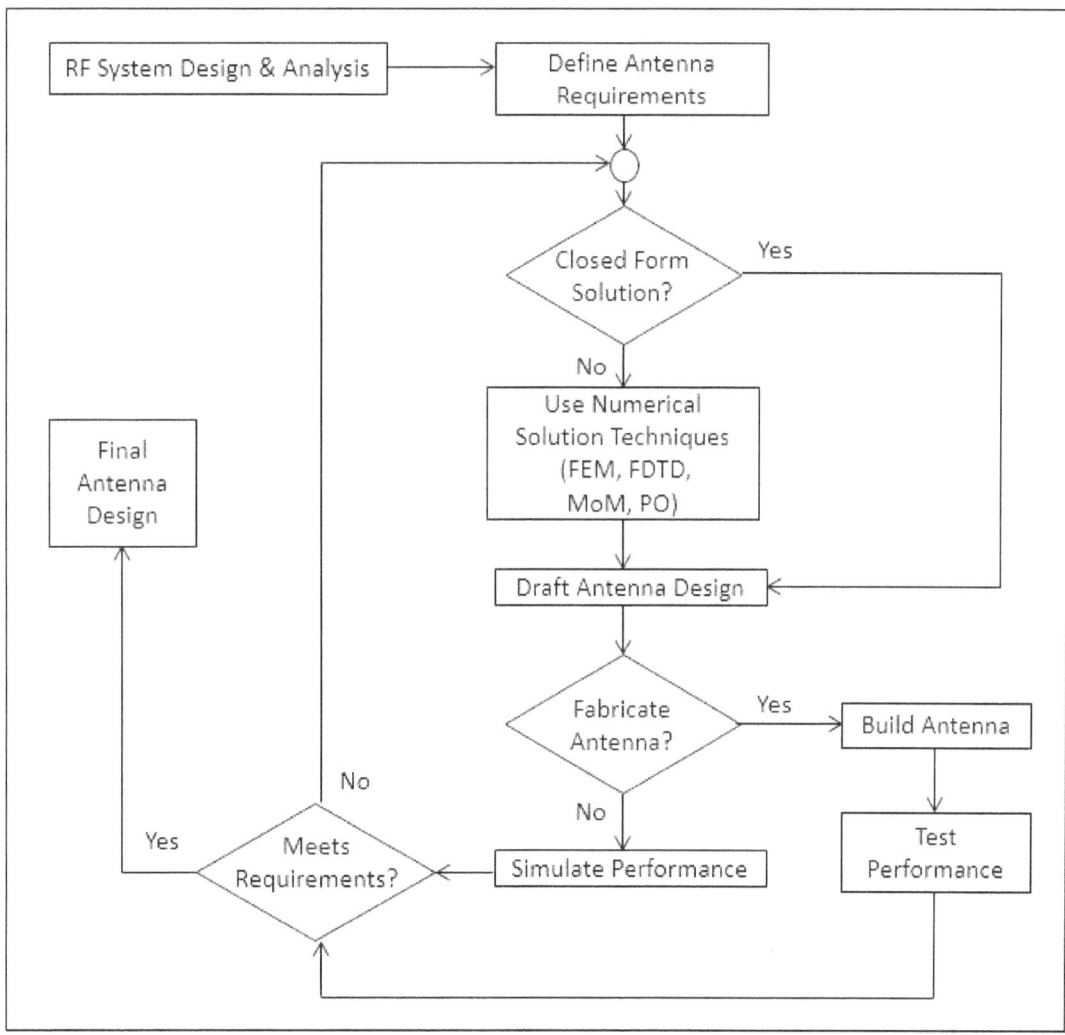

**Figure 9.1.a**

First and foremost, the technical requirements for the antenna must be defined. These are derived from the analysis and design of the proposed RF system. The various antenna parameters, which we discussed in **Chapter 6**, will come into play here, such as the *frequency used*, *bandwidth*, *return loss*, intended *gain* and *polarization*, *axial ratio*, *EIRP*, *beamwidth*, *physical structure* of the antenna, *size, weight and power* considerations.

Next step is to figure out if a *closed form solution* is available for the target antenna requirements. A closed form solution requires the target antenna parameter values to be input to a set of known design equations, which are then solved to arrive at the physical dimensions of the antenna. For example, for a half wave dipole antenna, the only parameter we need is the frequency used, from which we calculate wavelength, and half of wavelength gives us the length of dipole antenna element.

If a closed form solution is available, then the next step is to simply make a draft of the physical design of antenna. If a closed form solution is not available, then numerical computation methods are used to find a solution. Some of the techniques are *finite element method* (FEM), *finite difference time domain* (FDTD), *method of moments* (MoM) and *physical optics* (PO). Some of the commercially available antenna design software solutions incorporate one or more of these numerical methods.

Either using closed form solution method or numerical methods, leads to a draft design of the proposed antenna. The next step is to decide, whether to physically fabricate a working prototype of the antenna, so that its performance can be tested against the original design parameters. This decision depends on the actual cost of fabricating the prototype. For large and complex antenna structures, the cost of building the prototype can be prohibitive, especially when several design iterations are required before the right antenna prototype can be realized, which meets the design requirements. In order to offset the cost of building, testing and reworking a prototype, commercial electromagnetic software can be used to simulate the performance of the draft antenna design.

The ultimate goal of either of these methods, that is, build a prototype to test performance, or simulate performance with software, is to meet the original design requirements. If the requirements are not met, we must iterate through all the steps starting from finding a solution to testing the performance of proposed design.

Once the design and test cycle is out of the iteration loop, the design requirements are met and the antenna performance is validated, and the final antenna design is ready for production.

## 9.2 Vector Network Analyzer (VNA)

The *vector network analyzer* (VNA) is used to make the actual performance measurements on the antenna. VNAs are typically used to measure impedance, gain, return loss, field patterns and several other parameters, and they come in different flavors with a range of capabilities. **Figure 9.2.a** shows a 2-port and a 4-port VNA.

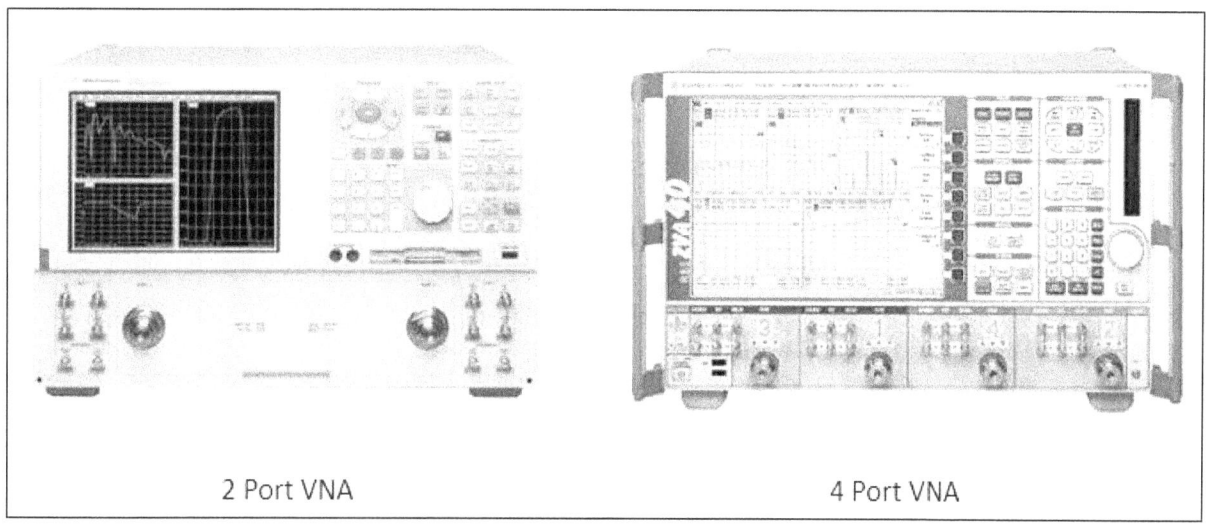

**Figure 9.2.a**

## 9.3 Impedance Measurement

*Impedance* of an antenna can be measured by using a 2-port network. **Figure 9.3.a** shows a set up to measure impedance of a half wave dipole antenna. One half wave dipole antenna is connected to the *transmit* (Tx) port of the VNA, and the other is connected to the *receive* (Rx) port. Each of the 2 ports of VNA has a typical impedance of 50 ohms. The VNA transmits RF signal at the designated frequency of the antenna from the Tx port and receives the same signal over the Rx port, thereby completing the circuit for measurement of antenna impedance. The objective here is to measure the incident and reflected RF voltage on each port, calculate the *s-parameters* for the RF circuit and then calculate impedance from the *s-parameters*.

**Figure 9.3.b** shows a representation of 2-port network and formulas for the *s-parameters*, which are essentially ratios of reflected and incident voltages from different ports. Also shown is the *scattering matrix*, which is a matrix representation of all *s-parameters*. So, the task for VNA here is to measure the reflected and incident voltages as defined in the figure, and determine s-parameter values from these ratios.

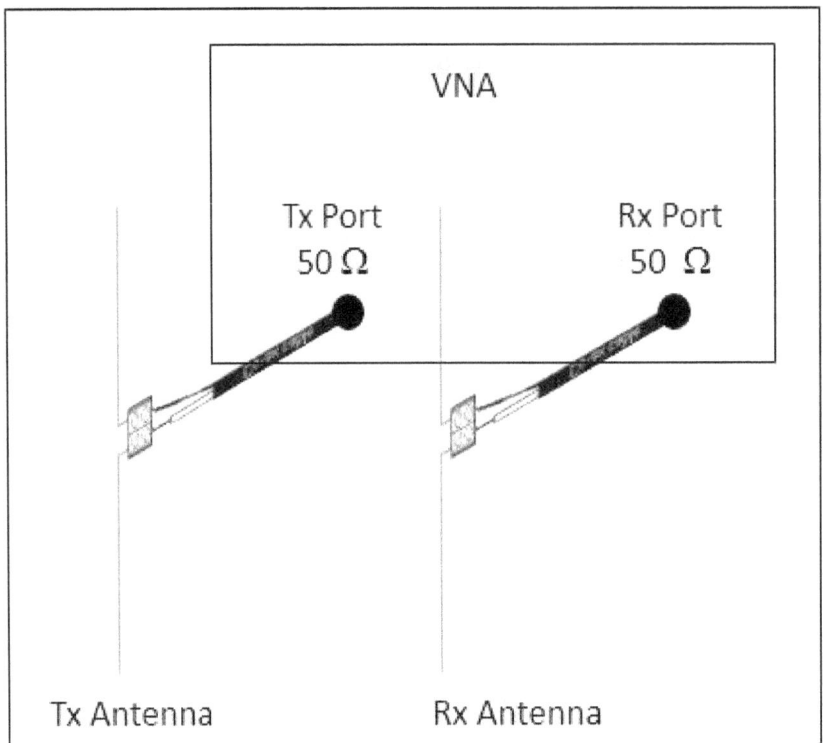

**Figure 9.3.a**

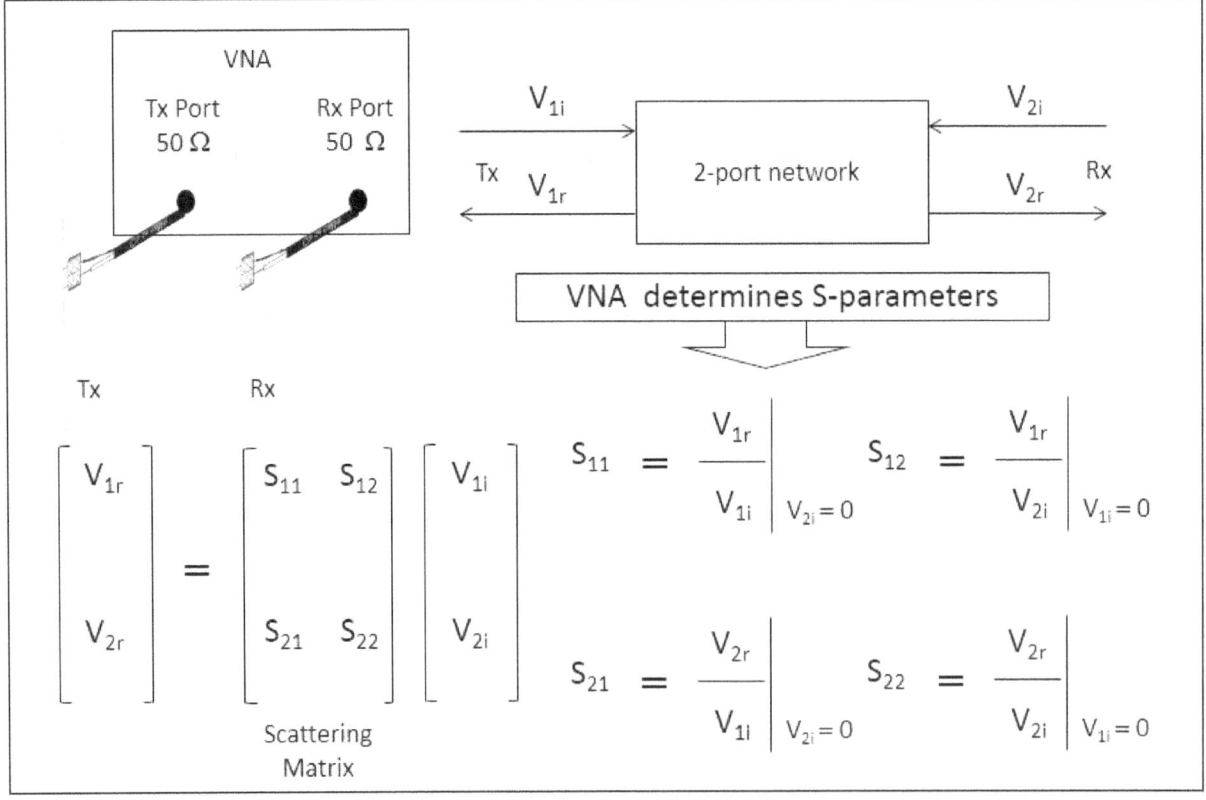

**Figure 9.3.b**

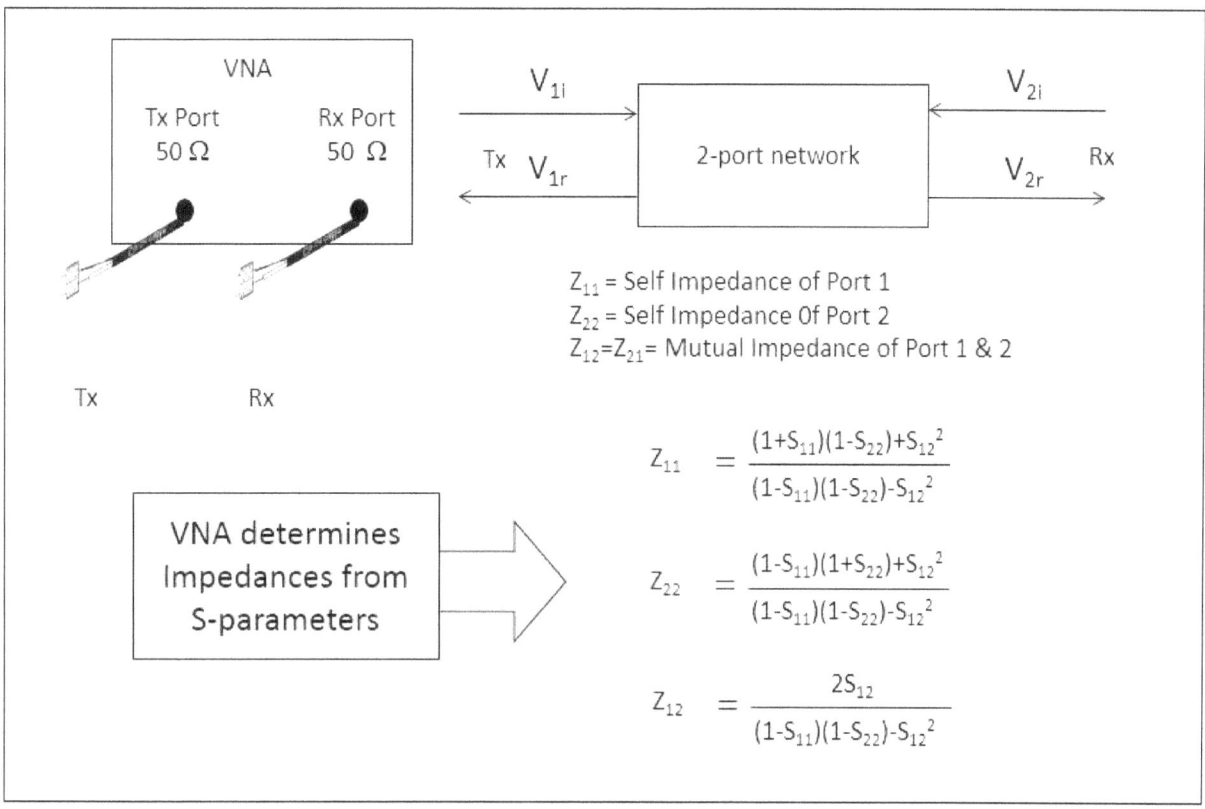

$Z_{11}$ = Self Impedance of Port 1
$Z_{22}$ = Self Impedance Of Port 2
$Z_{12}=Z_{21}$= Mutual Impedance of Port 1 & 2

VNA determines Impedances from S-parameters

$$Z_{11} = \frac{(1+S_{11})(1-S_{22})+S_{12}^{2}}{(1-S_{11})(1-S_{22})-S_{12}^{2}}$$

$$Z_{22} = \frac{(1-S_{11})(1+S_{22})+S_{12}^{2}}{(1-S_{11})(1-S_{22})-S_{12}^{2}}$$

$$Z_{12} = \frac{2S_{12}}{(1-S_{11})(1-S_{22})-S_{12}^{2}}$$

**Figure 9.3.c**

**Figure 9.3.c** shows the formulas for self impedance of Port 1 (Tx), self-impedance of Port 2 (Rx) and the mutual impedance of Port 1 and Port 2. The VNA determines these impedance values from the *s-parameters* calculated earlier from the ratios of measured reflected and incident voltages on different ports.

## 9.4   Gain and Field Pattern Measurements

In this section, we will study various techniques that are used to measure the gain and field patterns of an antenna in the *far field* as well as the *near* field.

## 9.4.1     Open Area Test Site (OATS)

An *open area test site* (OATS) is used for *far field* measurements of *gain* and *field pattern* for an antenna. **Figure 9.4.1.a** shows the illustration of an OATS. The *antenna under test* (AUT) is used as a *receive* (Rx) antenna and is placed on a tower with a position controller. The AUT is positioned in *line of sight* (LOS) of the reference *transmit* (Tx) antenna, so that the *electromagnetic* (EM) wave received by the Rx antenna is a plane wave, which is a condition for the *far field region* of the antenna.

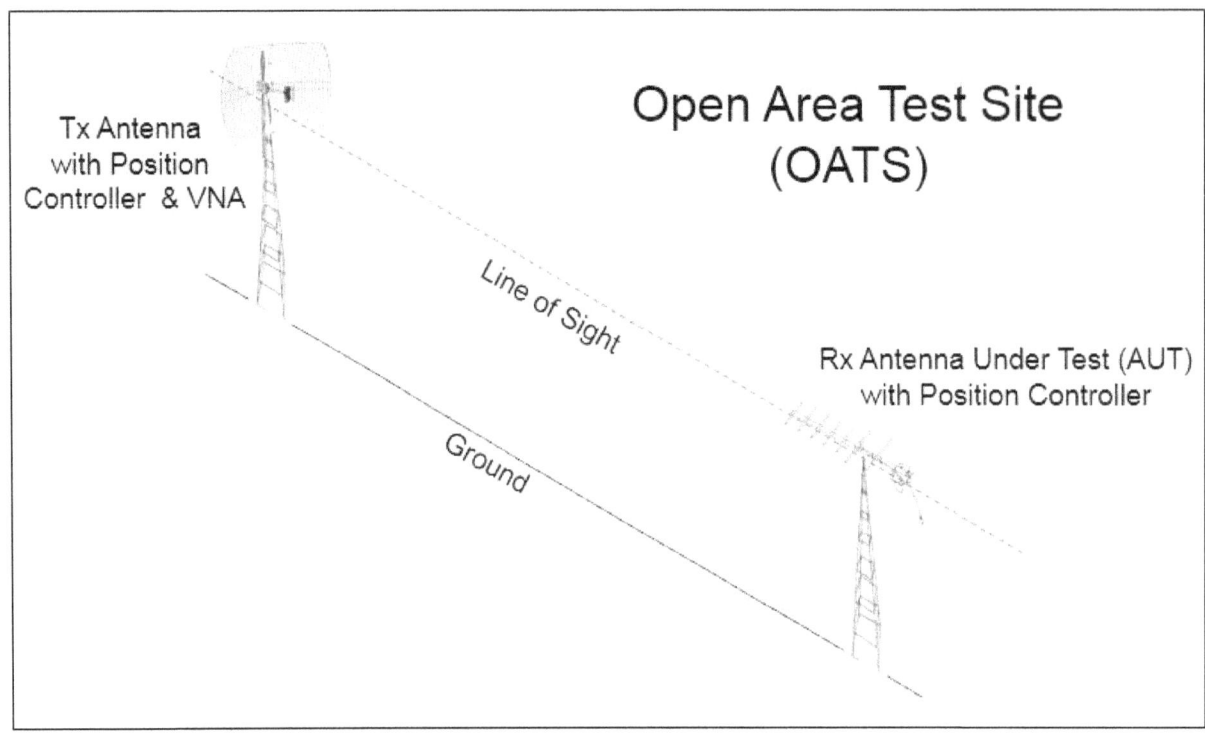

**Figure 9.4.1.a**

The transmit (Tx) antenna is housed with a position controller and a VNA. The Tx antenna is used to illuminate the Rx antenna in a controlled and consistent manner. The Rx antenna is moved through different angles and VNA is used to take measurements of gain and field strength in the far field, in E plane as well as H plane.

## 9.4.2    Indoor Anechoic Chamber

The *indoor anechoic chamber* can also be used for far field measurements. It simulates the outdoor conditions by having an electromagnetic shield around a large room to avoid *electromagnetic interference* (EMI), and electromagnetic absorbent material on the walls, floor and ceiling to avoid reflections. The Tx and Rx antennas are placed inside the chamber and connected to a VNA to make the antenna measurements. **Figure 9.4.2.a** shows an illustration of the indoor anechoic chamber.

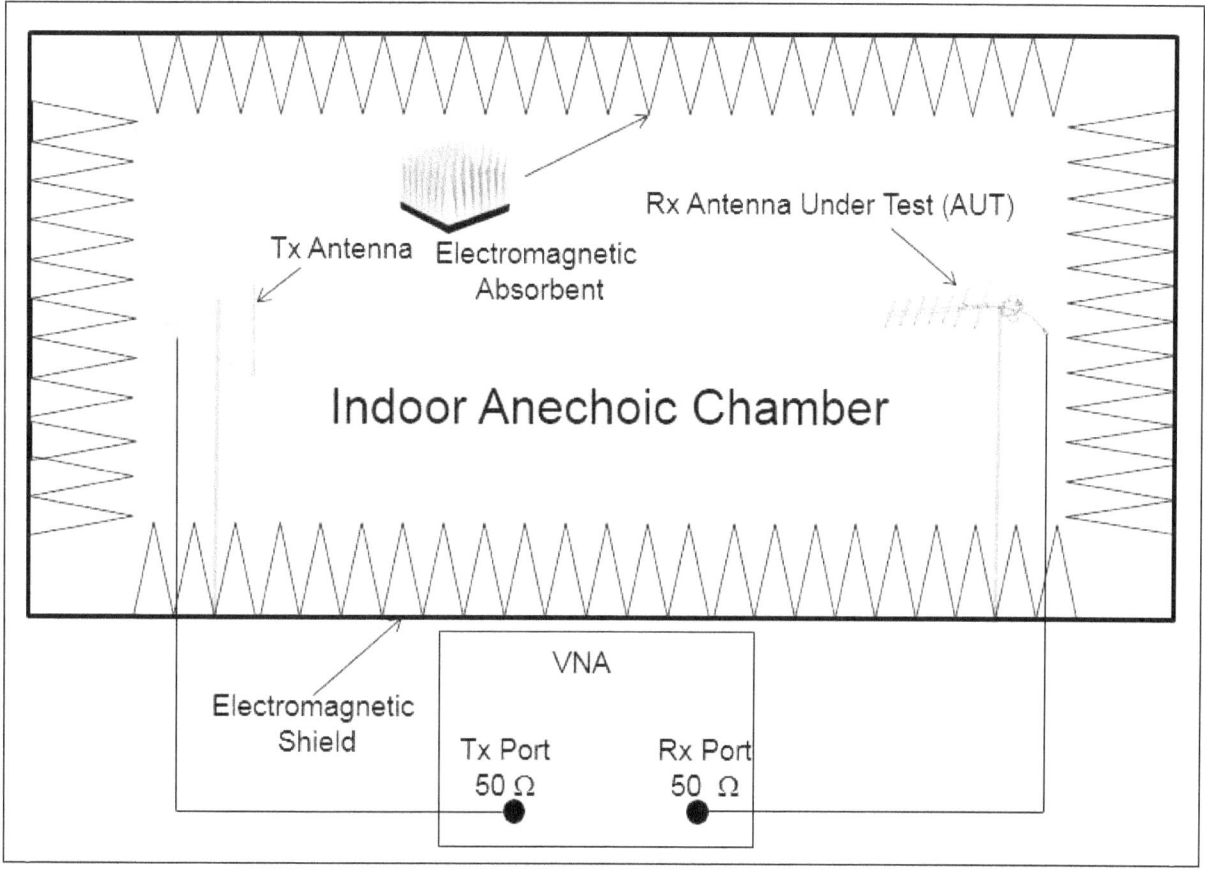

**Figure 9.4.2.a**

## 9.4.3     Indoor Compact Range

Another option is to use *indoor compact range*, which is smaller than *indoor anechoic chamber*. The *indoor compact range* uses a reflector to generate a plane *electromagnetic* (EM) wave that illuminates the Rx *antenna under test* (AUT). The Tx antenna here is usually a calibrated horn antenna that illuminates the reflector, as shown in **Figure 9.4.3.a**.

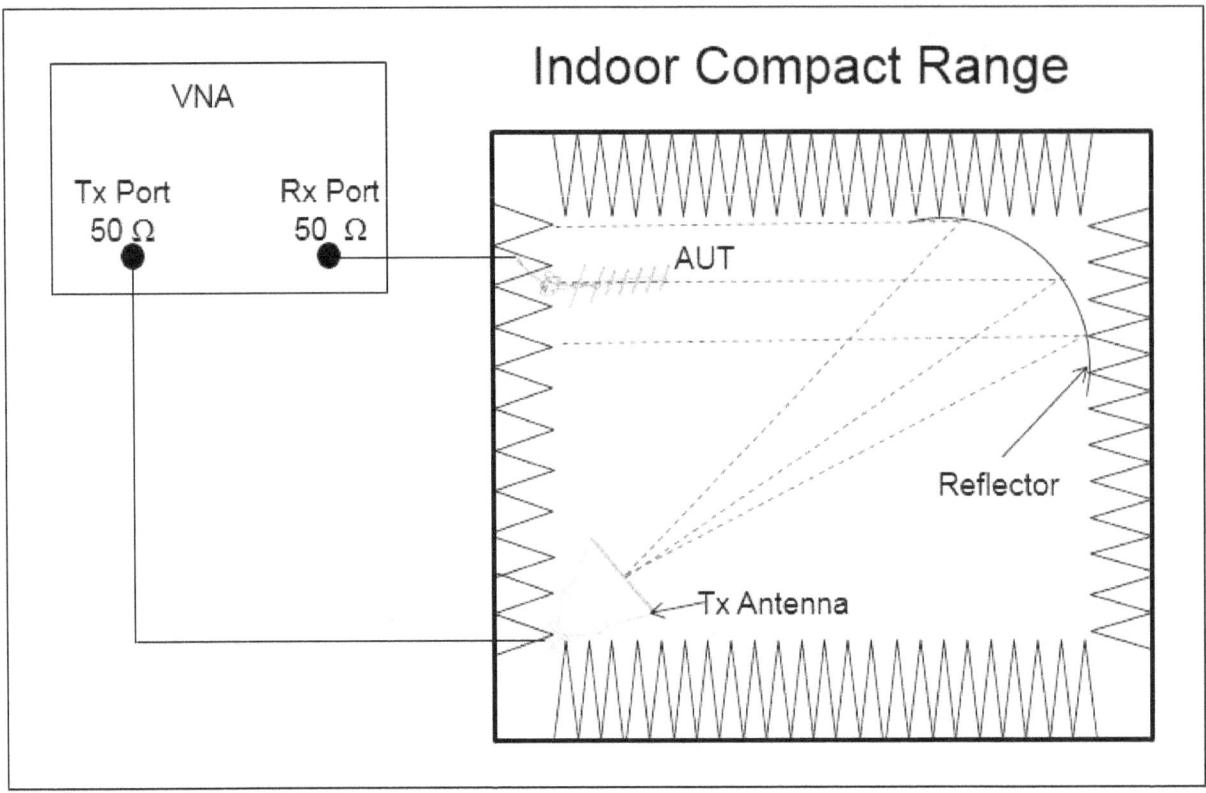

**Figure 9.4.3.a**

## 9.4.4    Procedure for Far Field Measurements

The various steps involved in far field measurement techniques that we have discussed so far, are listed sequentially in **Figure 9.4.4.a**.

The first step involves calibration of the range with a standard gain horn antenna at the receiver. Next, the standard gain Rx horn antenna is removed and instead the *antenna under test* (AUT) is mounted on the Rx tower.

The AUT Rx antenna is then moved in the azimuth and elevation planes to take measurements at different positions. Typically, the *magnitude* and *phase* of the received RF signal is measured at different positions of azimuth and elevation.

After data collection is completed, measurement and control software is used to determine the field patterns in the E and H plane.

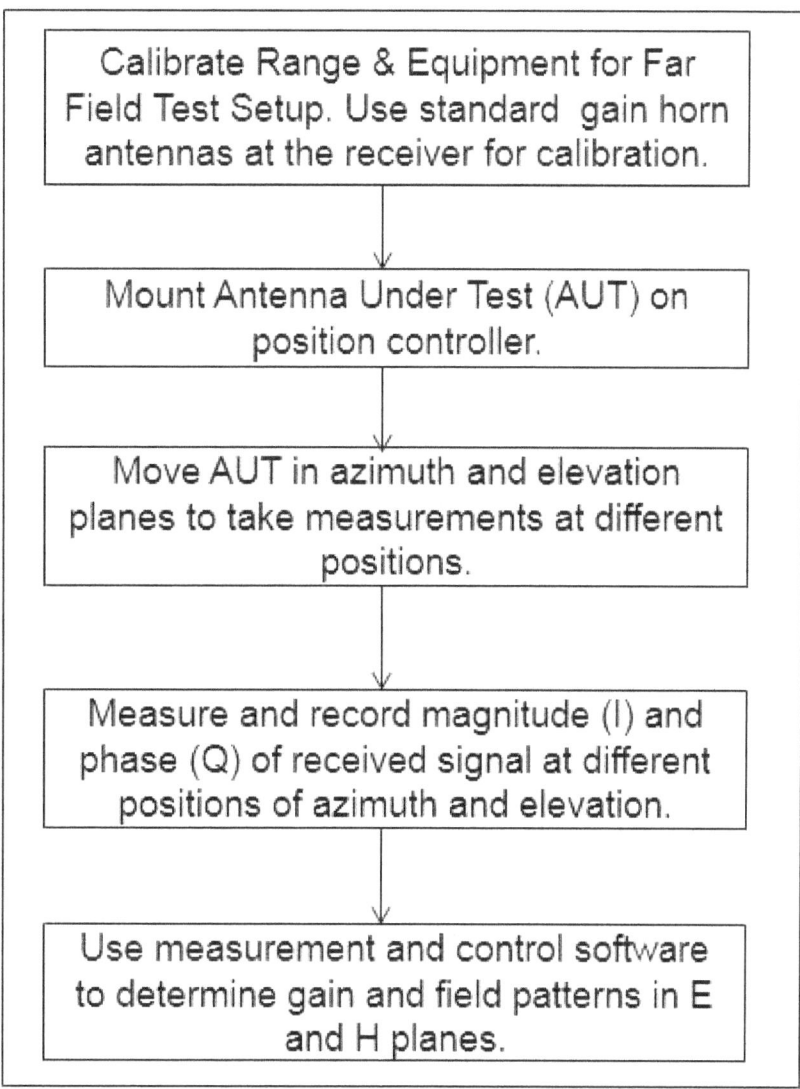

**Figure 9.4.4.a**

## 9.4.5    Near Field Measurements

If an *open area test site* (OATS), or an indoor anechoic range or compact chamber are not available, then the alternate way is to make *near field measurements*. The gain and field pattern measurements in the near field region of an antenna are also necessary if the antenna size is very large or if it is very small. As shown in **Figure 9.4.5.a**, in this case, the *antenna under test* (AUT) is a transmit antenna, and a small measurement probe is moved over the aperture of the antenna to make measurements. The probe is usually in the form of a small horn antenna.

The data collected by the measurement probe is then computed to determine the *far field pattern* by using one of these techniques-*aperture integration* (AI), *geometrical theory of diffraction* (GTD), *fast fourier transforms* (FFT). However the *far field* pattern computed from near field measurements is accurate only over a limited range.

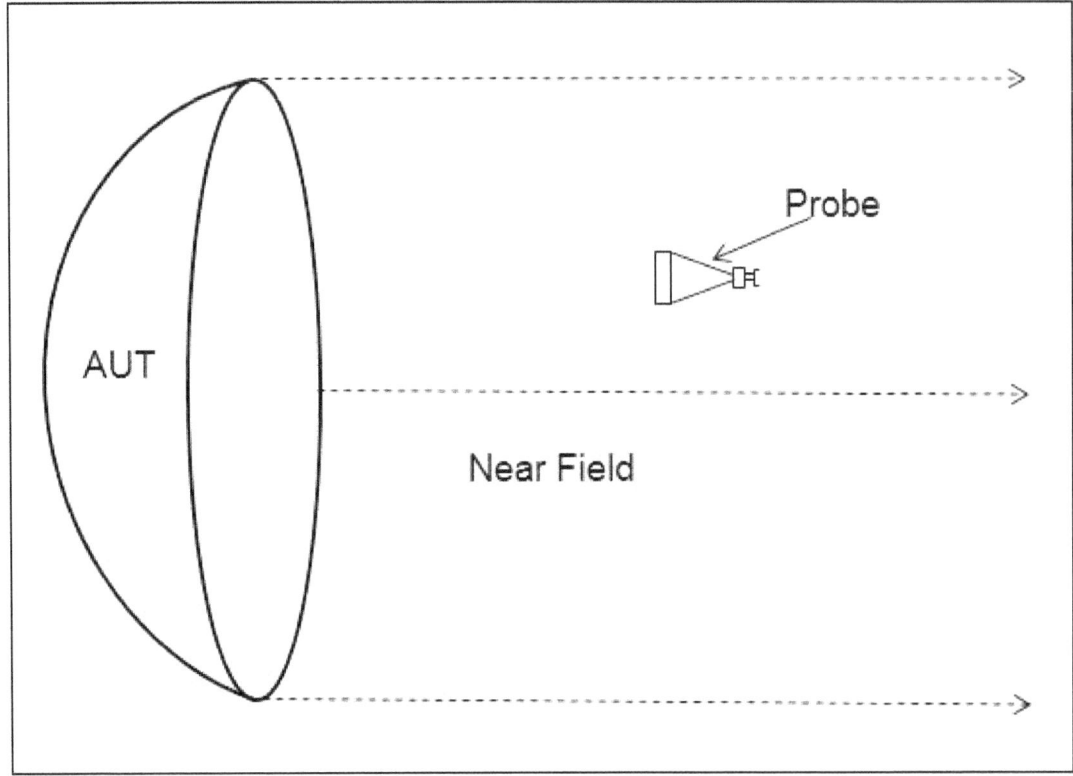

**Figure 9.4.5.a**

The near field measurements can be made over a *planar surface* as shown in **Figure 9.4.5.b**. In this case the probe moves along x and y axes. This method is best for high gain antennas.

**Figure 9.4.5.c** shows near field measurements made over a *plane-polar* surface. In this case, the AUT rotates around z-axis and the probe moves along y-axis. This method is also best for high gain antennas.

**Figure 9.4.5.d** shows near field measurements made on *cylindrical surface* around the antenna. In this case the probe moves along z-axis and rotates around z-axis on a cylindrical surface.

**Figure 9.4.5.e** shows near field measurements made over a *spherical surface* around the antenna. In this case the probe moves along spherical coordinates. Although the figure shows a parabolic antenna, this technique is best suited for low gain omni-directional antennas.

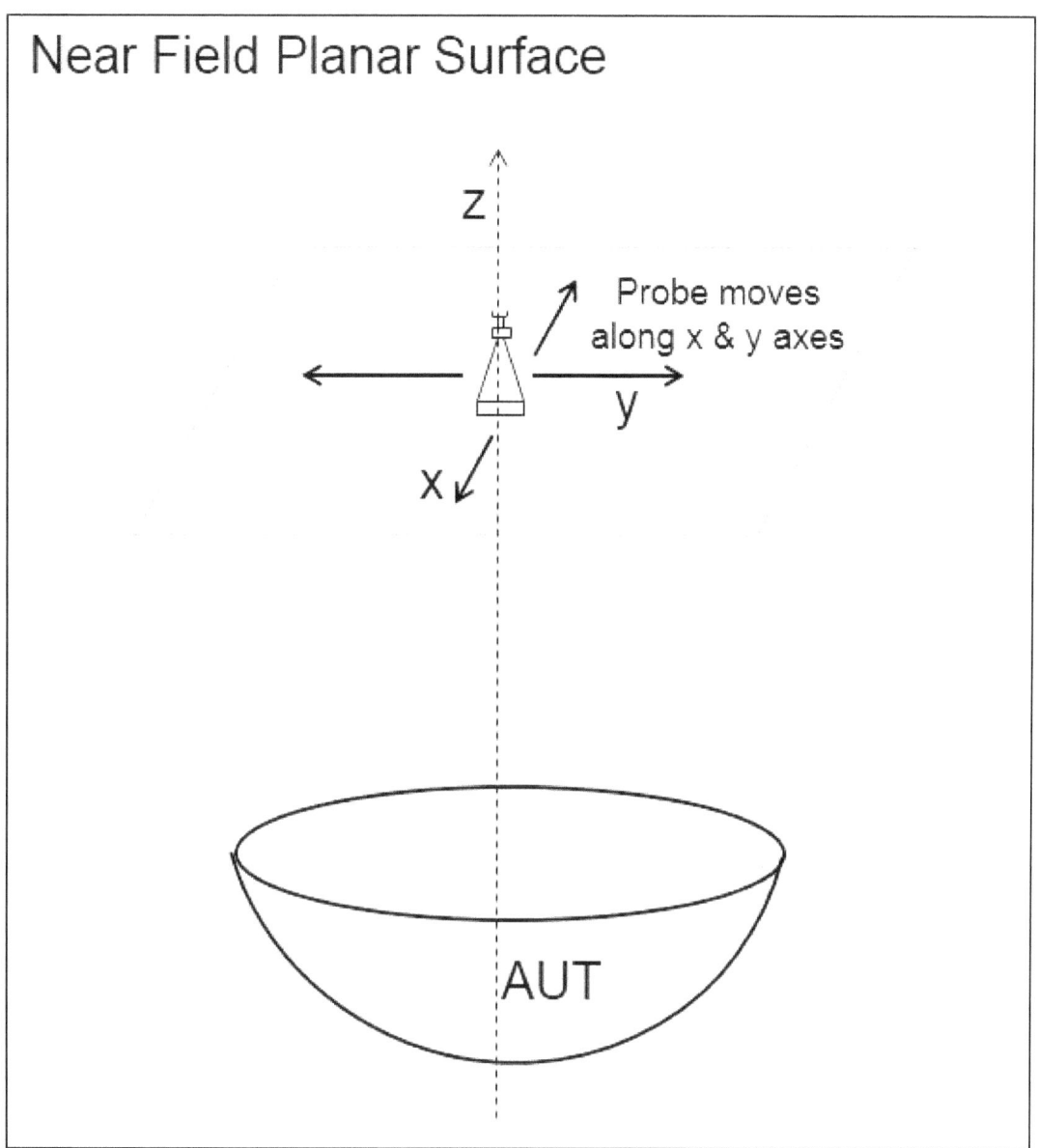

**Figure 9.4.5.b**

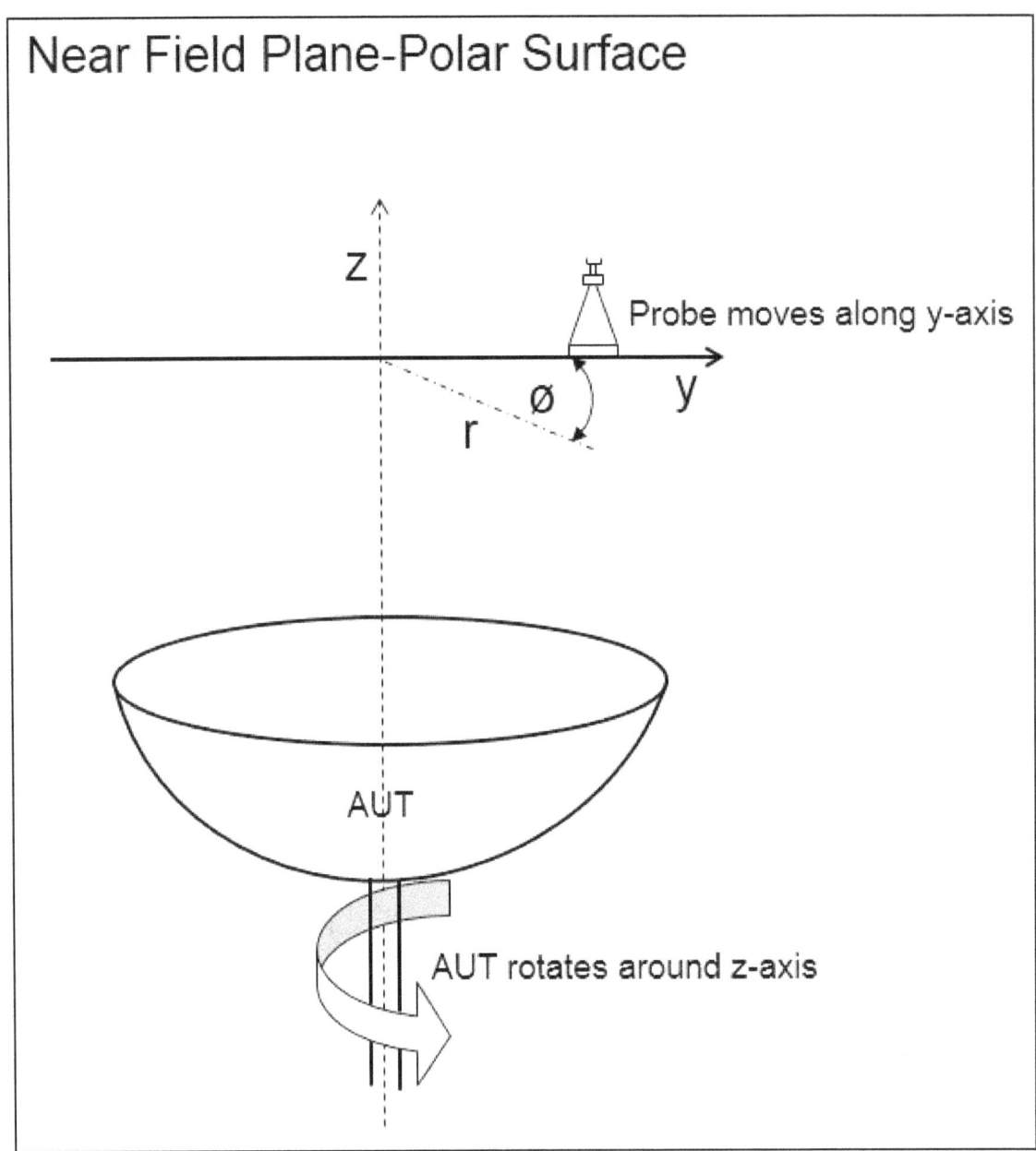

**Figure 9.4.5.c**

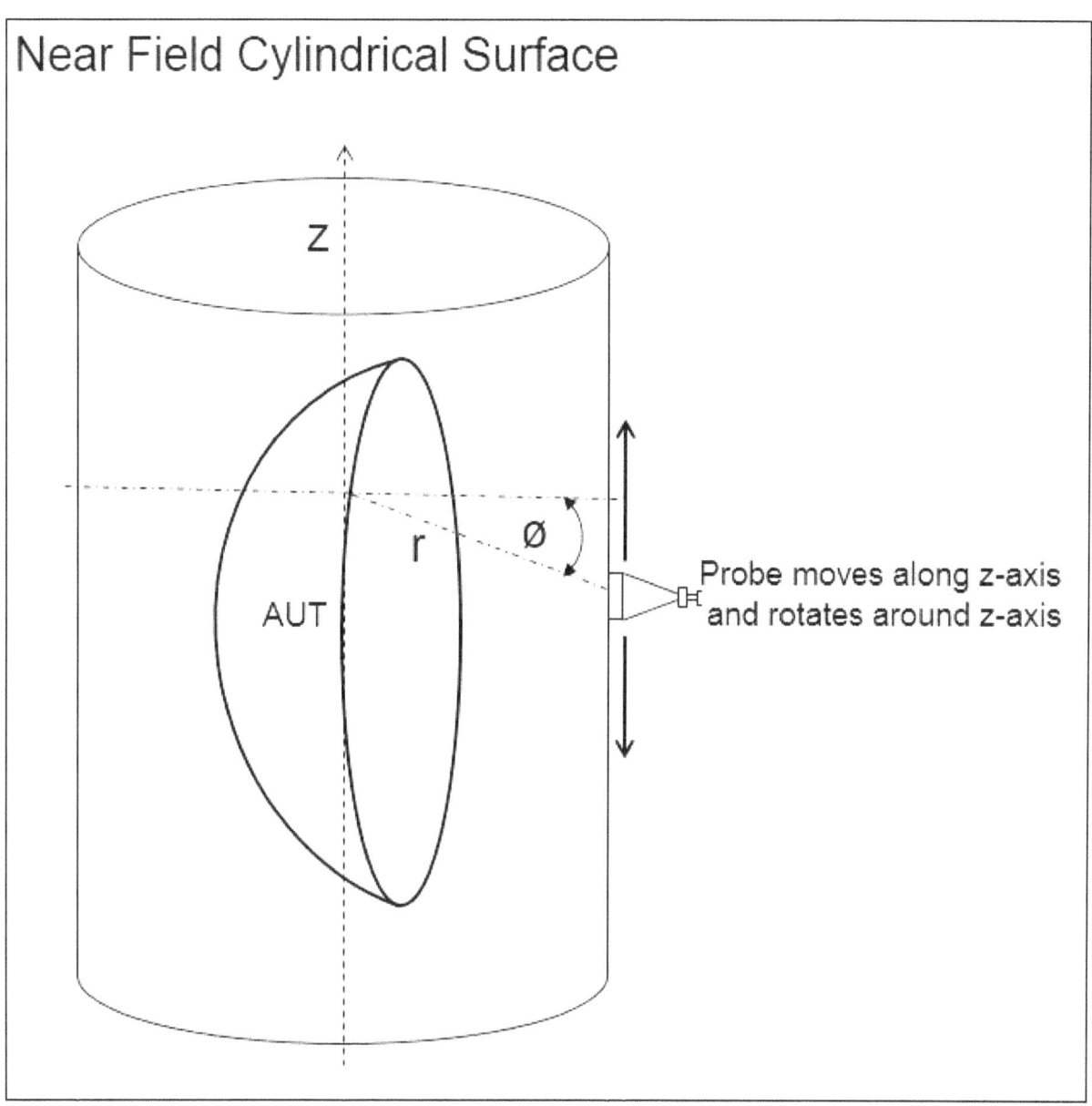

**Figure 9.4.5.d**

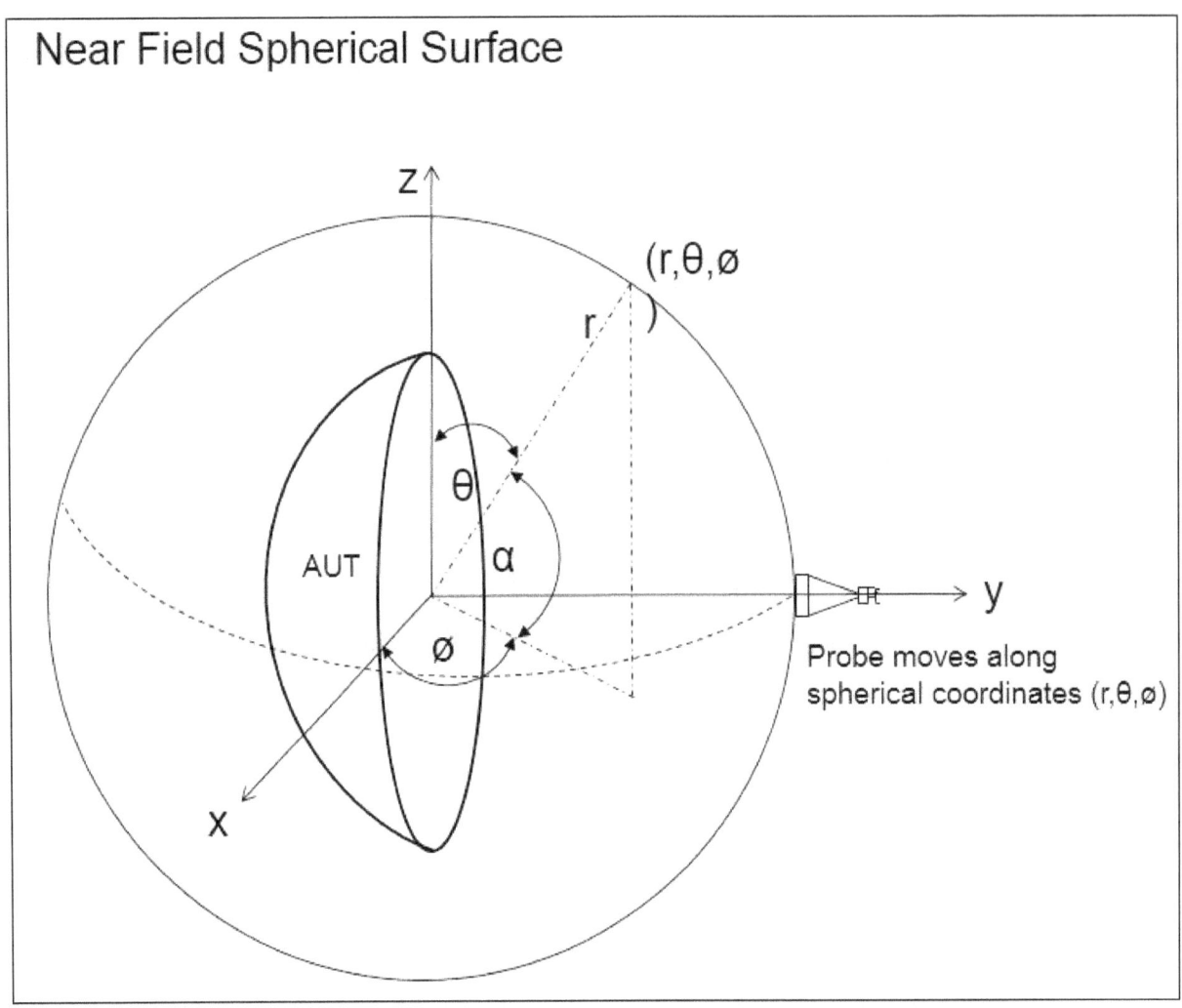

**Figure 9.4.5.e**

# 10. Basics of Link Budget Analysis

In this chapter we will get familiar with the basics of link budget analysis. It is an essential procedure for the determination of the performance of a wireless communication link.

Link budget analysis involves calculation of the ratio of received *signal power* to the *noise power* at the receiver. For meaningful reception of a radio signal at the receiver, the received signal power must be higher than noise power; otherwise, the signal will be so weak that it will be indistinguishable from noise at the receiver. Link budget analysis helps us predict if for a given wireless link, we are going to have meaningful reception of radio signal or not. The results of this analysis help us tweak the design of a given wireless link so that its performance complies with the desired requirements and performance thresholds.

## 10.1 Generic Wireless Link

**Figure 10.1.a** shows the parameters involved in link budget analysis of a generic wireless link.

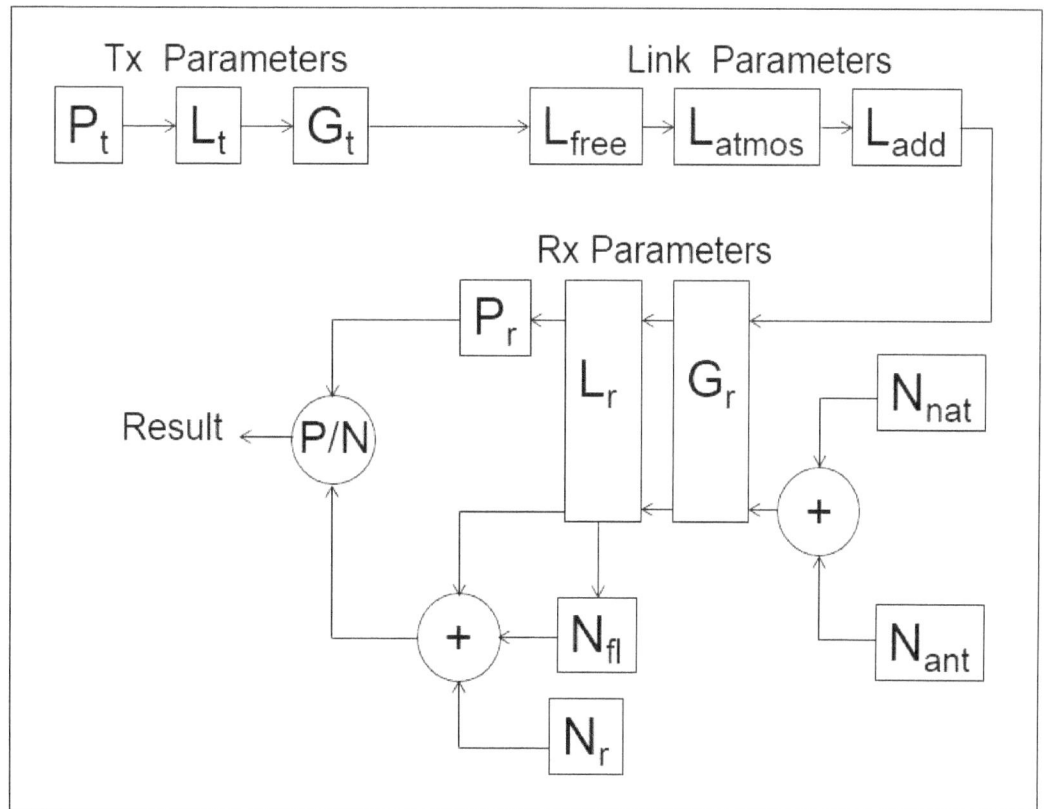

**Figure 10.1.a**

There are three categories of parameters that we have to deal with. These are transmitter (Tx) parameters, radio link parameters and receiver (Rx) parameters. We will now explore each of these parameters in little more detail.

## 10.2 Transmitter Parameters

**Figure 10.2.a** shows the transmitter (Tx) parameters, which includes *transmitted power*, $P_t$, *transmission losses*, $L_t$ and *antenna gain*, $G_t$. The *effective isotropic radiated power* (EIRP) is calculated by subtracting the *transmission losses* from *transmitted power* and adding the *antenna gain*, where all parameters are expressed in decibel (dB).

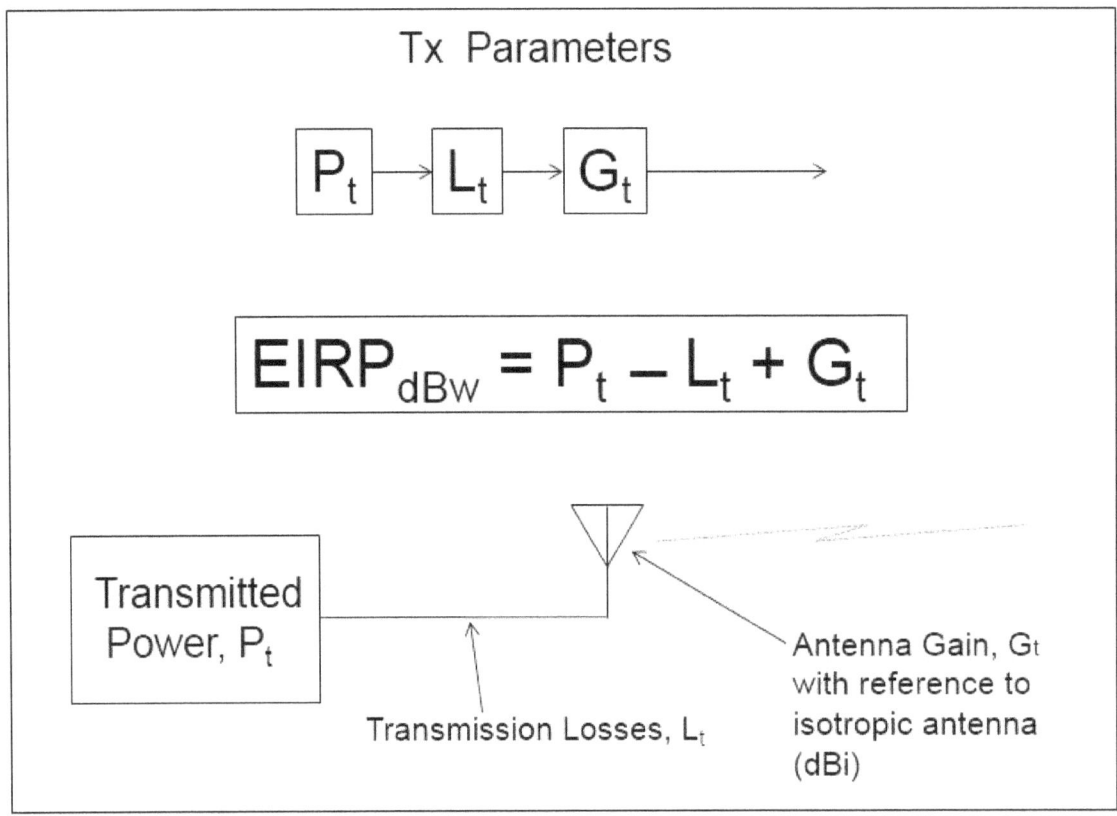

**Figure 10.2.a**

## 10.3 Link Parameters

**Figure 10.3.a** shows the link parameters, which includes *free space loss*, $L_{free}$, *atmospheric loss*, $L_{atmos}$, *additional losses* due to *reflection, refraction and scattering*, $L_{add}$. All these losses are added up to determine the *path loss*, PL.

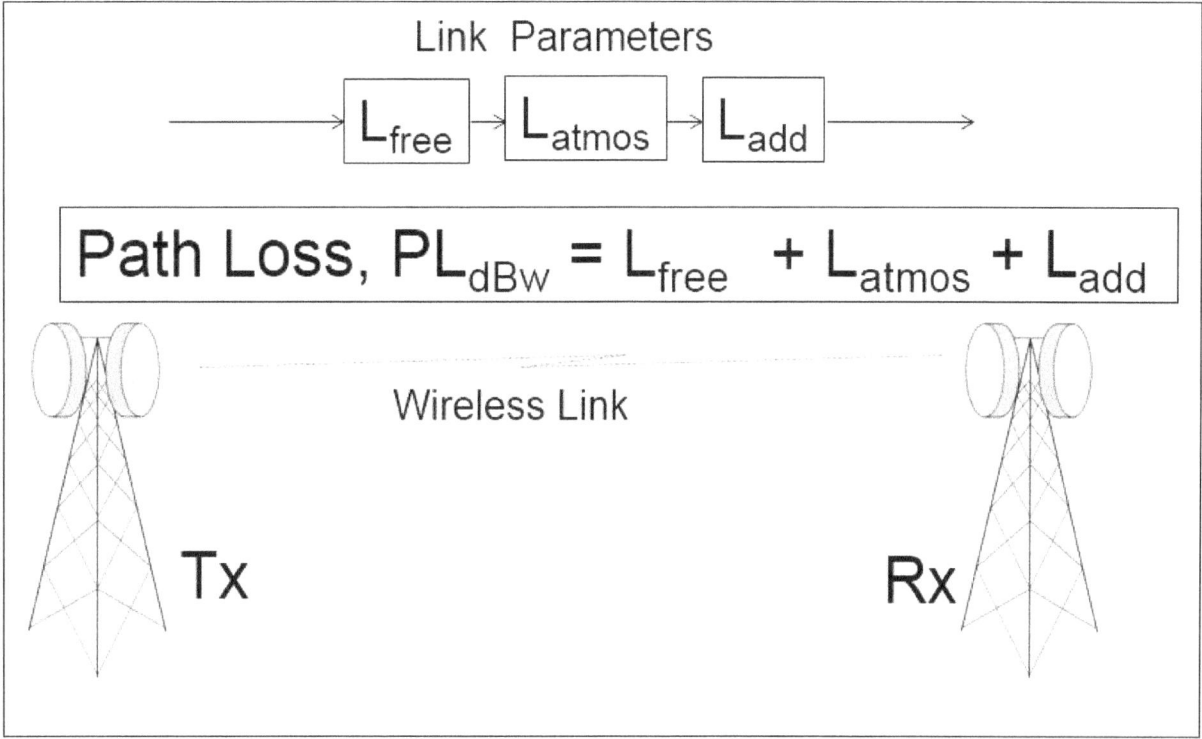

**Figure 10.3.a**

## 10.4 Receiver Parameters

**Figure 10.4.a** shows the receiver parameters, which includes receiver *antenna gain*, $G_r$, *transmission losses*, $L_r$, *received power*, $P_r$, *noise power from natural sources*, $N_{nat}$, *noise power from receiver antenna*, $N_{ant}$, *noise power from feeder*, $N_{fl}$, and *noise power from receiver equipment*, $N_r$.

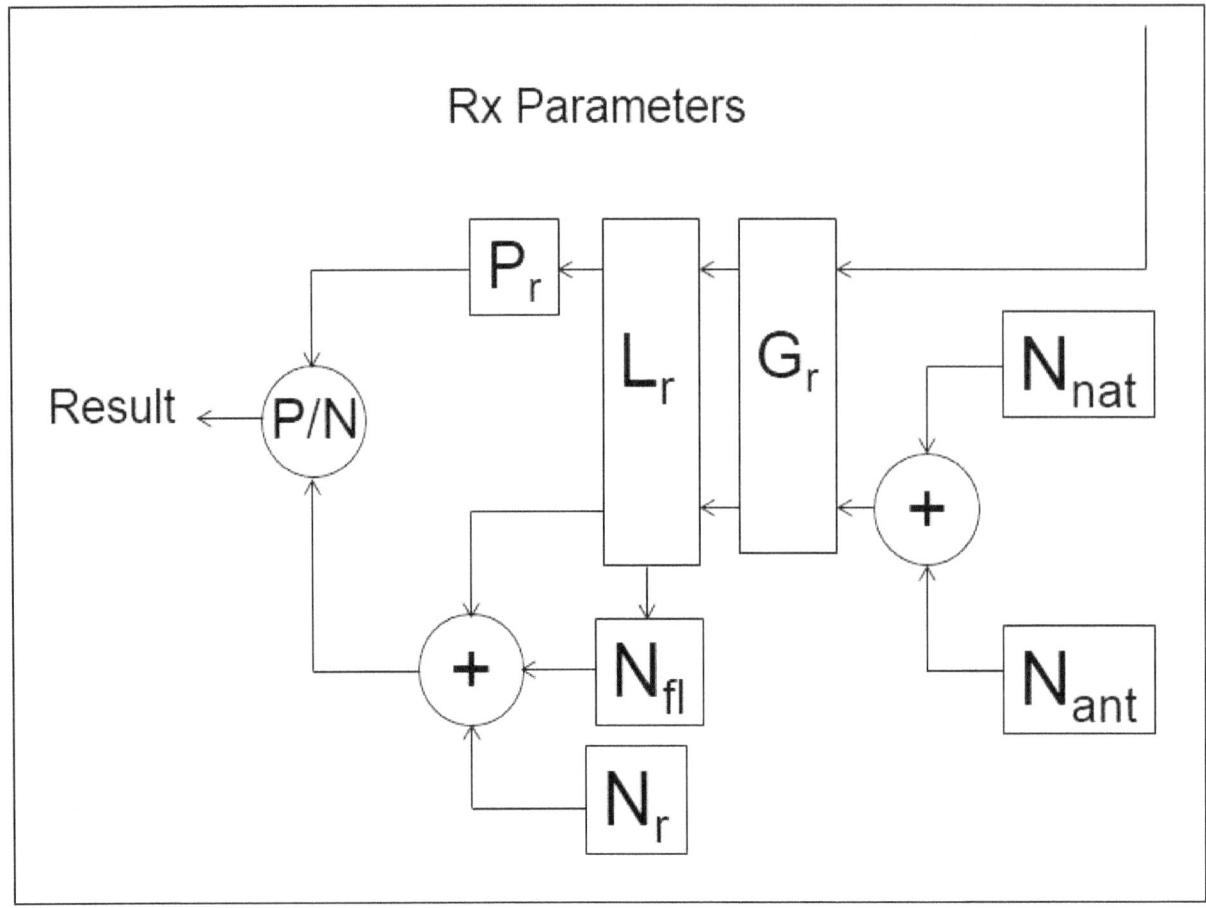

**Figure 10.4.a**

## 10.5 Signal Power at Receiver

**Figure 10.5.a** shows the formula to calculate *signal power* at the receiver.

Signal Power at Receiver

$$P_{r,dBw} = P_t - L_t + G_t - L_{free} - L_{atmos} - L_{add} + G_r - L_r$$

$$P_{r,dBw} = EIRP_{dBw} - PL_{dBw} + G_r - L_r$$

**Figure 10.5.a**

Notice that the *path loss* is subtracted from EIRP, because it reduces the *signal power*. Receiver *antenna gain* adds to the *signal power*, whereas receiver transmission losses reduce the net signal power at the receiver.

## 10.6 Noise Power

*Noise* is an unwanted random input to a wireless communications system. It is modeled as *additive white gaussian noise* (AWGN), for which the *power spectral density* is constant with respect to frequency.

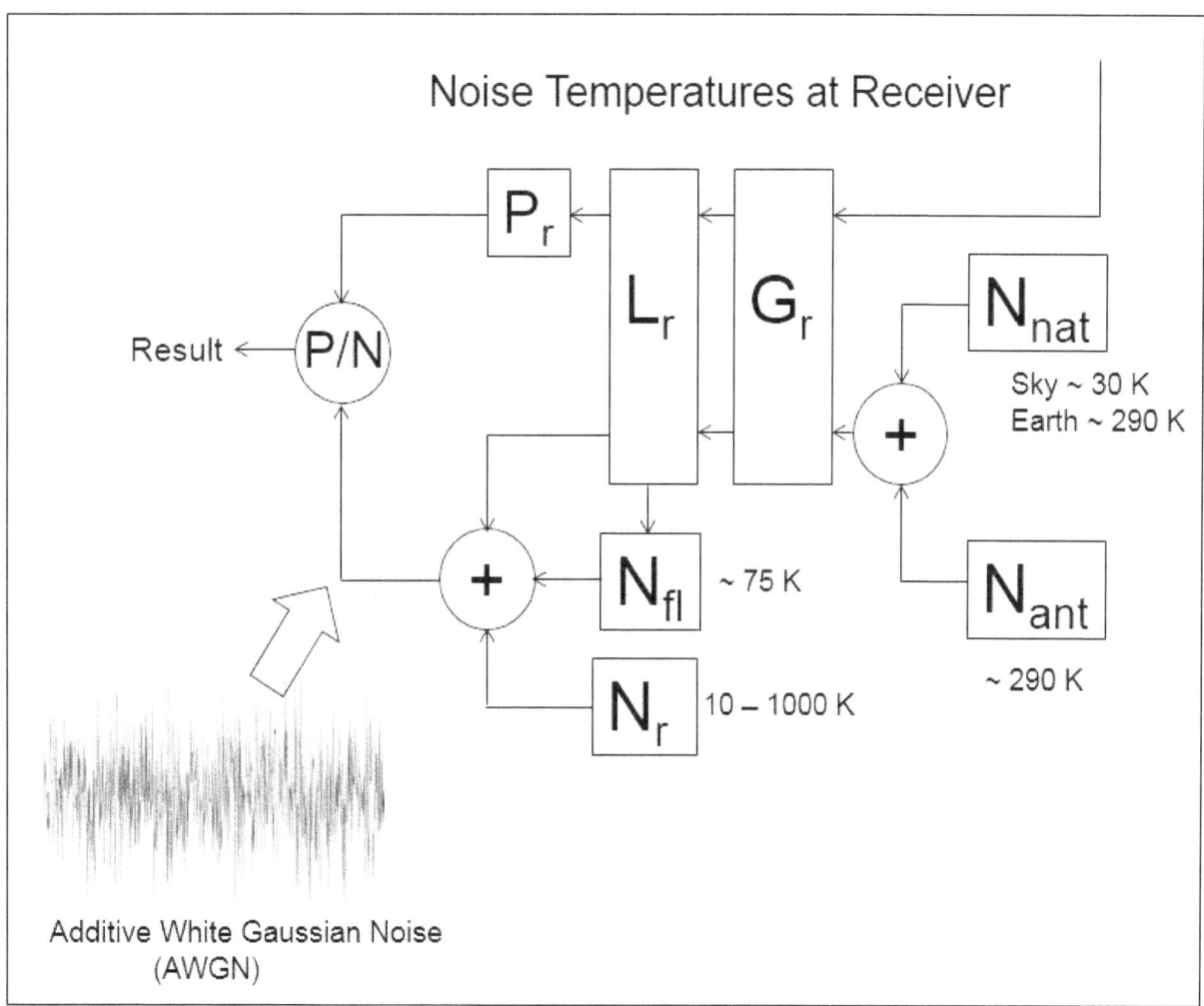

**Figure 10.6.a**

A physical object at absolute zero temperature, which is 0 K ( -273 deg C), does not generate noise, because nearly all molecular motion stops and there are no vibrations. However, real world objects have much higher temperature and therefore, generate noise. The antenna generates noise through its own temperature, which is typically assumed to be 290 K. It also picks up noise from earth, which is typically assumed to be at a temperature of 290 K, and from the sky, which is assumed to be at 30 K. The noise generated due to the feeder from receiver antenna is typically equivalent to a temperature of 75 K. The noise picked up from the cosmic background by deep space antennas is equivalent to a temperature of 2 K. The receiver equipment houses RF amplifiers, power supply, and other electronics that generate heat with a temperature of the order of 10 – 1000 K, and is a significant source of noise. (See **Figure 10.6.a**)

*Noise power* (N) is directly proportional to the *temperature* (T) and the frequency *bandwidth* (B) under consideration. The *Boltzmann's constant* (k) completes the mathematical relationship between N, T and B, as shown in **Figure 10.6.b**. The formula for calculation in *decibels* is shown in **Figure 10.6.c** along with an example of *noise power* calculation.

$$N = k \cdot T_e \cdot B$$
where k is Boltzmann's constant
$k = 1.38 \times 10^{-23}$ J / K
N is Noise Power in W
$T_e$ is equivalent Receiver Noise Temperature in K
B is Noise Bandwidth in Hz

Noise Power per Hz
$$N_o = k \cdot T_e$$

**Figure 10.6.b**

$$N_{dBw} = k_{dBw} + 10 \log T_e + 10 \log B$$
where k is Boltzmann's constant
$k = -228.6$ dBw/Hz.K
$T_e$ is equivalent Receiver Noise Temperature in K
B is Noise Bandwidth in Hz

Example: $T_e$ = 290 K, B = 1 MHz
$N = -228.6 + 10 \log 290 + 10 \log 10^6$
$N = -228.6 + 24.6 + 60$
$N = -144$ dBw

**Figure 10.6.c**

## 10.7 Noise Factor

The *noise factor*, F is a measure of noise added by an active device, such as an amplifier. **Figure 10.7.a** shows an amplifier terminated in 50 ohms resistance in both input as well as output. The resistance at the input of amplifier is assumed to be at a temperature of 290 K. It generates noise power $N_{in}$ that is input to the amplifier which has a gain of $G_r$. The input noise power gets amplified and appears at the output of amplifier as the product of gain $G_r$ and $N_{in}$, even when the amplifier is ideal and does not generate any internal noise. However, in real world, all amplifiers generate noise, and noise power at the output of an amplifier is greater by a ratio that is called *noise factor*, F. Considering F, the noise power at the output of a real amplifier, $N_{out}$ is given as product of F, $N_{in}$ and $G_r$.

**Figure 10.7.a** shows three alternate formulas for *noise factor*, F. It can be expressed as the ratio of *signal to noise ratio*, SNR at the input to the SNR at the output of amplifier. Alternately, it can also be expressed in terms of the ratio of amplifier temperature, $T_{re}$ to the temperature of resistance at the input, $T_0$.

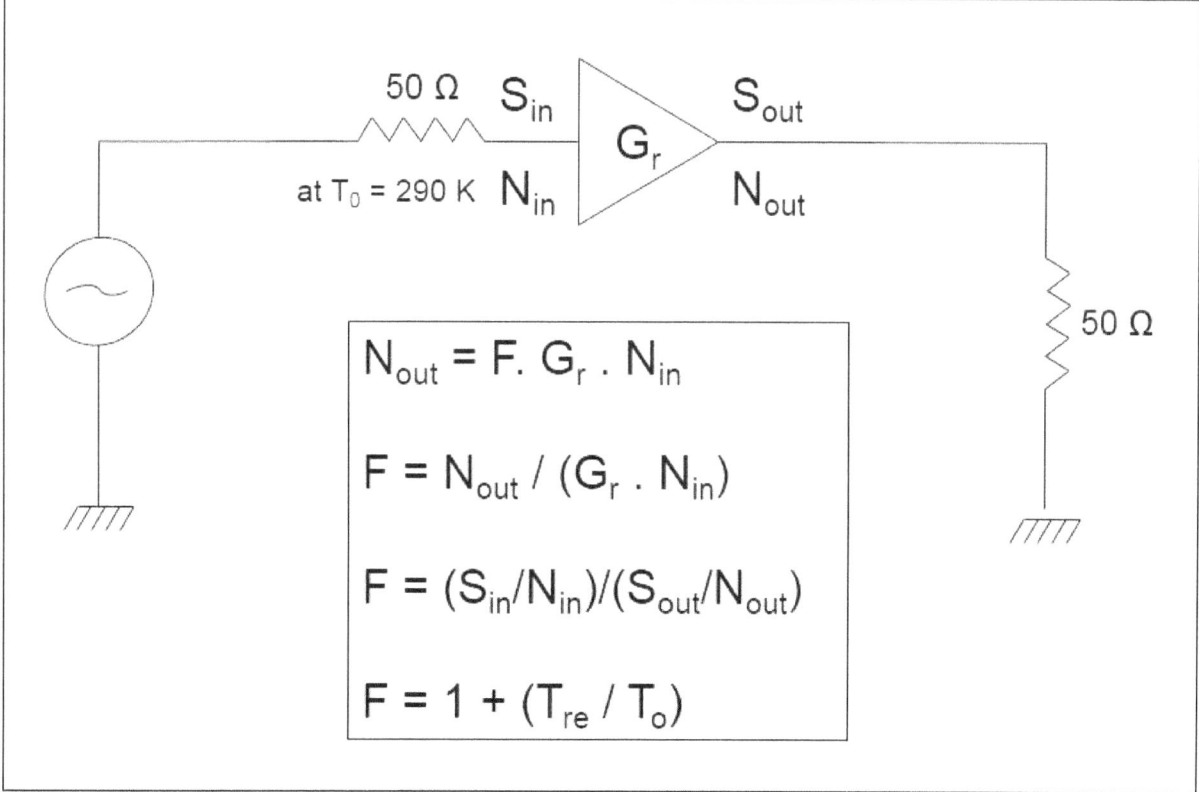

**Figure 10.7.a**

## 10.8 Noise Figure

The *noise figure*, NF is more practical term in use and is defined as the 10 log of *noise factor*, F. The *noise figure* is expressed in *decibels* (dB). It is common to find *noise figure* of a commercial RF amplifier specified in dB. **Figure 10.8.a** shows the formulas for *noise figure*.

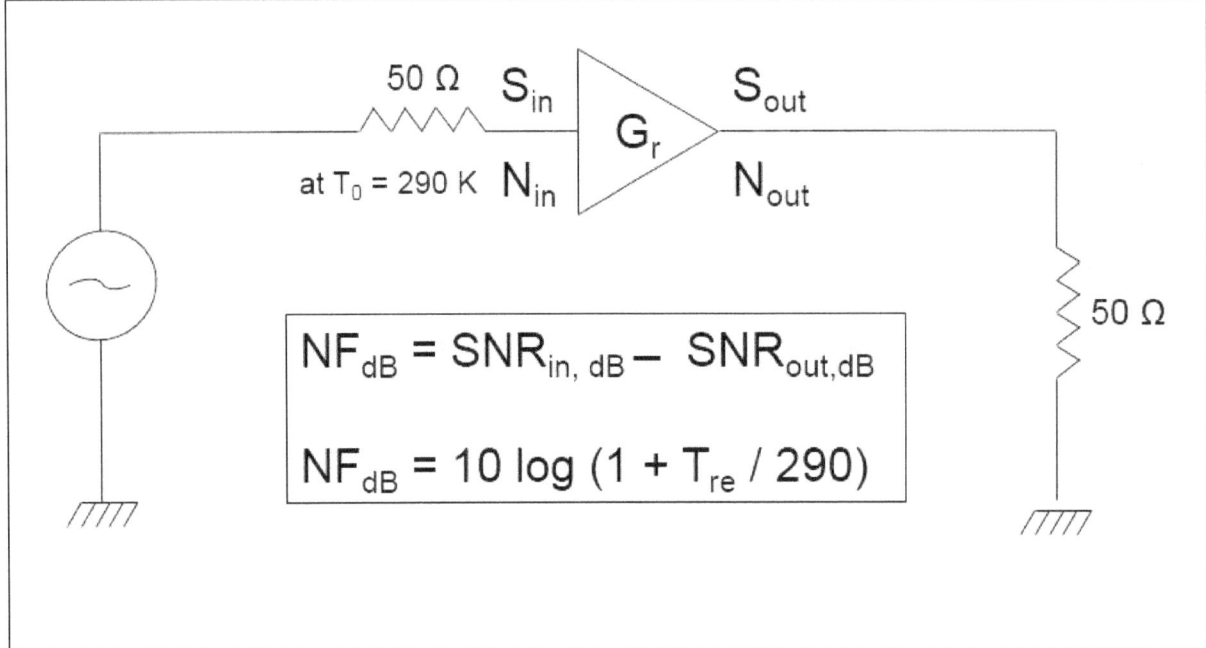

**Figure 10.8.a**

## 10.9 Noise Factor of Cascaded Sources

A typical radio equipment has a chain of cascaded devices. For example, a satellite receiver has a *low noise amplifier* (LNA) at the front end, followed by one or more *intermediate frequency* (IF) amplifiers, and finally a *mixer/demodulator* and a *power amplifier* for the output signal. Each of these devices contributes *noise power* proportional to its noise temperature, and the noise power is amplified with the *gain* of each amplifier stage.

**Figure 10.9.a** shows a representation of cascaded sources of noise, with each stage having a *temperature* $T_n$, *gain*, $G_n$ and *noise factor* $F_n$. Also, shown is the *Friis* formula to calculate the *noise factor* of the cascaded sources.

It is obvious from this formula that the noise factor of the 1st noise source in the chain is the most significant contributor to the total noise factor, F of the cascade. Therefore the first stage of a well designed RF system must have the best noise performance of all the stages and

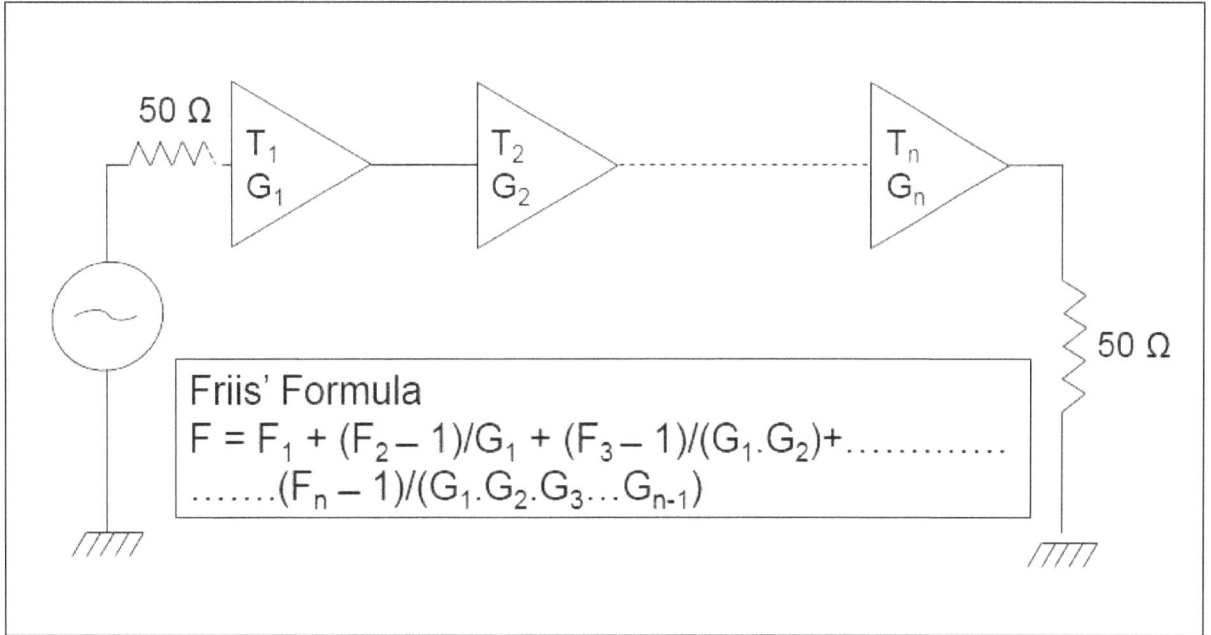

Friis' Formula

$$F = F_1 + (F_2 - 1)/G_1 + (F_3 - 1)/(G_1.G_2)+\ldots\ldots\ldots$$
$$\ldots\ldots(F_n - 1)/(G_1.G_2.G_3\ldots G_{n-1})$$

**Figure 10.9.a**

the highest gain. For example, in a satellite communication system, the presence of a well designed *low noise amplifier* (LNA) at the front end, which is usually mounted on the parabolic antenna, provides the best *signal to noise ratio* (SNR), and makes it possible to detect even the weakest of RF signals received from the communication satellite.

# 10.10    SNR at Receiver

Knowing the received signal power and noise power at the receiver helps to calculate the *signal to noise ratio* (SNR) at the receiver. **Figure 10.10.a** shows the formula for SNR. Notice that SNR in dBw is simply received *signal power*, Pr, minus the *noise power*, N, at the receiver.

Signal to Noise Ratio

$$SNR_{dBw} = P_{r,dBw} - N_{dBw}$$

where

$$P_{r,dBw} = P_t - L_t + G_t - L_{free} - L_{atmos} - L_{add} + G_r - L_r$$
$$N_{dBw} = -228.6 + 10\log T_e + 10\log B$$

**Figure 10.10.a**

## 10.11    Microwave Link Example

We will now walk through an example to calculate SNR for a microwave radio link with a frequency of 1 GHz and a distance of 30 km between the transmitter and receiver. The transmitter is assumed to have a power of 5W. The transmitting antenna has a gain of 34 dBi and receiving antenna has a gain of 34 dBi.

> # Step 1: Calculate EIRP of Transmitter
>
> ## Tx Parameters
> Transmitted Power, $P_t$ = 5 W
>  = 10 log 5 = 7 dBw
> Waveguide Loss, $L_t$ = 3 dB
> Transmitter Antenna Gain, $G_t$ = 34 dBi
>
> $$EIRP_{dBw} = P_t - L_t + G_t$$
> $$= 7\ dBw - 3dB + 34\ dBi$$
> $$= +38\ dBw$$

**Figure 10.11.a**

Step 1 in **Figure 10.11.a** involves calculation of *effective isotropic radiated power* (EIRP) of transmitter from the transmitted power, waveguide transmission losses and the gain of transmitter antenna.

Step 2 in **Figure 10.11.b** calculates the *path loss*, (PL), for the radio link between the microwave transmitter and receiver over a distance of 30 km. The *free space loss* and *atmospheric loss* are calculated. Additional losses due to reflection, refraction and scattering are assumed to be 0 dB.

In step 3 of **Figure 10.11.c**, the gain of receiver antenna and receiver feeder losses are determined from the technical specifications of the receiver.

Step 4 in **Figure 10.11.d** calculates the received signal power from EIRP, path loss, receiver gain and receiver feeder losses.

## Step 2: Calculate Path Loss, PL

## Link Parameters

Calculate Free Space Loss, $L_{free}$

$L_{free}$ (dBw) = 32.45 + 20 log f (MHz) + 20 log d (km)

f = 1 GHz , d = 30 km

$L_{free}$ (dBw) = 32.45 + 20 log 1000 + 20 log 30

$L_{free}$ (dBw) = 122 dBw

Calculate Atmospheric Loss, $L_{atmos}$

for f = 1 GHz, specific attenuation = 0.0025 dB/km

For d = 30 km,

$L_{atmos}$ = 0.0025 x 30 = 0.075 dBw

Assume, $L_{add}$ = 0 dBw

$$PL_{dBw} = L_{free} + L_{atmos} + L_{add}$$
$$= 122 \text{ dBw} + 0.075 \text{ dBw} + 0 \text{ dBw}$$
$$\sim 122 \text{ dBw}$$

**Figure 10.11.b**

## Step 3: Determine Receiver Gain and Feeder Losses

## Rx Parameters

Receiver Gain $G_r$ = 34 dBi

Receiver Feeder Losses, $L_r$ = 5 dB

**Figure 10.11.c**

---

## Step 4: Calculate Received Signal Power, $P_r$

$$P_{r,dBw} = EIRP_{dBw} - PL_{dBw} + G_r - L_r$$

$$= 38 \text{ dBw} - 122 \text{ dBW} + 34 \text{ dBi} - 5 \text{ dBw}$$

$$= -55 \text{ dBw}$$

---

**Figure 10.11.d**

Step 5 in Figure **10.11.e** calculates the total equivalent noise temperature at the receiver from the estimated values of noise temperature from natural sources, antenna temperature, noise temperature due to feeder and the receiver noise temperature. Notice that the receiver noise temperature is calculated from the *noise figure* (NF) value of the receiver, which is determined from the technical specifications of receiver.

Step 6 in Figure **10.11.f** calculates the noise power at the receiver from the equivalent noise temperature value determined in Step 5.

Finally, Step 7 in **Figure 10.11.g** calculates the signal to noise (SNR) ratio from the calculated values of received signal power in Step 4, and noise power at the receiver, N.

# Step 5: Calculate Noise Temperature at Receiver, $T_e$

## Rx Parameters

Assume antenna side lobes see half ground (290 K) and half sky (30 K)

Average Noise Temperature from natural sources, $T_{nat}$ = 160 K

Assume Antenna Noise Temperature, $T_{ant}$ = 290 K

Assume Feeder Noise Temperature, $T_{fl}$ = 75 K

Assume Receiver Noise Figure, NF = 6 dB

$NF_{dB} = 10 \log (1 + T_{re} / 290)$

$6 = 10 \log (1 + T_{re}/290)$

$0.6 = \log (1 + T_{re}/290)$

$10^{0.6} = 1 + T_{re}/290$

Effective Receiver Noise Temperature, $T_{re}$ = 864 K

$$T_e = T_{nat} + T_{ant} + T_{fl} + T_{re}$$

$$= 160 \text{ K} + 290 \text{ K} + 75 \text{ K} + 864 \text{ K} = 1389 \text{ K}$$

**Figure 10.11.e**

| Step 6: Calculate Noise Power at Receiver, N |
|---|
| For $T_e$ = 1389 K , B = 10 MHz<br>$N_{dBw}$ = $-$ 228.6 + 10 log $T_e$ + 10 log B<br>$\qquad$ = $-$ 228.6 + 10 log 1389 + 10 log $10^7$<br>$\qquad$ = $-$ 127.17 dBw |

**Figure 10.11.f**

| Step 7: Calculate SNR at Receiver |
|---|
| $SNR_{dB}$ = $P_{r,dBw}$ $-$ $N_{dBw}$<br>$\qquad$ = $-$ 55 dBw $-$ ($-$ 127.17 dBw)<br>$\qquad$ = 72.17 dBw |

**Figure 10.11.g**

## 10.12    Noise Floor

The noise power that was calculated in **Section 10.11**, establishes the *noise floor* for the receiver in question. The *noise floor* of a receiver is essentially the sum of noise power from all sources. It represents the noise power that must be overcome by the received signal power, for the received signal to have any meaningful value. In other words, received signal power must be greater than the *noise floor* of a receiver.

**Figure 10.12.a** is a bar chart representation of the link budget analysis for the microwave link example, and indicates all *transmit, link* and *receive* parameters that were used, including their mathematical relationships. Notice the *noise floor* and *received signal power values* calculated for the given example.

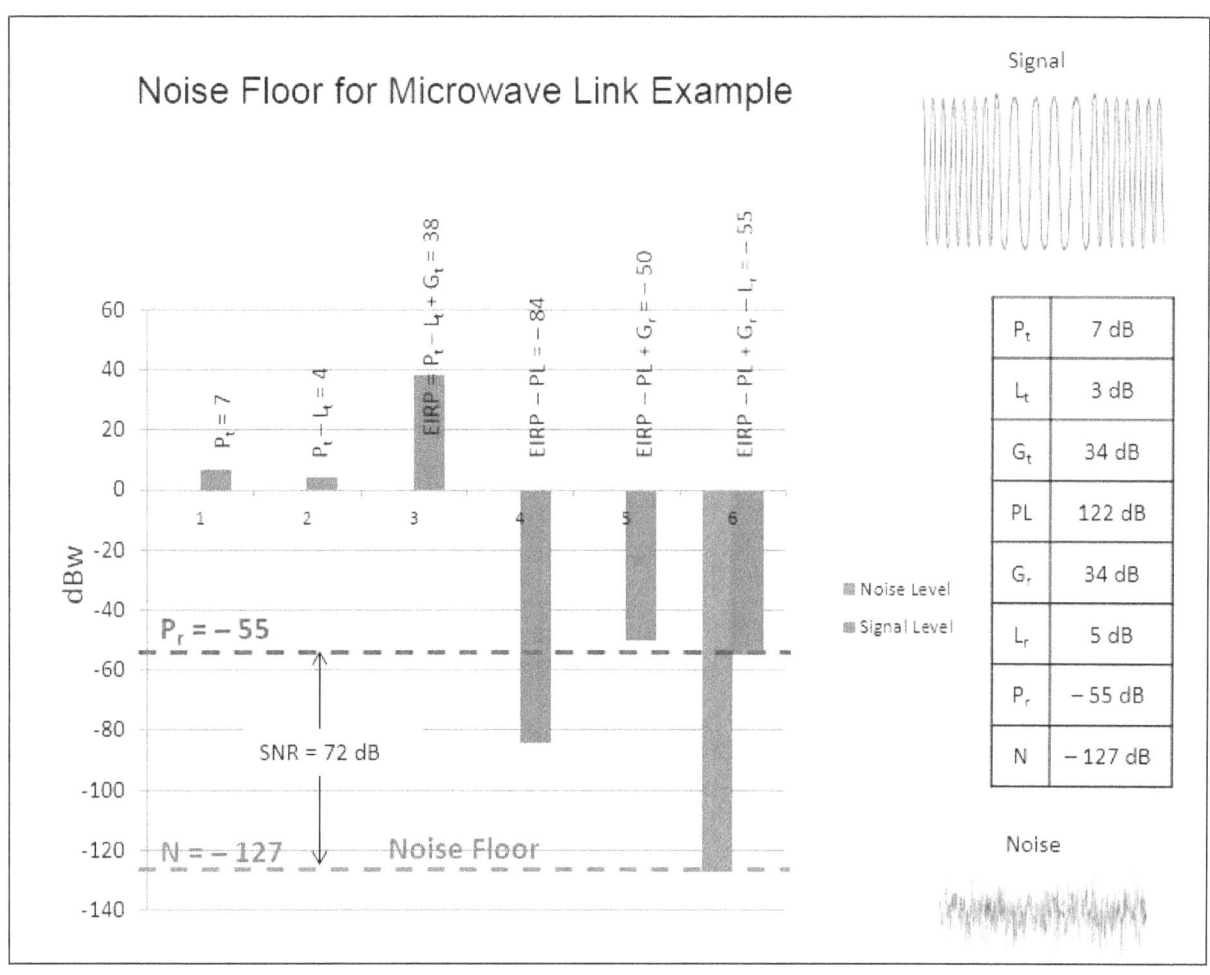

**Figure 10.12.a**

## 10.13   Fade Margin

In the link budget example of **Section 10.11**, a received signal power level of -55 dB is established. However, this level of received signal power level cannot be maintained over long periods of time for the microwave radio link. This is because the propagation path conditions may change because of changes in the environment. There is a distinct probability of drop in signal power level for brief periods of time, which may be as high as 30 dB or more. This phenomenon is called *fading*, and it must be accounted for when sizing or dimensioning a wireless communication link.

For a digital wireless communication link, there is a need to meet the acceptable *bit error rate* (BER) performance requirement. In order to account for *fading* effects, there is a need to determine what percent of the time the BER performance requirement can be met. This is called *time availability* of the radio link. If the time availability of a radio link is 99%, then the link meets BER performance requirement 99% of the time, this implies that the radio link is not available 1% of the time.

| Time Availability (%) | Required Fade Margin (dB) |
|---|---|
| 90.000 | 8 |
| 99.000 | 18 |
| 99.900 | 28 |
| 99.990 | 38 |
| 99.999 | 48 |

**Table 10.13.a**

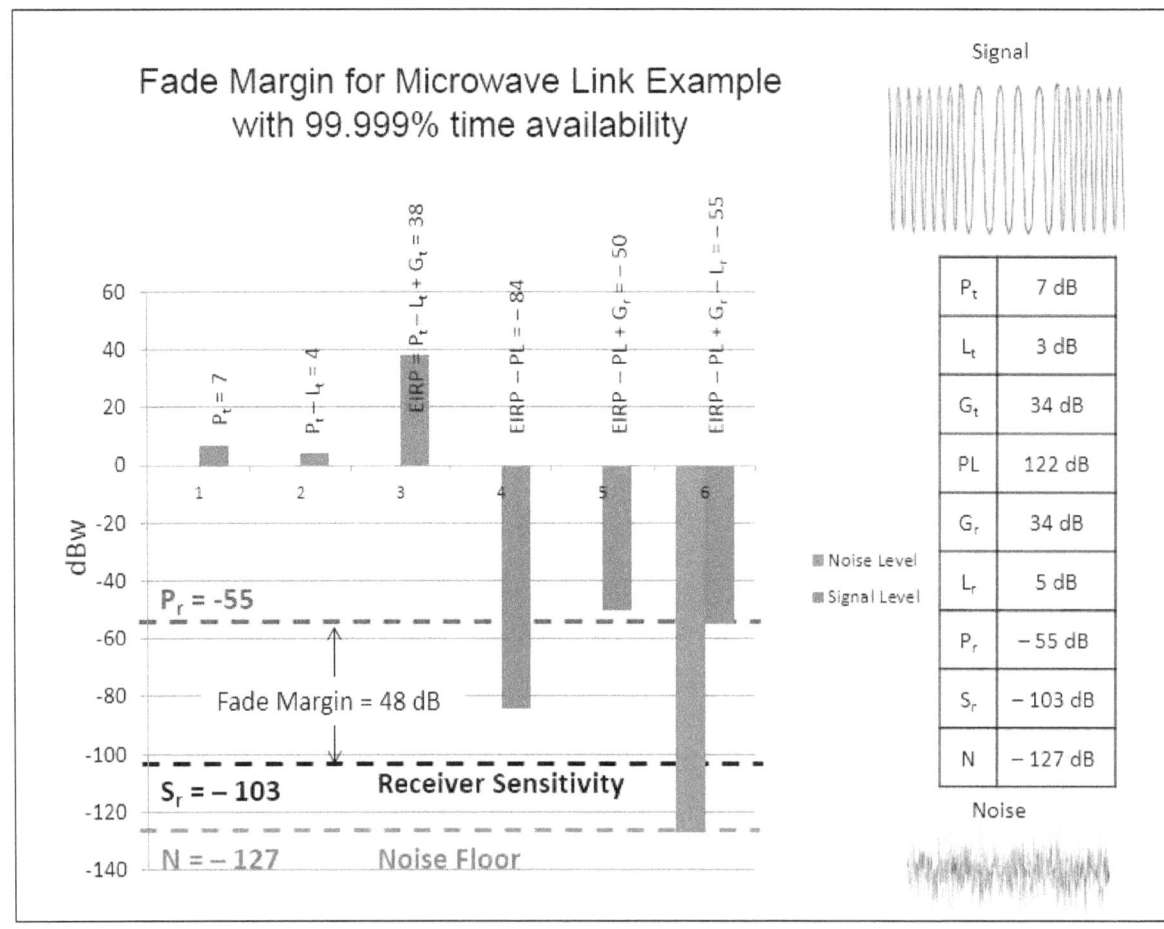

**Figure 10.13.b**

To improve *time availability* of a radio link, there is a need to increase *fade margin* and determine how much additional signal power is needed to overcome the effects of fading.

One approach is to assume the worst case scenario and use *Rayleigh distribution*. Using this approach, the recommended fade margins as shown in **Table 10.13.a** can be used.

For the microwave link example, if we assume 99.999% *time availability*, then from **Table 10.13.a**, the required *fade margin* is 48 dB. This is shown on the bar chart representation of link budget in **Figure 10.13.b**. Notice that with a fade margin of 48 dB, the receiver must have a *sensitivity* of -103 dB for signal power. This means that the receiver must be capable of detecting the presence of signal even if the signal power fades by 48 dB to a level of -103 dB.

# 10.14    $E_b / N_o$

$E_b / N_o$ is an important ratio for digital radio links. It is a measure of *signal to noise (SNR) ratio per bit* of data transmitted on a radio link. It is useful when comparing BER performance of digital modulation schemes in a radio link. See **Figure 10.14.a** for the formulas that can be used for calculations.

**Figure 10.14.b** shows a calculation for microwave link example. The $E_b / N_o$ comes out to be 34.16 dB for a data rate of 1 Mbps at the calculated received power level of -103 dB accounting for fade margin and equivalent receiver temperature of 1389 K.

**Figure 10.14.c** shows how BER vs $E_b / N_o$ curves can be used to determine if at a specific digital modulation technique, if the target bit rate satisfies the $E_b / N_o$ requirement. The example here shows that for QPSK modulation and BER of $10^{-8}$ the $E_b / N_o$ requirement is 12 dB. This means that the energy per bit must be at least 12 dB. Recalculating $E_b / N_o$ for the microwave link example reveals that even if we increase the bit rate to 100 Mbps for the radio link, the $E_b / N_o$ is calculated to be 14.16 dB, which is higher than the minimum required ratio of 12 dB. Hence, 100 Mbps bit rate can be supported for the given microwave radio link.

$$E_b = P_r - 10 \log R$$

where

$E_b$ is Energy per bit

$P_r$ is Received Signal Power Level in dB

R is bit rate in bps

---

$$N_0 = k_{dBw} + 10 \log T_e$$

where

$N_0$ is Noise Power Spectral Density (Noise Power per Hz)

$k = -228.6$ dBw/Hz.K is Boltzmann's Constant

$T_e$ is Equivalent Receiver Noise Temperature in K

---

$$E_b / N_o = P_r - 10 \log R + 228.6 - 10 \log T_e$$

where

$P_r$ is Received Signal Power Level in dB

R is bit rate in bps

$T_e$ is Equivalent Receiver Noise Temperature in K

**Figure 10.14.a**

## Calculate $E_b/N_o$ for Microwave Link Example

For $P_r = -103$ dB (accounting for fade margin)
R = 1 Mbps, $T_e$= 1389 K

$$E_b/N_o = P_r - 10 \log R + 228.6 - 10 \log T_e$$
$$= -103 - 10 \log 10^6 + 228.6 - 10 \log 1390$$
$$= 34.16 \text{ dB}$$

**Figure 10.14.b**

From BER vs $E_b/N_o$ curves
For BER = $10^{-8}$ , Modulation = QPSK
Required $E_b/N_o$ = 12 dB

Available $E_b/N_o$ = 34 dB at 1 Mbps

Recalculating $E_b/N_o$ at 100 Mbps
$$E_b/N_o = P_r - 10 \log R + 228.6 - 10 \log T_e$$
$$= -103 - 10 \log (10^8) + 228.6 - 10 \log 1390$$
$$= 14.16 \text{ dB}$$

**Figure 10.14.c**

## 10.15    Figure of Merit

The *figure of merit* is the ratio of *antenna gain* to *antenna temperature* as shown in **Figure 10.15.a**, and is measured in dB/K. It is an important measure for satellite communication links.

Figure of Merit = G / T
where
G is antenna gain in dB
T is antenna temperature in K

**Figure 10.15.a**

**Figure 10.15.b** shows how in order to satisfy the INTELSAT earth station requirement for *figure of merit* (G/T), the noise temperature of antenna must be brought down to 85.13 K, which is actually -187.87 deg C. This is achieved by cryogenically cooling down the antenna, and also by having *low noise amplifiers* (LNA) at the front end.

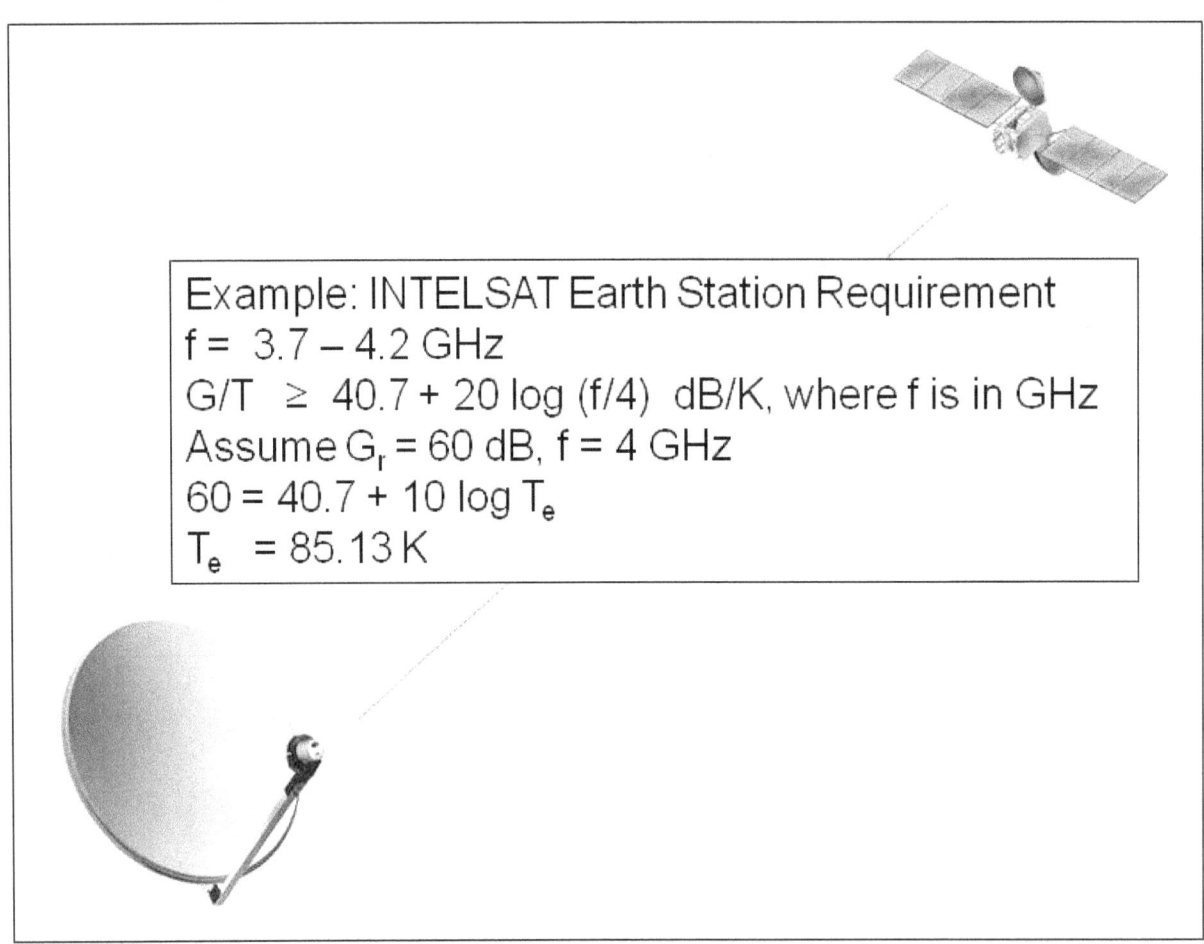

Example: INTELSAT Earth Station Requirement
$f = 3.7 - 4.2$ GHz
$G/T \geq 40.7 + 20 \log (f/4)$ dB/K, where f is in GHz
Assume $G_r = 60$ dB, $f = 4$ GHz
$60 = 40.7 + 10 \log T_e$
$T_e = 85.13$ K

**Figure 10.15.b**

# 11. RF Site Surveys

In this chapter, we will get familiar with various steps involved in RF site survey for indoor and outdoor wireless networks.

## 11.1 Indoor Wireless Network

**Figure 11.1.a** shows the steps for site survey for an indoor wireless network, such as a

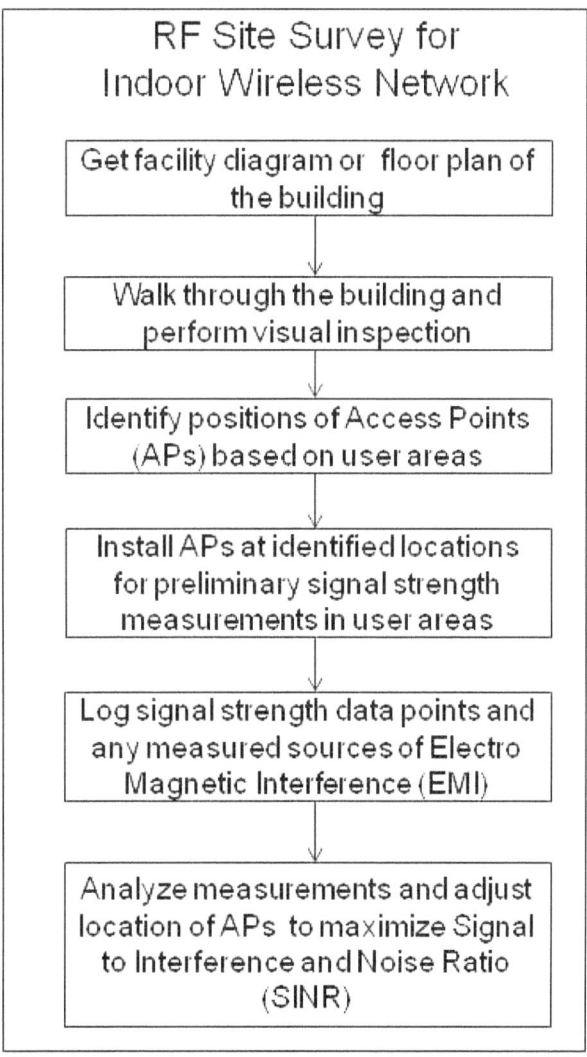

**Figure 11.1.a**

WLAN based on IEEE 802.11 standard or a *femto cell* site for 2G, 3G or 4G LTE telecommunication standards. It really begins with access to the building layout and the floor plan to facilitate a walk through and visual inspection of the user areas. The user density at specific locations in the building and the floor area to be covered by the RF signal determines the number of *access points* (AP), and their distribution throughout the building. It is a good practice to install APs at key locations and measure the preliminary signal strength in the target user areas. Also needed is measured data about any sources of *electromagnetic interference* (EMI) from electrical equipment installed in the building or in nearby area or from other nearby radio transmitters in adjacent frequencies. The collected data about received signal strength and interference is analyzed and the location of APs is adjusted to maximize the *signal to interference and noise ratio* (SINR) in the user areas.

Notice the difference between SINR and SNR. Whereas SINR measures the ratio of *signal to (interference + noise)*, SNR only measures the ratio of *signal to noise*. SINR also accounts for deterioration in the quality of received signal because of adjacent frequencies from other nearby transmitters and EMI. Therefore, SINR is a more objective measure of signal quality in real world situations.

# 11.2 Outdoor Wireless Network

**Figure 11.2.a** shows steps for site survey for an outdoor wireless network, such as a micro or a macro cell site for 2G, 3G or 4G LTE or IEEE 802.16 WiMax standard. The steps are similar to those for indoor wireless network, but different means are used because of nature of the problem. It is expensive to relocate cell towers, therefore the cell site must be chosen carefully after survey of the terrain in the coverage area, and before the installation of the tower and the antennas. Thereafter, drive tests are performed through the user areas and signal strengths and interference levels are measured and logged. The measured data is analyzed and used to make recommendations about the transmit power level, antenna tilt, cell splitting and cell sectoring to improve signal quality in the target user areas.

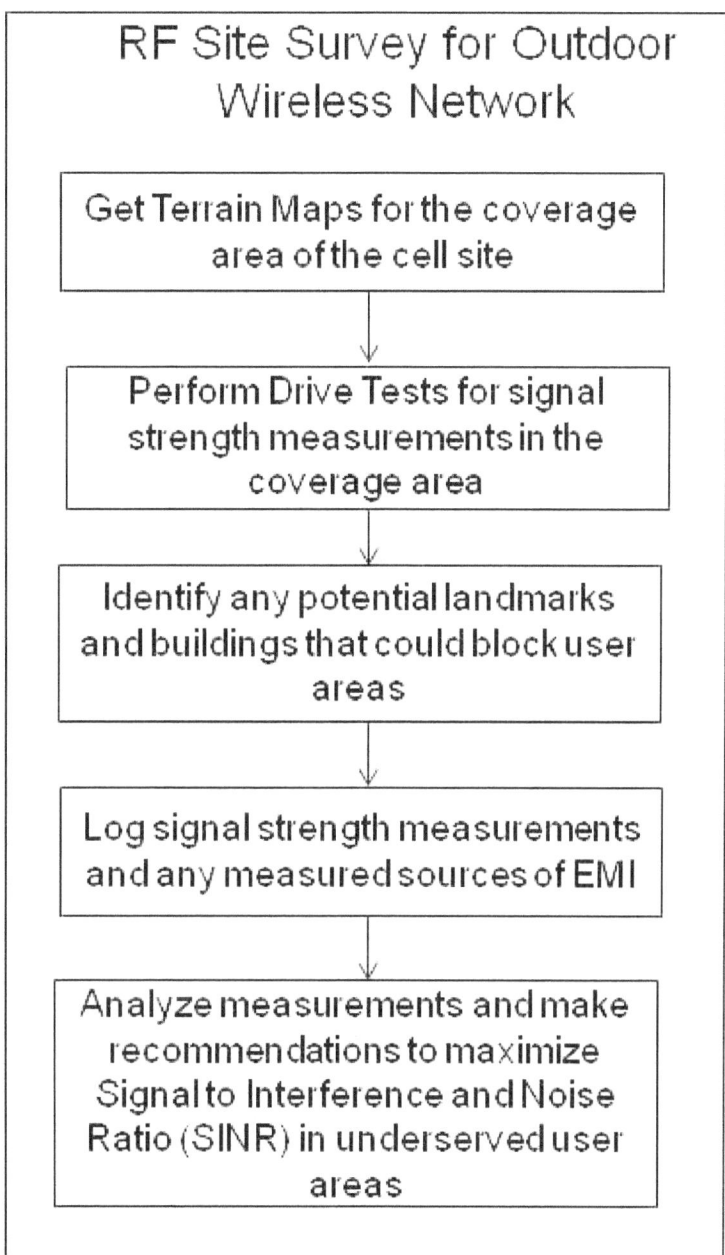

**Figure 11.2.a**

# 12.  Antenna Diversity

In this chapter we will learn about various antenna diversity techniques that can be used to overcome the effects of *fading* and *inter symbol interference* (ISI) in RF propagation.

## 12.1 Multipath Propagation

Before we begin to study antenna diversity techniques, let us first understand the motivation and need behind antenna diversity. In radio links, unlike wired links, the radio signal does not travel over a single physical path from the transmitter to the receiver. It is common for the radio receiver to receive multiple copies of the radio signal that have travelled over paths of different lengths, and therefore arrive at different times. **Figure 12.1.a** shows a mobile radio receiver, which receives three radio signals travelling over three different paths of propagation.

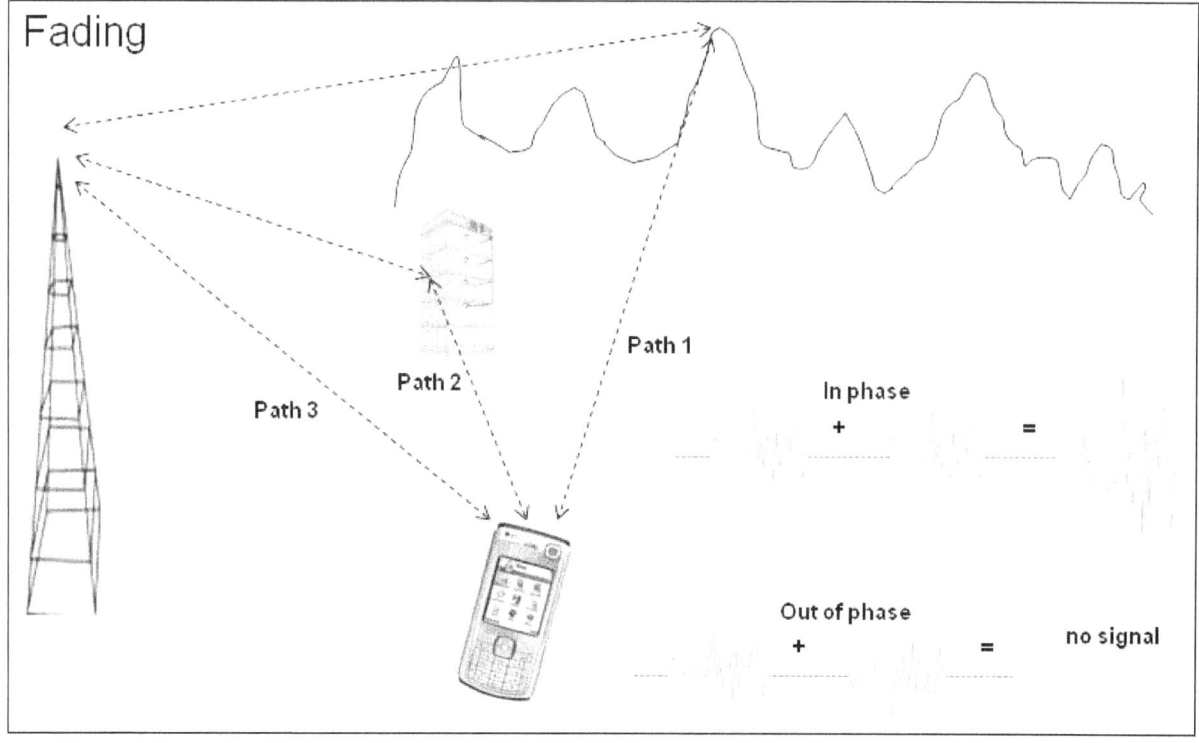

**Figure 12.1.a**

Path 1 has radio waves reflected from the top of a hill. Path 2 involves reflection from the side of a tall building and Path 1 is *line of sight* (LOS). The radio signals will arrive at the receiver with different amplitudes and phases. If the signals are 180 deg out of phase, they will cancel out each other, resulting in no signal at the receiver. If the signals are in phase, they will add up, resulting in a strong signal at the receiver. When the mobile receiver moves around, the phases and amplitudes of radio signals received over different propagation paths vary in a random manner.

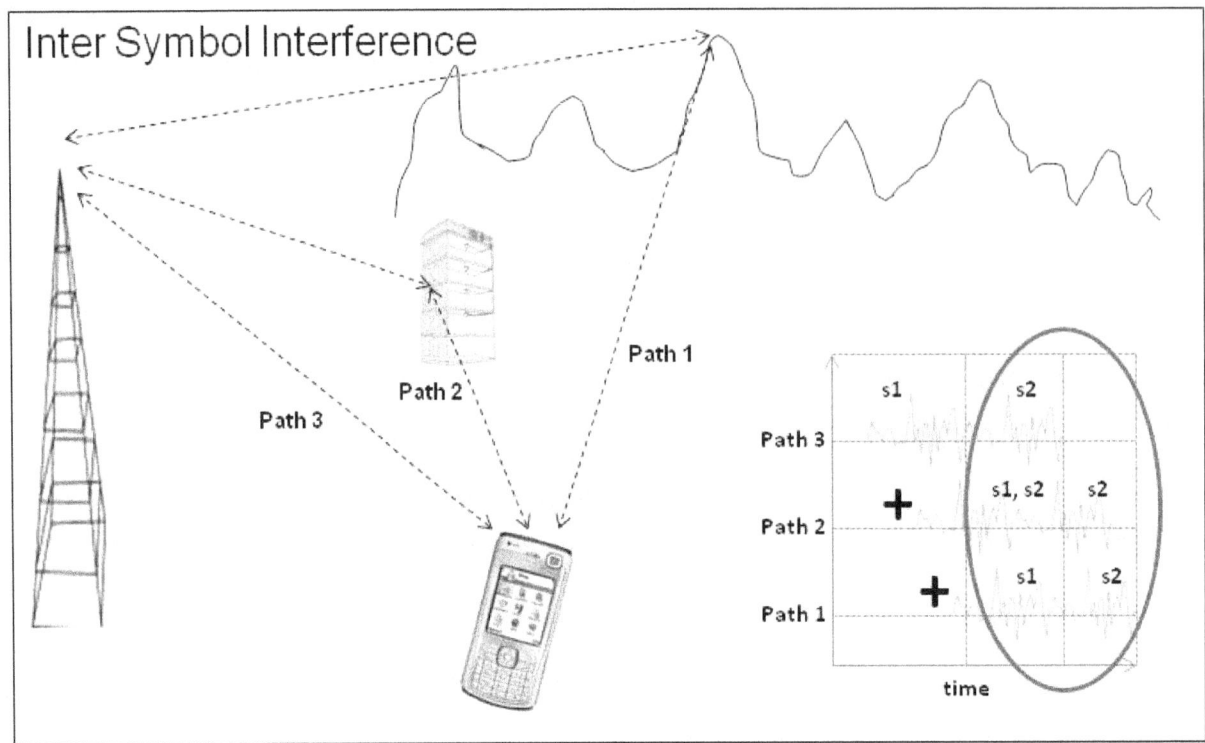

**Figure 12.1.b**

Since there are in reality several propagation paths, especially in a dense urban environment, the signals from different paths rarely cancel out each other completely. However it is common to have a drop in signal level of about 20 – 30 dB, and this drop in signal level causes *fading*.

For digital wireless communication links, multipath propagation also results in *inter symbol interference* (ISI). The receiver sees the adjacent symbols received over different paths as superimposed, which appears as useless or corrupted data. **Figure 12.1.b** shows symbols s1 and s2 received over LOS Path 3. Path 2 has a time delay for s1 and s2. Path 1 has even greater time delay for s1 and s2. The symbols received from the three different paths get superimposed, for example, s1 and s2 get corrupted due to superposition and hence the data contained in these symbols loses its integrity and becomes meaningless.

## 12.2 Frequency Diversity

One method to mitigate the effects of *fading* is to deploy *frequency diversity*. It involves transmission of same information over different RF carriers, each of which fades independently of the other.

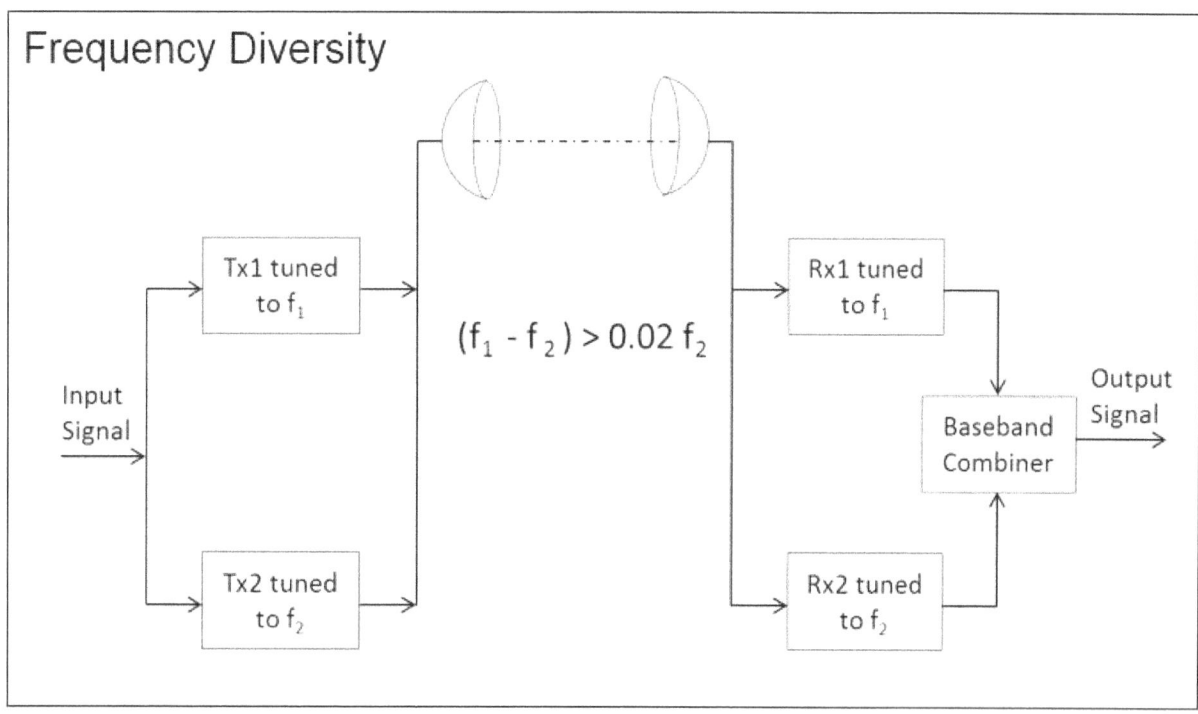

**Figure 12.2.a**

**Figure 12.2.a** shows a *frequency diversity* RF system, which has 2 transmitters, Tx1 and Tx2, tuned to carrier frequencies f1 and f2, and 2 receivers Rx1 and Rx2, tuned to carrier frequencies f1 and f2 respectively. *Frequency diversity* uses the fact that fading period differs for frequencies that are 2-5% apart. The input signal is simultaneously transmitted over frequencies f1 and f2, and also received over f1 and f2. A baseband combiner at the receiver combines the 2 signals received over f1 and f2. The probability that both signals will fade simultaneously is considerably smaller, because each signal fades independently of the other. In this way, the fading effects are mitigated by having a redundant copy of information on a second frequency.

Frequency diversity ensures higher reliability for radio propagation path. In addition, it provides a simple and full redundancy of radio equipment, and therefore avoids a single point of failure for the transmitter or the receiver equipment.

Despite its benefits, frequency diversity has few disadvantages. It is more complex, almost doubles the cost of equipment and needs more allocation of frequency spectrum which is an expensive resource.

## 12.3 Space or Receive Diversity

The *space or receive diversity* technique uses the well-known fact that radio waves travel along multiple paths of propagation. So, if multiple antennas are deployed, which are spaced sufficiently apart, it is possible to receive multiple independently fading signals. This method of diversity also reduces the probability that multiple received signals will fade simultaneously.

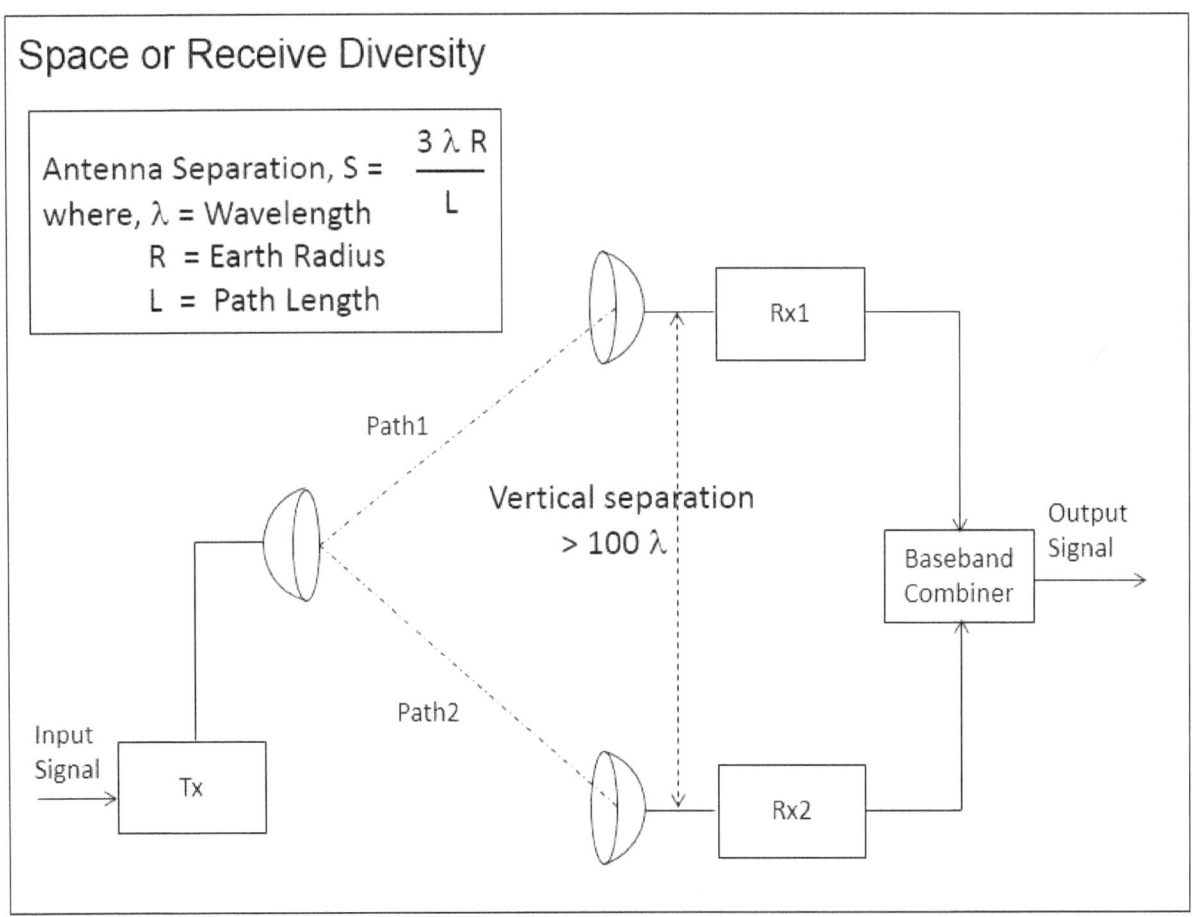

**Figure 12.3.a**

**Figure 12.3.a** shows a space diversity radio system, which uses 2 receivers Rx1 and Rx2 with their receiving antennas separated by at least 100 times the wavelength, λ. Notice that the radio signal has only one frequency, and 2 antennas are used to receive the radio signals from 2 different directions. It is required that the 2 signals be independently fading, so sufficient separation between the receiving antennas is very important to realize the full benefits of space diversity. The figure also shows a formula to estimate the antenna separation if the frequency and length of propagation path are known.

## 12.4 Time Diversity

The *time diversity* technique involves multiple transmissions of data packets, with each transmission separated from the next repeat transmission by a time interval, T. In order to reap the benefits of transmission redundancy, each transmission of a packet must *fade* independently of the next transmission. A quantity called *coherence time,* τ of a radio channel is defined, which is the duration of time for which impulse response of the radio channel is expected to remain constant. In other words, the propagation characteristics of the channel will not change for the duration of *coherence time*.

Once the coherence time for the channel is known, it is easy to figure out the time interval, T between consecutive repeat transmissions of a packet. T must be greater than the *coherence time* for the channel to ensure that each repeat transmission has independent *fading* characteristics.

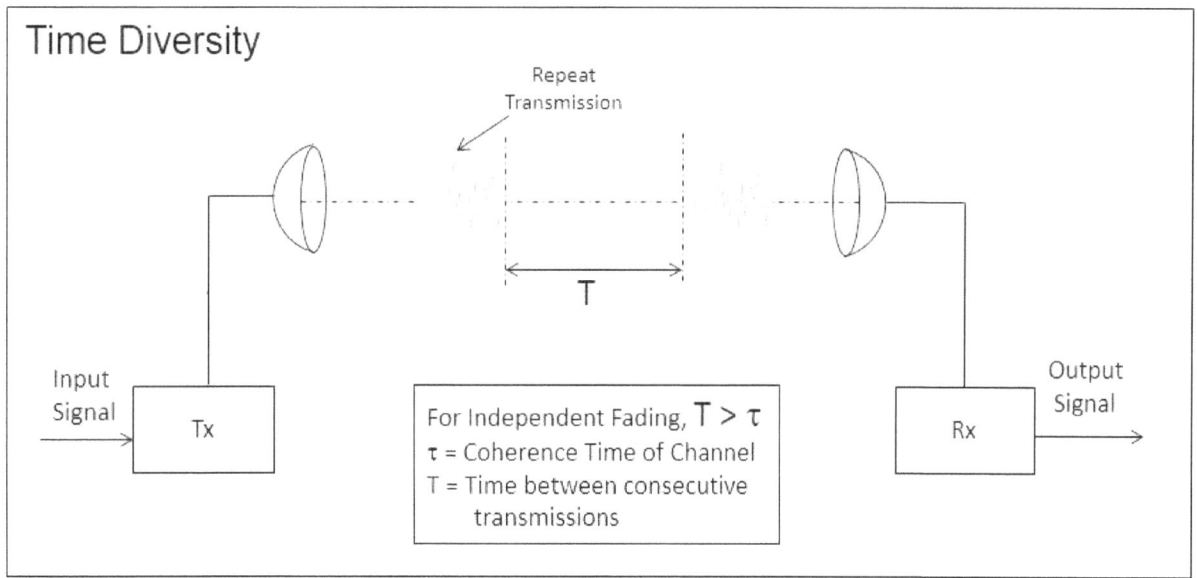

**Figure 12.4.a**

**Figure 12.4.a** shows a radio system using time diversity. Also shown is the criteria for independent fading.

## 12.5 Polarization Diversity

The *polarization diversity* technique uses redundant transmission of packets over the horizontal and vertical polarizations of the radio wave, H and V. **Figure 12.5.a** shows how the two versions of the signal, horizontal and vertical, are transmitted. There is very little correlation between horizontal and vertical components of the radio wave and they are independently fading.

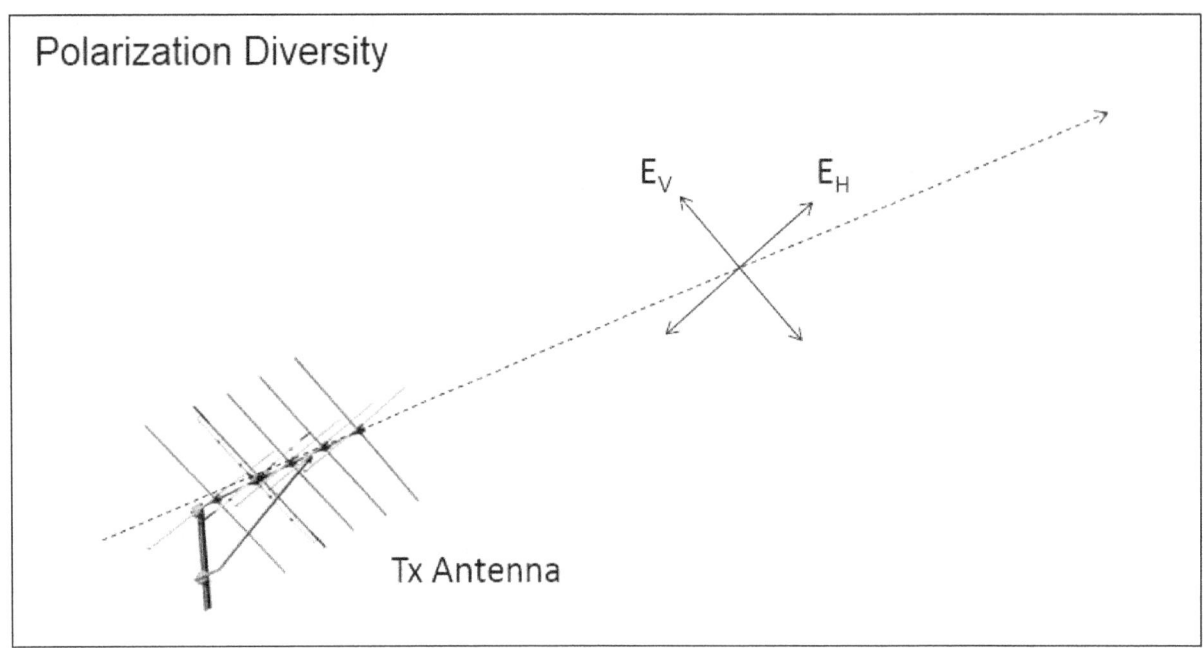

**Figure 12.5.a**

*Polarization diversity* technique has useful implications for hand held wireless terminals. The actual orientation of a hand held terminal is usually somewhere between horizontal and vertical, so if it uses polarization diversity, it will not completely lose the signal due to fading. Either the horizontal or the vertical component will compensate for fading, because both components will not fade at the same time.

# 12.6 Pattern (Angle) or Transmit Diversity

In *transmit diversity* technique, multiple antennas are deployed on the transmitter, so that each antenna transmits its own independent beam, which can be spatially distinguished by the receivers. If each independent radio beam carries a copy of the same radio signal, which is transmitted over different coordinates of space and time, then we have built in redundancy, because not all beams will fade at the same time.

The *space time block codes* are used to insert time delays in the transmission paths of the signal, so that individual signals can be separated. These time delays can range between 50 and 200 ns.

The benefits of pattern, angle or transmit diversity are, reduced effects of *fading* and *inter symbol interference* (ISI) due to multipath propagation. See **Figure 12.6.a** for illustration.

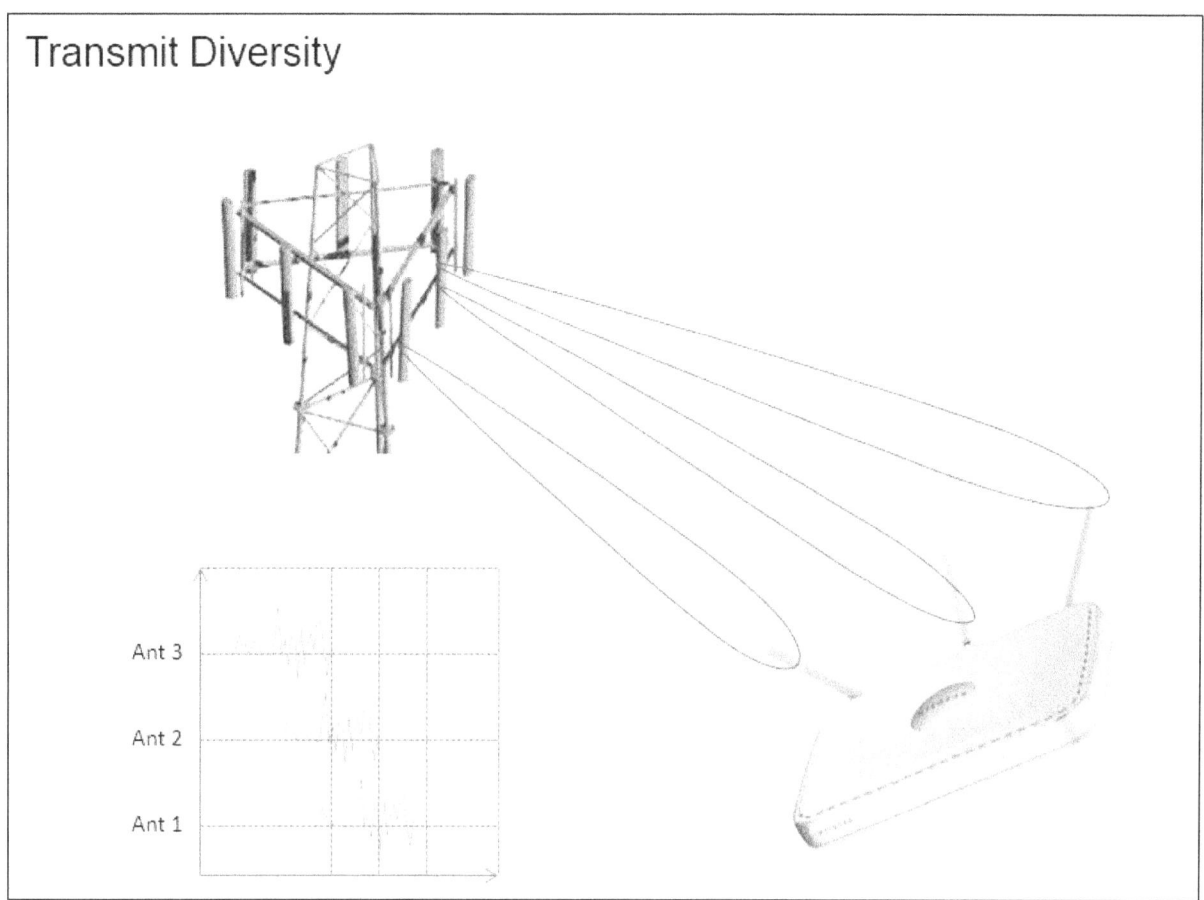

## Transmit Diversity

Ant 3

Ant 2

Ant 1

**Figure 12.6.a**

# 12.7 Diversity Combining Techniques

In this section, we will study various diversity combining techniques that are used to combine multiple radio signals.

## 12.7.1    Pre-detection Combining

In *pre-detection combining* technique, diversity combining happens before the detection of baseband signal in the receiver, that is, at the *intermediate frequency* (IF) stage. **Figure 12.7.1.a** shows an illustration of pre-detection combining in a radio receiver. The input signal 1 and input signal 2 enter their own respective mixers and after IF amplification, are fed into IF hybrid combiner, before sending the combined signal to the detector stage. Notice that the two input signals are actually copies of the same original signal, each of them having travelled on an *independently fading* path of propagation. Multiple copies of the signal reduce the probability of fading of the combined signal that is fed to the baseband signal detector.

This technique uses only a single baseband signal detector stage, however, there are multiple mixer and IF amplifier stages, one for each incoming signal. Also, an IF hybrid combiner is needed that adds to the cost and complexity of the system.

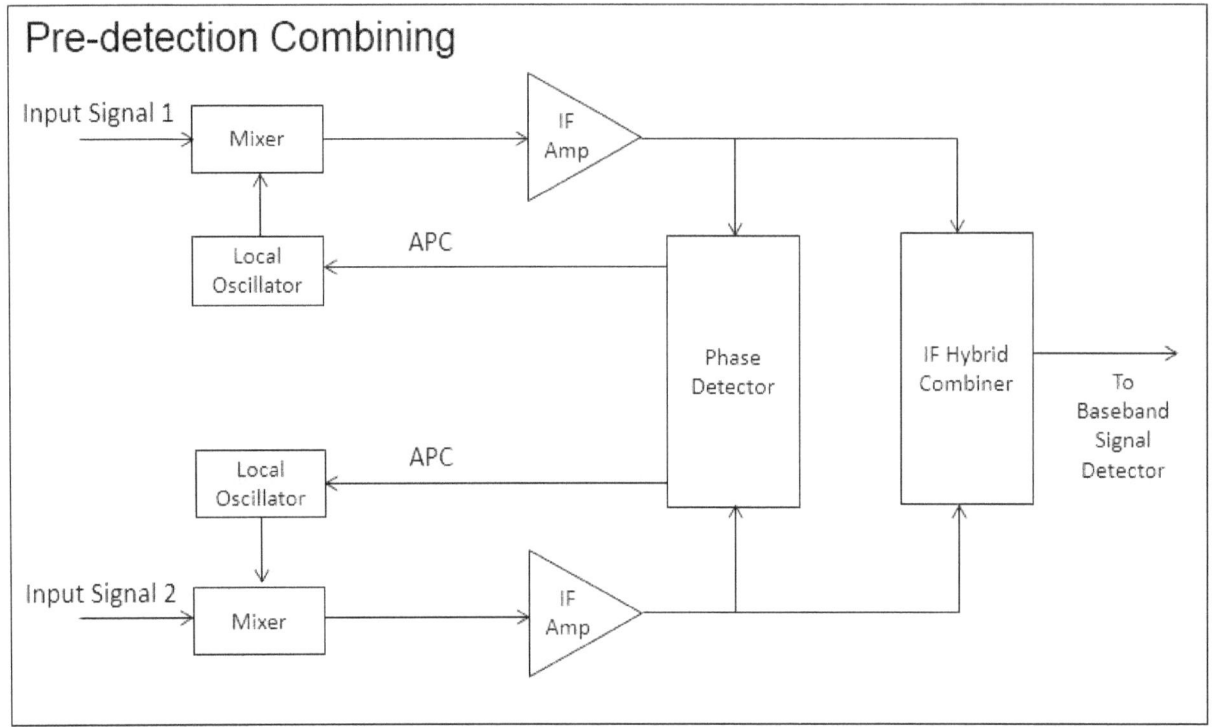

**Figure 12.7.1.a**

## 12.7.2    Post-detection Combining

In post-detection combining, the multiple received radio signals are combined after the detection stage. **Figure 12.7.2.a** illustrates the technique, where baseband signal 1 and baseband signal 2 are first extracted from the incoming radio signals by their respective detectors, before their combining at the output of baseband amplifiers.

In post detection combining too, there are multiple IF amplifier stages, one for each incoming radio signal, however, there are additional baseband signal detector stages too, one for each incoming radio signal input. This adds to the cost of the system, however, the added benefit is a simplified baseband signal combiner.

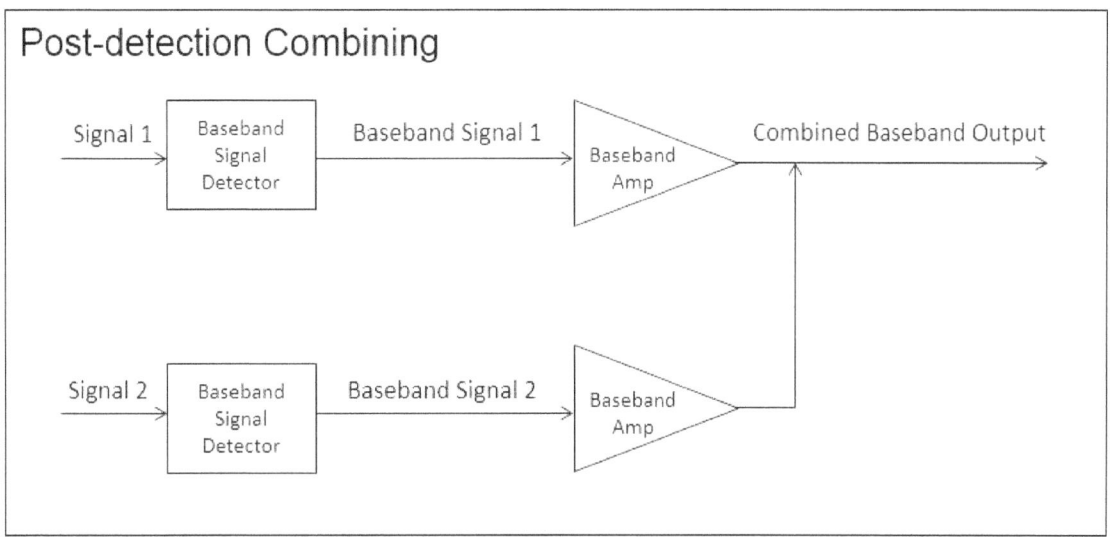

**Figure 12.7.2.a**

## 12.7.3   Selection Combining

The *selection combining* technique uses an electronic switch to receive the radio signal which has maximum *signal to noise ratio* (SNR).

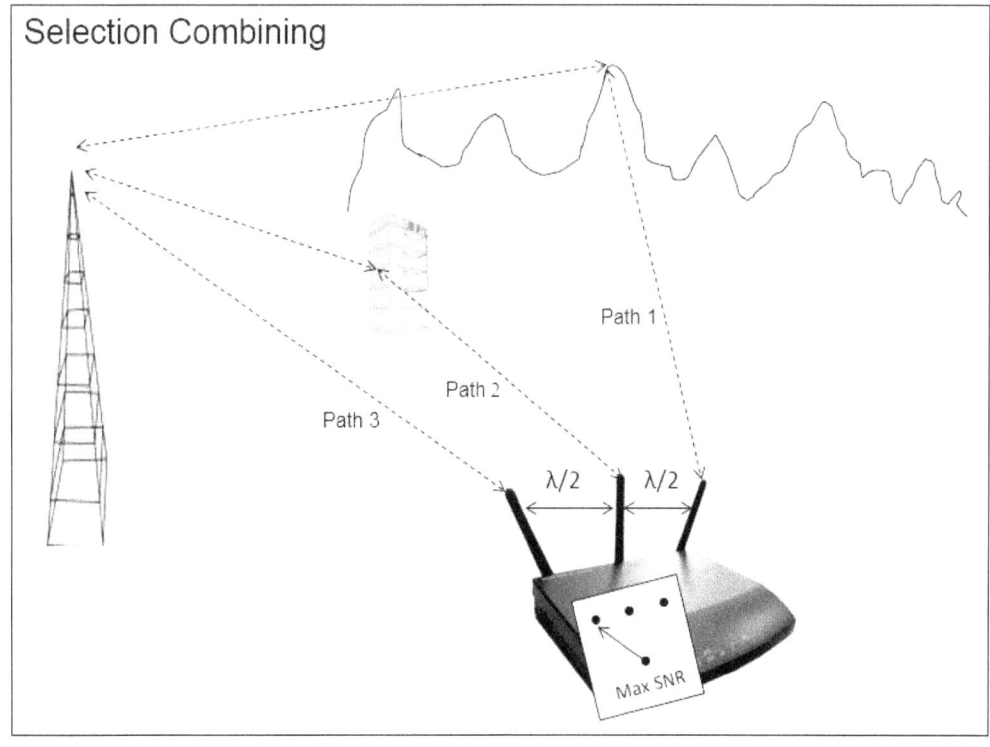

**Figure 12.7.3.a**

**Figure 12.7.3.a** shows three copies of the radio signal received over Path 1, Path 2 and Path 3. The three receiving antennas are spaced apart at least half wavelength to avoid mutual coupling between the antennas. An electronic switch inside the receiver dynamically selects the strongest signal, and hence the name *selection combining*.

## 12.7.4     Equal Gain Combining

The *equal gain combining* technique uses equal weighted summation of all received signals in amplitude and phase. **Figure 12.7.4.a** shows how the combined signal consists of

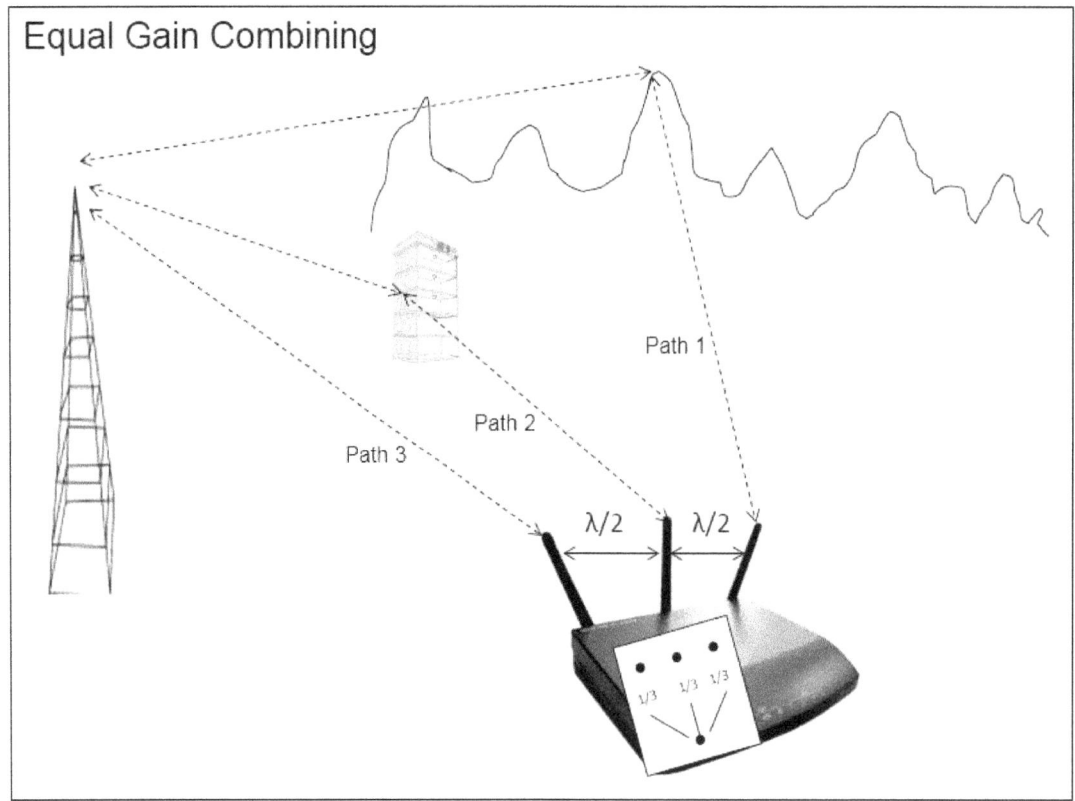

**Figure 12.7.4.a**

equal portions of each of the three received radio signals.

## 12.7.5     Maximal Ratio Combining

The *maximal ratio combining* technique uses a weighted combination of all the received signals based on their *signal to noise ratio* (SNR). The stronger signal has more weight than the weaker one. **Figure 12.7.5.a** shows how the received signals are combined in maximal ratio. The signal received on Path 3 is the strongest, so it takes a bigger share, that is, half of the combined signal. The signal received on Path 2 is less strong, and takes a smaller share, that is, one third of the combined signal. The signal received on Path 3 is the weakest and takes smallest share, that is, one sixth of the combined signal.

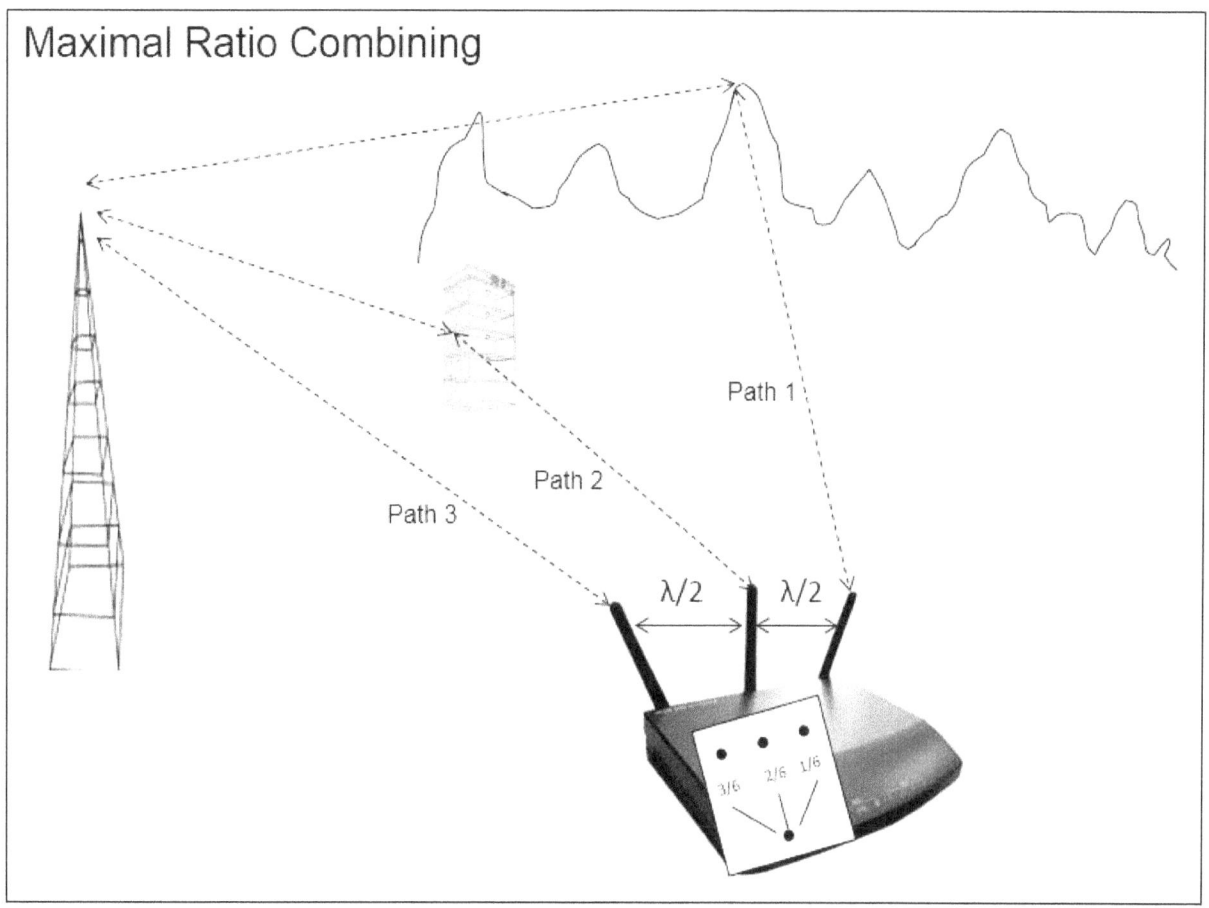

**Figure 12.7.5.a**

## 12.8 Diversity Gain

The term *diversity gain* defines the gain achieved by deploying any antenna diversity technique. It represents relative increase in *signal to noise ratio* (SNR), due to diversity. It is really the SNR for a diversity combined signal to the original signal when there is no diversity.

**Figure 12.8.a** shows an example of diversity gain for the space diversity scheme. The original signal in the absence of space diversity propagates on Path 1, and its signal to noise ratio is represented by $SNR_1$. After space diversity is deployed in the system, there is a second signal that propagates on Path 2 and its signal to noise ratio is represented by $SNR_2$. With space diversity, the two diverse signals are combined, which results in a combined signal with signal to noise ratio of $SNR_{comb}$. So, in this example, *diversity gain* is the ratio of $SNR_{comb}$ to $SNR_1$.

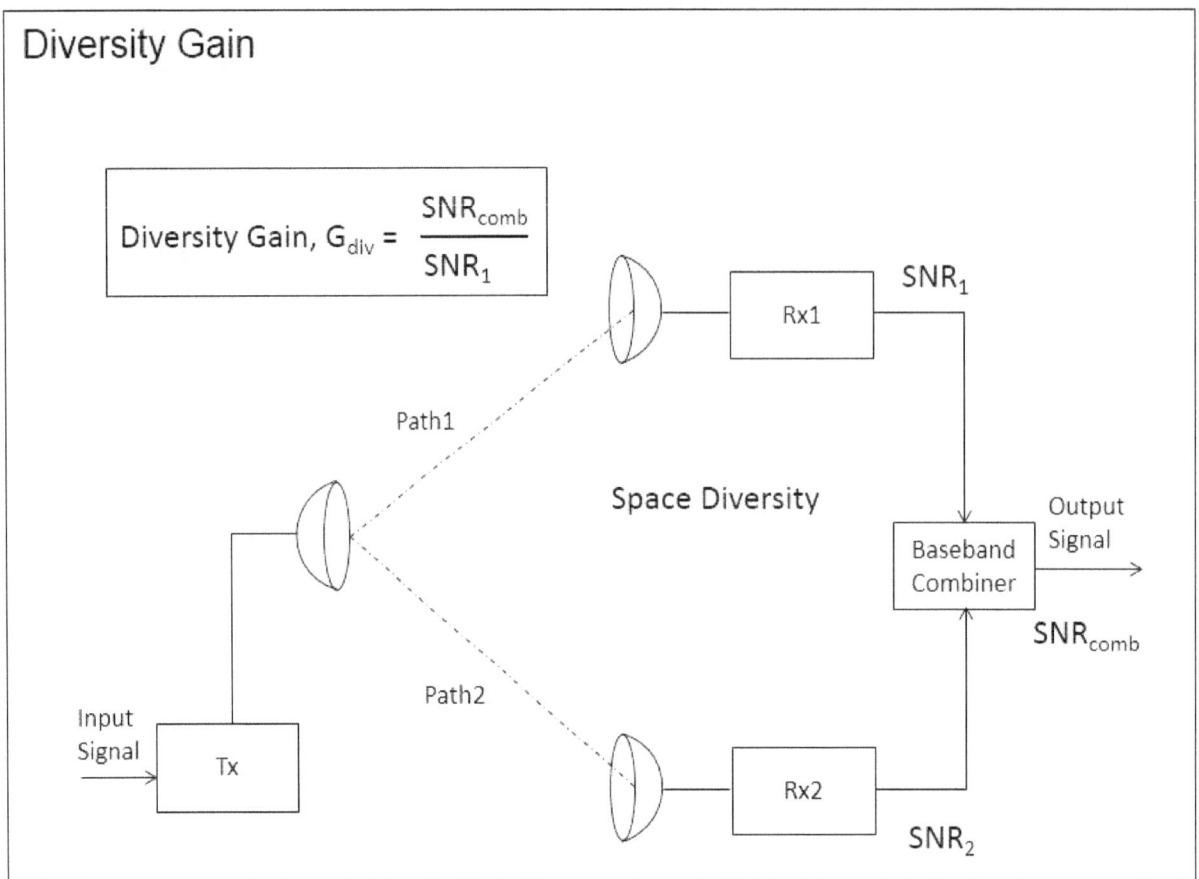

Figure 12.8.a

# 13. System Capacity and Multiple Access

For any wireless communications system, frequency spectrum is a scarce physical resource. This is because the use of wireless communications has grown tremendously around the world. A commercial carrier of wireless services often finds it challenging to find an available frequency spectrum, which may have to be leased at a premium from licensees of the spectrum or from government regulatory authorities.

Frequency spectrum is used to create multiple physical wireless channels to serve multiple users in a given geographical region. Since frequency spectrum is expensive, it is useful to figure out the *system capacity* of a wireless communications system. The *system capacity* of a wireless communications system defines how many users can be served by the system simultaneously, by independently allocating a physical wireless channel to each user. So, our emphasis here is to explore different techniques which are used to maximize the *system capacity* of a wireless communications system, given the constraint of expensive frequency spectrum. In other words, we want to maximize the number of simultaneous users of a wireless communications system.

## 13.1 Frequency Reuse

The reuse of available frequency spectrum is a widely used technique to increase the system capacity of a wireless communications system. The geographical region in which the served wireless users exist, is sub divided into smaller regions of 4-8 miles size called *cells*. The users in each *cell* are served by a *base station*, which has a typical antenna height of 50 – 300 ft and a typical ERP from 40 – 100 W. The antenna height and ERP of base station is adjusted in such a way that the radio signal is significantly powerful only within the region of a cell.

**Figure 13.1.a** illustrates a geographical region sub divided into hexagon shaped cells numbered 1,2,3,4…. The hexagon shape is just for illustration. In reality the coverage area of a cell may be a more irregular shape and depends on how the base station antenna elements are aligned. Each of the cells has its own base station antenna at the center of the cell.

In fact the numbers 1,2,3,4…in this example represent the unique frequency channels assigned to each cell. It should be obvious from the figure that adjacent cells are assigned different frequency channels. For example, a cell that is assigned frequency channel 2, is surrounded by cells with frequency channels 1, 3 and 4. In this way, this frequency allocation scheme allows reuse of frequency channels for distant cells, which are not adjacent. Notice that in this figure there are only four available frequency channels 1,2,3,4. However, these four channels are serving 19 cells, each of size 4-8 miles, and there is potential to expand coverage simply by deploying more base stations in new cells and by reusing the available four frequency channels.

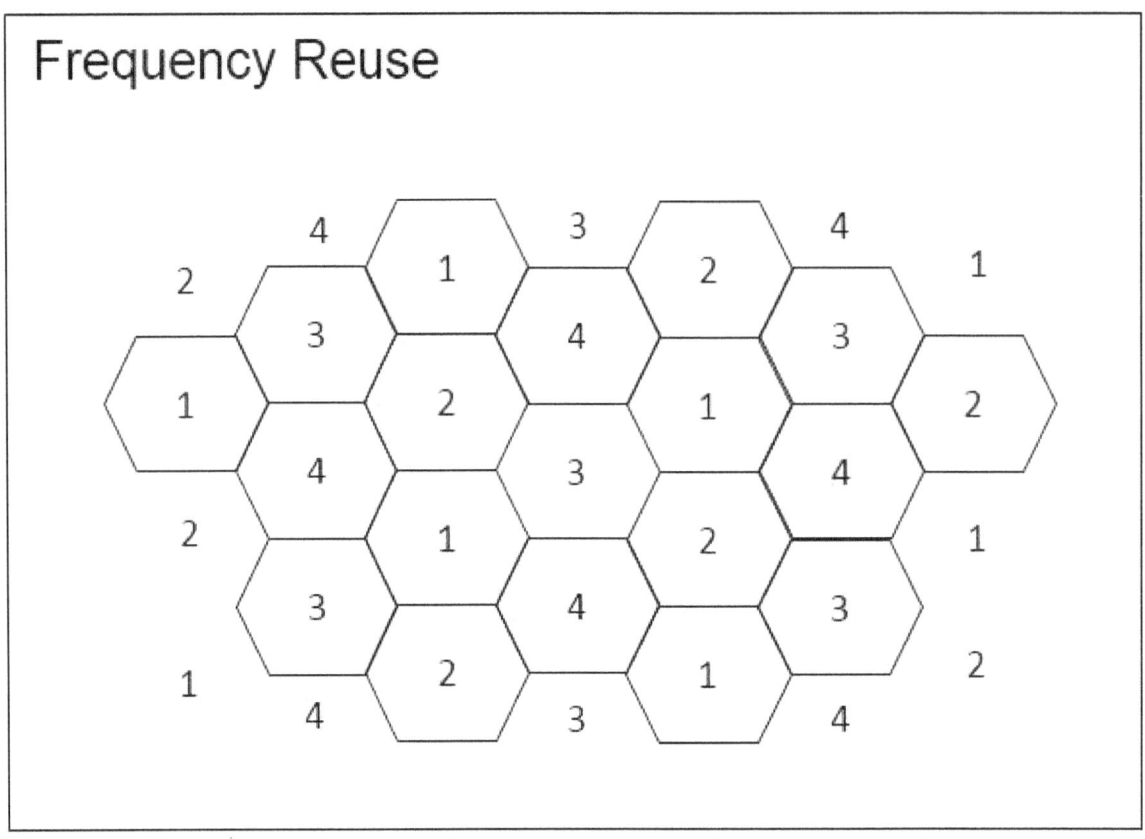

**Figure 13.1.a**

There is an additional benefit of frequency reuse scheme. It minimizes *co-channel interference* (CCI), because of the fact that adjacent cells are on different frequencies. CCI will be a problem if the hand set were to receive 2 different signals from 2 base stations at the same frequency.

## 13.1.1    Frequency Reuse Parameters

We will now understand some frequency reuse parameters that are useful when planning frequencies for a geographical region divided into cells.

N, is the number of frequency channel sets into which the whole available frequency spectrum is divided. For example, for **Figure 13.1.1.a**, the value of N is 4, since only 4 frequency channel sets are used. Notice that only specific values of N leads to regular repeat patterns, such as, such as N = 3,4,7,9,12,13,… Notice that for each cell, no adjacent cell has the same frequency channel set assigned.

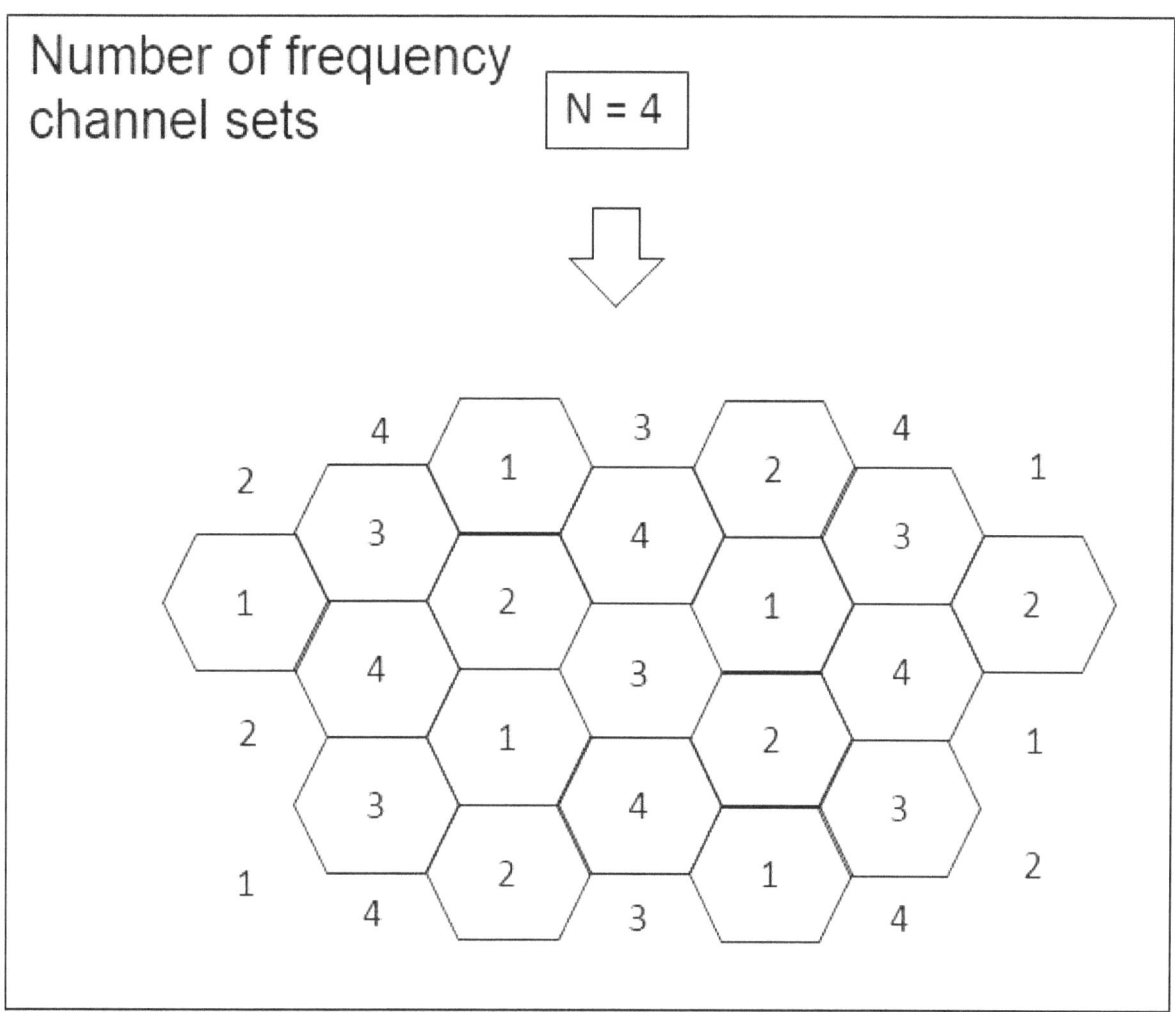

**Figure 13.1.1.a**

*Frequency Reuse Factor* (FRF) is the reciprocal of N, the number of available frequency channel sets, as illustrated in **Figure 13.1.1.b**. This implies that each cell uses only 1/N of the available frequency spectrum.

*Reuse distance ratio*, Q, is the ratio of *co-channel cell spacing*, D and the *radius* of cell, R, as shown in **Figure 13.1.1.c**. *Co-channel cell spacing* is the distance between the centers of 2 cells that are allocated same frequency channel set. When planning frequency reuse, D and R can be proportionally reduced, thus reducing the size of cell and adding to *system capacity*. However, as the cell size is reduced, there is an increase in *co-channel interference* (CCI), which is a limiting factor to the size of a cell.

There is a mathematical relationship between Q and N as shown in **Figure 13.1.1.d**. For given values of N, Q can be calculated. For example, for N = 4, the value of Q is calculated to be 3.46, and for N = 7, Q is 4.58.

# Frequency Reuse Factor

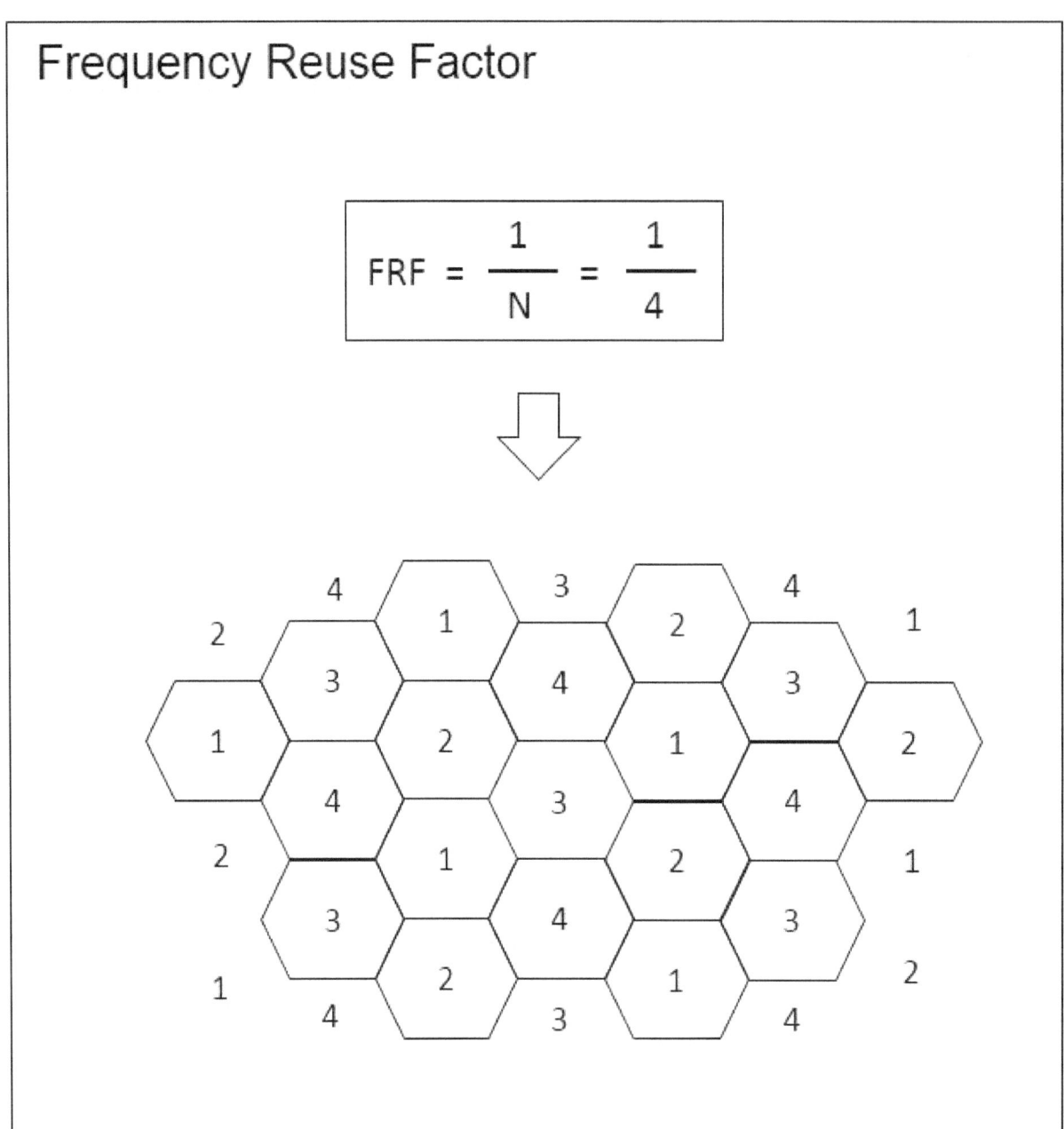

**Figure 13.1.1.b**

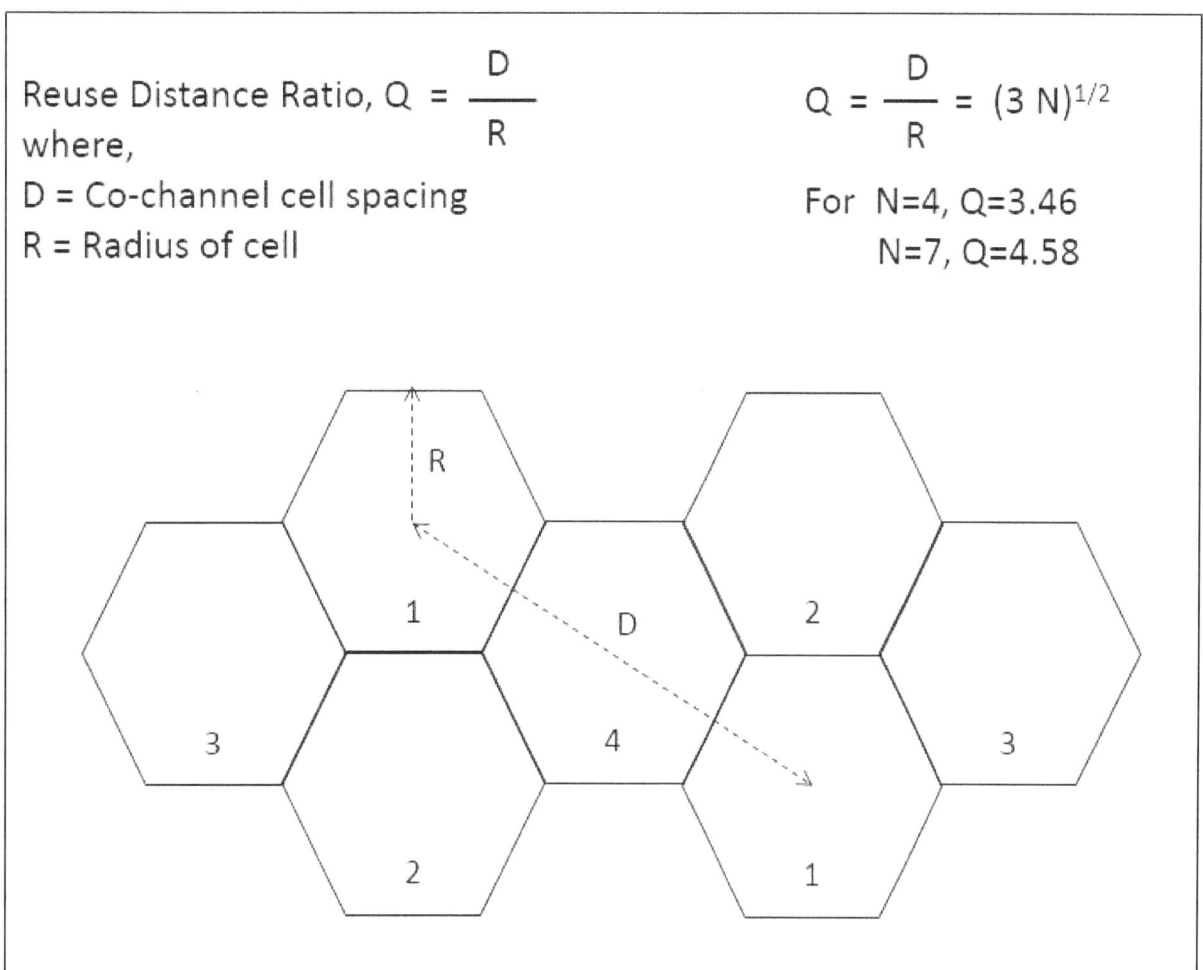

Reuse Distance Ratio, Q $= \dfrac{D}{R}$

where,

D = Co-channel cell spacing

R = Radius of cell

$Q = \dfrac{D}{R} = (3\,N)^{1/2}$

For N=4, Q=3.46

N=7, Q=4.58

**Figure 13.1.1.c**

The *carrier to interference ratio* (CIR or C/I), a.k.a. *signal to interference ratio* (SIR), is the ratio of average received *carrier or signal power* to average received *co-channel interference power*. In order for the received signal to be meaningful for a handset inside a cell, the CIR must be at least 18 dB or a numerical ratio of 63:1. This is an important design objective when planning the cell size.

**Figure 13.1.1.d** shows a relationship between CIR, reuse distance ratio, Q, and the number of frequency channel sets. Notice that if N increases, the co-channel spacing, D also increases. This results in an improved CIR, because the cells with same allocated frequency set are farther apart, resulting in reduced co-channel interference.

Let us assume that we have the same available frequency band as before, which should be divided into N frequency channel sets. If we choose a larger value for N, the frequency reuse factor, which is 1/N, is reduced. This results in each cell with a smaller share of the frequency spectrum, and hence lower *system capacity* for each cell. There will be lesser number of physical wireless communication channels available for the users in each cell.

Therefore, the choice of optimum number of frequency channel-sets, N, is really a trade-off between *system capacity* and *carrier to interference ratio* (CIR).

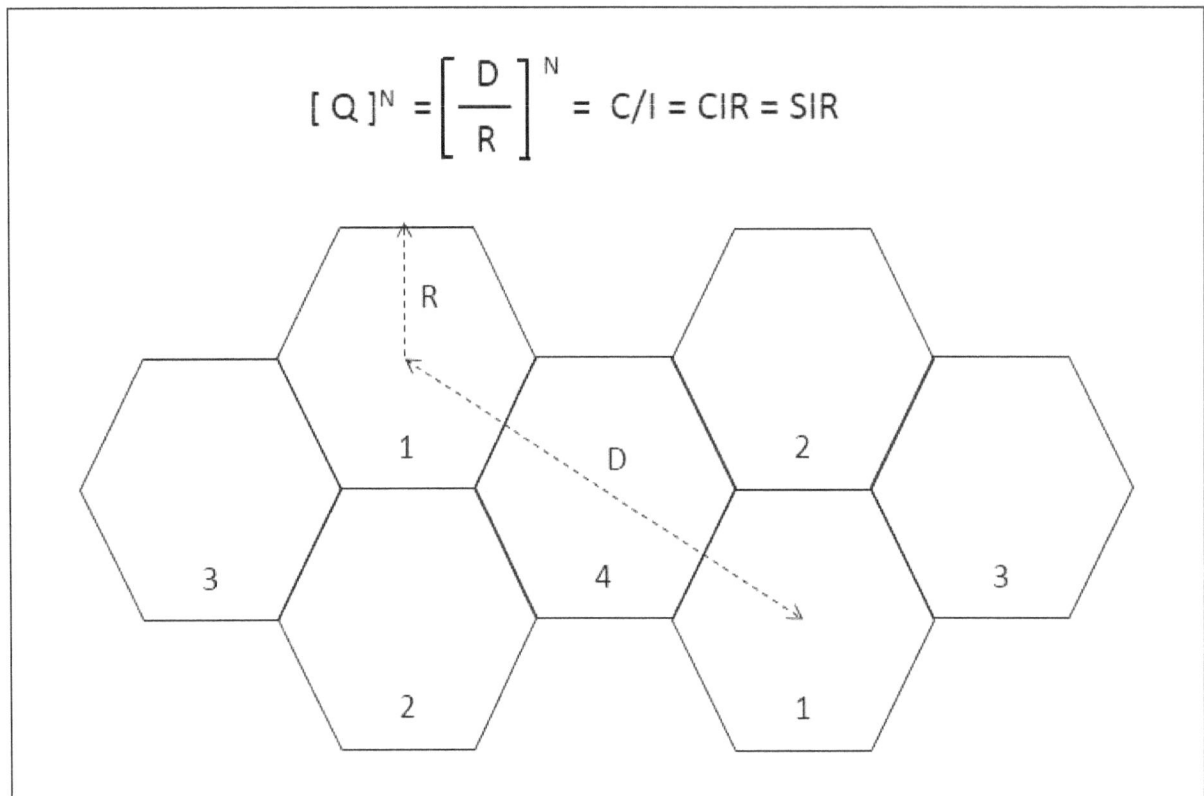

$$[\,Q\,]^N = \left[\,\frac{D}{R}\,\right]^{N} = C/I = CIR = SIR$$

**Figure 13.1.1.d**

## 13.2 Cell Splitting

In dense urban areas, where user density is very high, the technique of *cell splitting* can be used to add *system capacity*. In this technique, congested cells are subdivided into smaller cells each with its own base station. Smaller cells need lower power transmitters and lower height antennas, and therefore permit greater frequency reuse. **Figure 13.2.a** shows the illustration.

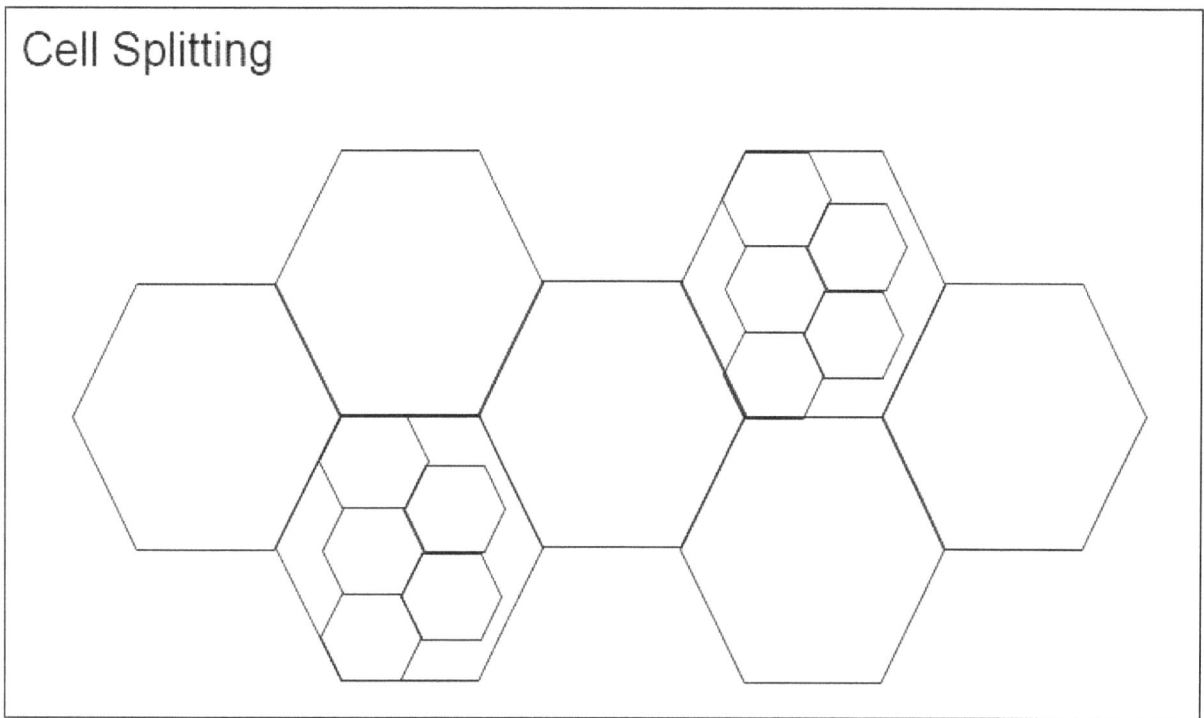

**Figure 13.2.a**

## 13.3 Cell Sectoring

Another technique to add system capacity is *cell sectoring*. Each cell is split into three sectors of 120 deg each or six sectors of 60 deg each as shown in **Figure 13.3.a**.

Each sector is served by a separate directional antenna and is assigned its own unique frequency channel set. It also helps reduce *co-channel interference (CCI)*. In a standard cell with no sectoring, CCI enters the cell from all directions. However, with 3 sectors, each sector has to contend with only one-third of the CCI, and with 6 sectors, each sector has to contend with only one-sixth of the CCI.

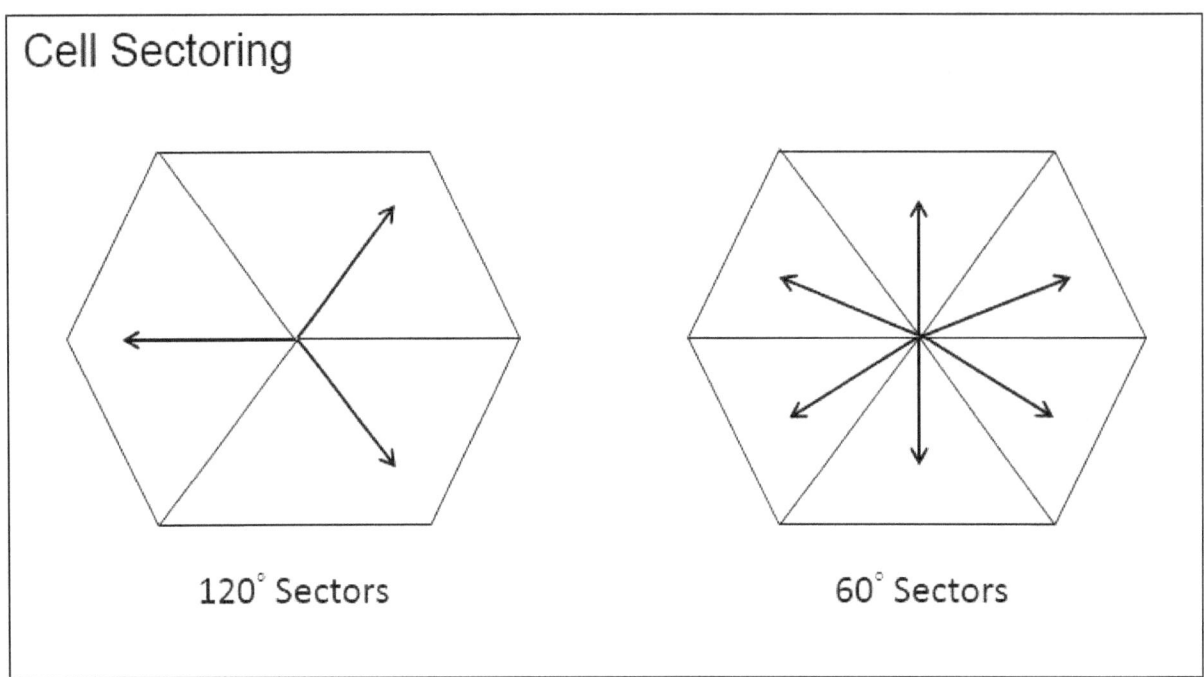

**Figure 13.3.a**

## 13.4 Flexible Spectrum Reuse

We will now study *flexible spectrum reuse* techniques as they are used in contemporary 4G wireless communications networks, such as LTE. *Long term Evolution* (LTE) is the wireless broadband communication standard, which aims to maximize *system capacity* by having a *frequency reuse factor* (FRF) of 1. This implies that all cells can use the full available frequency spectrum because the number of frequency channel sets, N is 1. However, as we have seen before, this will cause significant *co-channel interference* (CCI) or *inter cell interference* (ICI).

Therefore, an effective technique for *inter cell interference coordination* (ICIC) and *spectrum usage management* is needed to mitigate the effects of having FRF = 1, and realize the benefits of higher *system capacity*.

*Fractional frequency reuse* (FFR) and *soft frequency reuse* (SFR) are the techniques used in 4G LTE wireless broadband networks to overcome the effects of interference from adjacent cells with FRF = 1.

In both FFR and SFR, the cell is divided into two regions as shown in **Figure 13.4.a**, the *cell center region* and the *cell edge region*. The users are also classified into two categories, the *cell center users* and the *cell edge users*. The boundary between the cell center region and the cell edge region is set on the basis of the distance from the base station or the *signal to interference and noise ratio* (SINR) threshold. The set of sub carriers for center and edge region

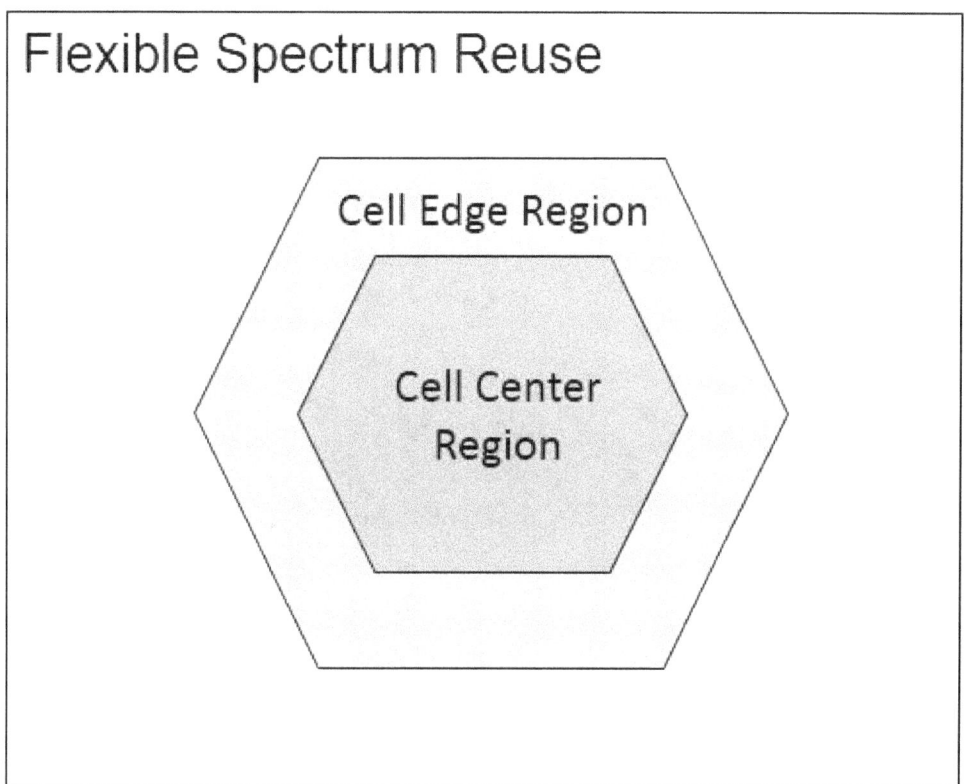

**Figure 13.4.a**

are selected randomly. The number of sub carriers in a cell region is proportional to the ratio of the number of users in the cell region to the total number of users inside the cell. The information received from the user such as *position information* (PI) and *channel quality indicator* (CQI) are used to dynamically allocate the subcarriers.

## 13.4.1    Fractional Frequency Reuse (FFR)

The *fractional frequency reuse* (FFR) technique splits the cell into inner or center band and outer or edge band as shown in **Figure 13.4.1.a.**

The cell center has limited signal power and a *frequency reuse factor*, FRF = 1. So, the users in the center band have access to full allocated frequency spectrum, but have access to limited signal power.

On the other hand, the cell edge has full signal power and a *fractional frequency reuse factor*, FRF = 2/3. So, the users in the cell edge region have access only to a fractional part of the frequency spectrum, 2/3$^{rd}$ in this case, but have access to full signal power. The remaining fraction of the spectrum in the cell edge region is shared by other users in the cell edge region of adjacent cells.

The allocation of signal power and frequency sub carriers to the cell center and edge regions is fixed at the time of radio network deployment.

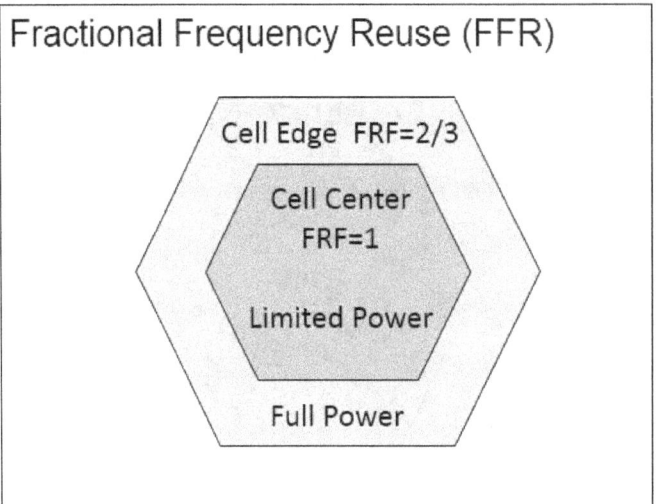

**Figure 13.4.1.a**

## 13.4.2    Soft Frequency Reuse (SFR)

The *soft frequency reuse* (SFR) techniques uses a *frequency reuse factor*, FRF = 1, for whole cell, including center region as well as cell edge region, as shown in **Figure 13.4.2.a**.

The frequency spectrum is sub divided into major sub carriers and minor subcarriers. The major sub carriers have higher signal power and are used by all cell users, in the center as well as edge region. Also, the major sub carriers are allocated in such a way that they are orthogonal in adjacent cells. This is done to mitigate the effects of co-channel interference.

In contrast the minor sub carriers have lower signal power and are used only by cell center users. The *power ratio* in SFR is defined as the ratio of transmit powers of major and minor subcarriers.

The *soft frequency reuse*, (SFR) technique is dynamic compared to the fixed *fractional frequency reuse* technique. The allocation of subcarriers in SFR can be dynamically changed to maximize *system capacity*. Also, SFR is found to have higher *spectral efficiency* in bps/Hz compared to FFR.

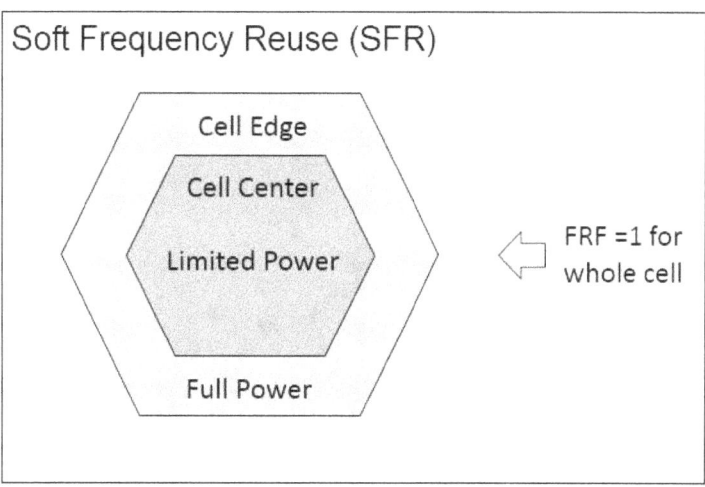

**Figure 13.4.2.a**

# 13.5 Small Cells

The 4G wireless broadband technologies such as LTE use *small cells* to add *system capacity*. The small cells have a coverage of less than 1 km and are served by low power compact base stations, which are easy to mount on the walls or ceiling. The *small cells* are widely used to add *system capacity* in dense urban areas and inside the buildings, where the signal coverage from a conventional base station is usually very poor.

The *small cells* constitute an overlay network over the conventional cellular network served by *macro cells*, which have a coverage of less than 30 km. The frequency sub-carrier allocation between the *small cells* and *macro cells* is controlled by the cellular operator. The overlay network of *small cells* provides a much higher *spectral efficiency* (bps/Hz) compared to other techniques.

**Figure 13.5.a** shows an arrangement of *small cells* in the macro cell center and macro cell edge region. The challenge for cellular operator in this overlay network of *small cells* and *macro cells* is to mitigate *co channel interference* (CCI). A common technique used is to assign orthogonal or non-overlapping frequency sub carriers between small cells and macro cells. **Figure 13.5.b** illustrates how for a macro user in the vicinity of a small cell, the signal from a small cell base station is interference. Similar situation exists for a small cell user inside a small cell region. For this user, the signal from macro cell base station is interference. So, it is obvious that the sub carriers must be allocated judiciously to avoid interference.

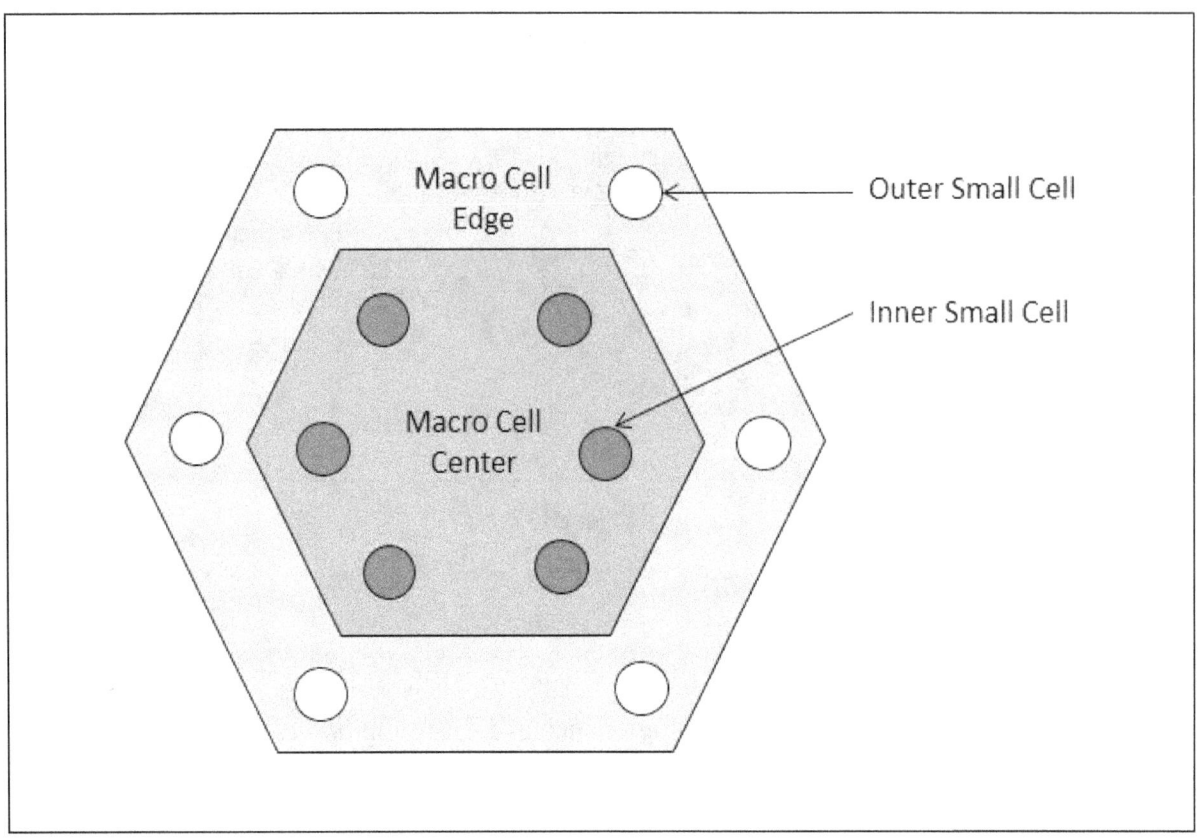

**Figure 13.5.a**

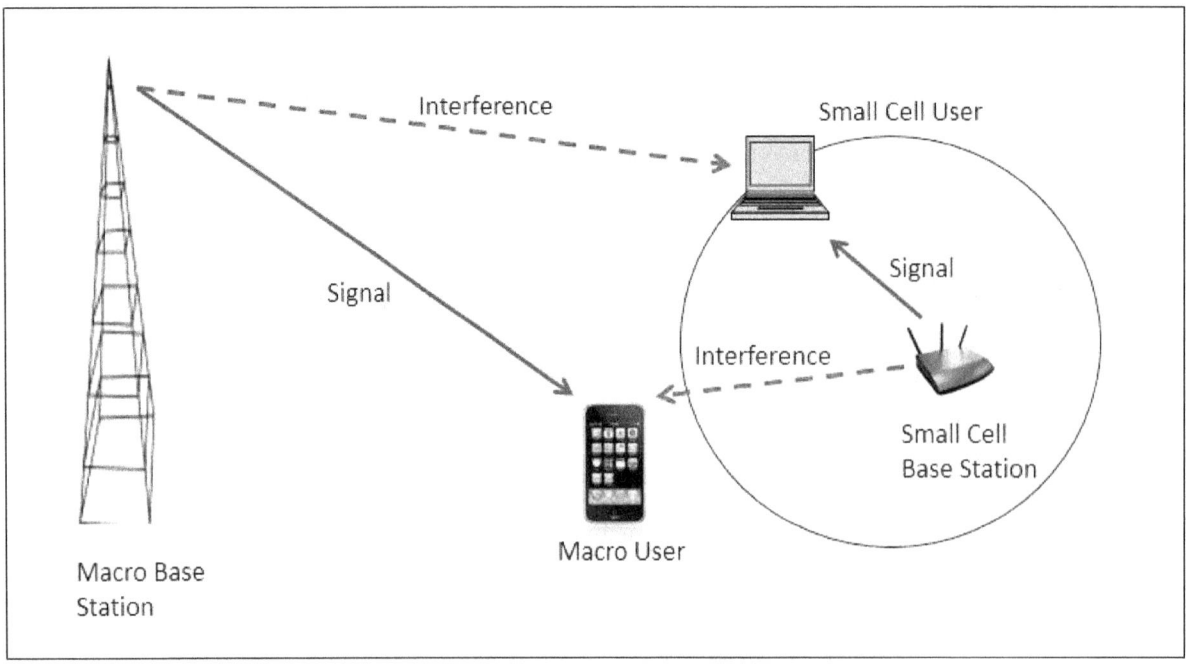

**Figure 13.5.b**

# 14. Base Station, WLAN and Handset Antennas

In this chapter we will learn some of the unique characteristics of antennas used for base stations, wireless LANs and handsets.

## 14.1 Cellular Base Station Antennas

The cellular base station antennas can be either omnidirectional or directional. The omnidirectional antennas are usually deployed in rural areas where user density is low and system

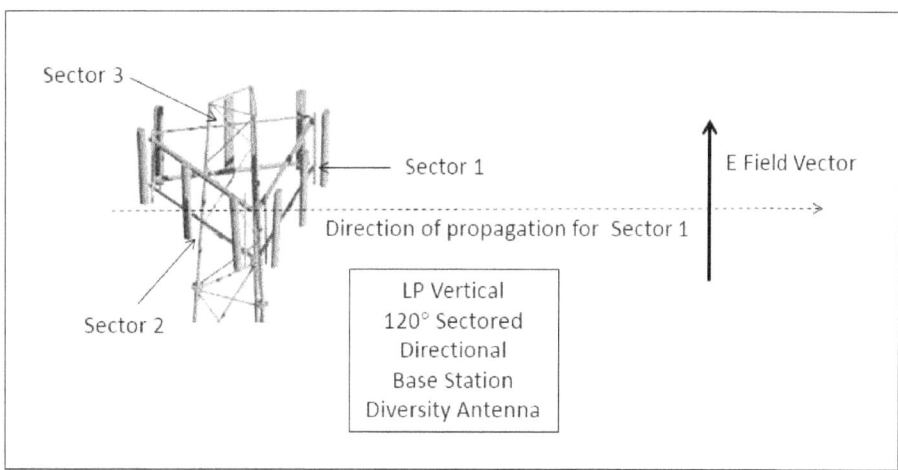

**Figure 14.1.a**

capacity is not much of a problem. In contrast, the urban areas have high user density and *cell sectoring* is used to add system capacity. Each sector is served by a separate set of antennas and frequency channels. **Figure 14.1.a** shows a directional base station antenna system serving three 120 deg sectors. The polarization is usually *linear polarized (LP) vertical* for commercial applications. The radiation is from broadside of the antenna or perpendicular to the antenna structure.

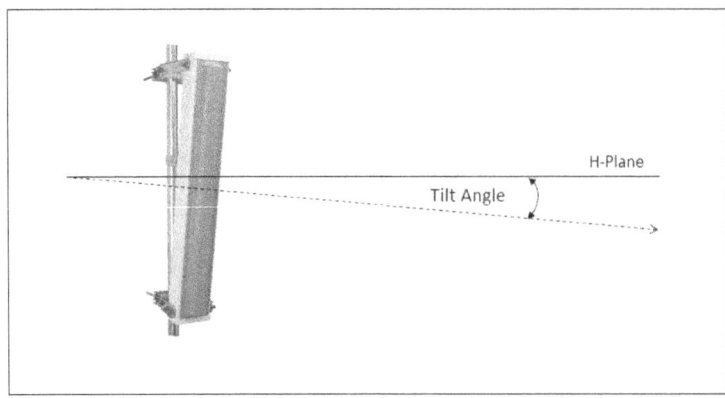

**Figure 14.1.b**

The base station antennas have moderate gain, typical values are from 5 dBi to 24 dBi, and they are often *tilted* as shown in **Figure 14.1.b**. The main radio beam from the antenna is tilted below the horizontal (H) plane, by an angle called the *tilt angle*. Tilted antennas help to focus the radio energy to regions of higher user density, and therefore help to reduce the effects of *propagation loss, delay spread* and *co channel interference* (CCI). It also helps to reduce the transmit power required to achieve acceptable signal strength in the served area of the cell.

If antenna tilt is *static*, it is preset during deployment of the antenna. However, to maximize the benefits of antenna tilt, often *remote electrical tilt* (RET) is used. RET is implemented with a stepper motor attached to the antenna to control its physical movement through the required *tilt angle*. RET allows the tilt angle to be controlled remotely for different hours of the day based on user density predictions. For example, during rush hours in the cell region, the tilt angle can be higher to provide stronger signal around the center of the cell.

## 14.2 Wireless LAN Antennas

The antennas used for wireless LANs come primarily in three different shapes as shown in **Figure 14.2.a**, the *parabolic reflector*, the *flat panel* and the *indoor omnidirectional* antenna.

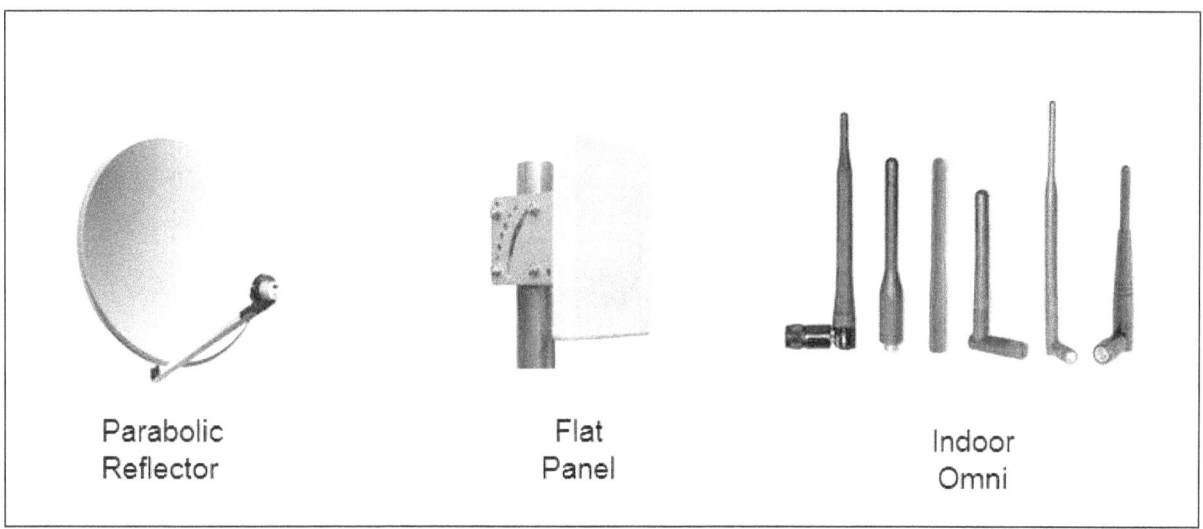

**Figure 14.2.a**

The *parabolic reflector* antennas are used for outdoor wireless LANs to provide coverage in long narrow regions or to provide radio links between buildings. They have comparatively higher gain in the range of 20-24 dBi.

The *flat panel* antennas can be used for large and, wide outdoor as well as indoor regions. They have gain in the range of 15-20 dBi.

The *indoor omnidirectional* antennas are used in wireless access points and routers, primarily in the indoor regions. They have a lower gain of about 6 dBi.

## 14.3 Handset Antennas

The handset antennas have additional challenges. These antennas are packed inside the handset, cellphone, or other mobile hand held equipment, along with several other peripherals, which are conductive and lossy, and that lowers the efficiency of these antennas. Therefore, trade-offs have to be made between design, placement and performance of the handset antennas.

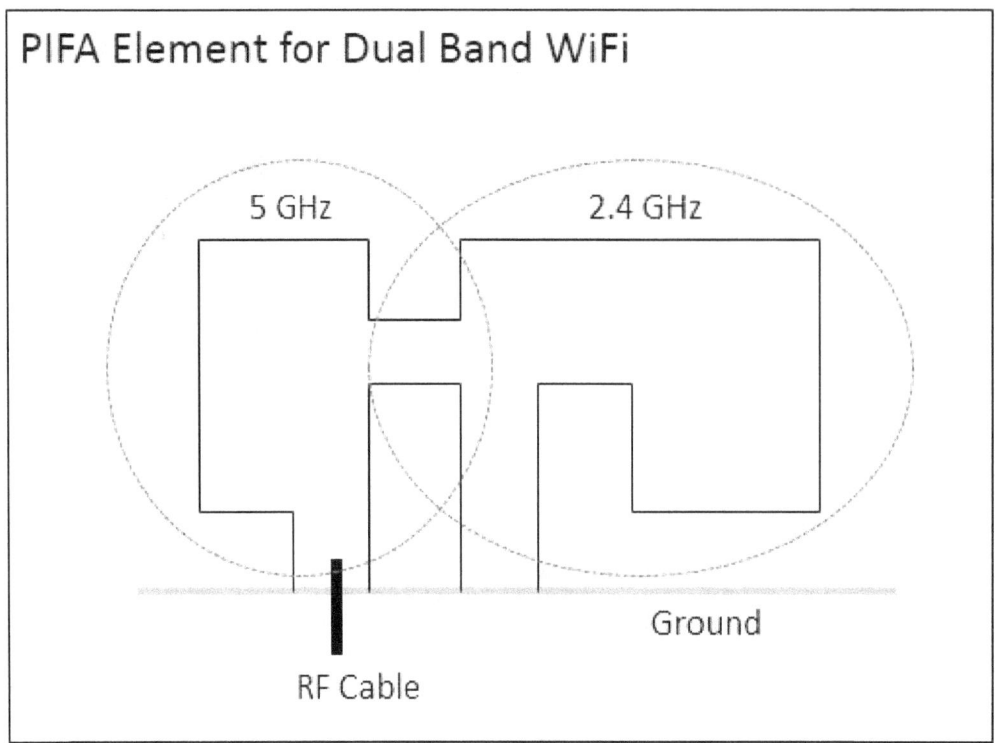

**Figure 14.3.a**

The performance parameters, such as *operating bandwidth*, *impedance matching*, *gain*, *radiation patterns*, *efficiency* and *specific absorption* rate are important for handset antennas. The human body effects also need to be considered.

The handset antennas are low gain and omnidirectional, and because of mobility, they have a random orientation.

**Figure 14.3.a** shows a multiband compact handset antenna design for 5/2.4 GHz dual band WiFi application, called a *planar inverted F antenna* (PIFA). The elements have inverted F shape and there are two parts, the left part is for 5 GHz and the right part is for 2.4 GHz. The planar structure of this antenna allows it to be fabricated on the inside surface of handset enclosure.

# 15. MIMO

In this chapter, we will learn about *multiple input multiple output* (MIMO) technology. In simple terms, MIMO implies multiple transmit and receive antennas as shown in **Figure 15.a**.

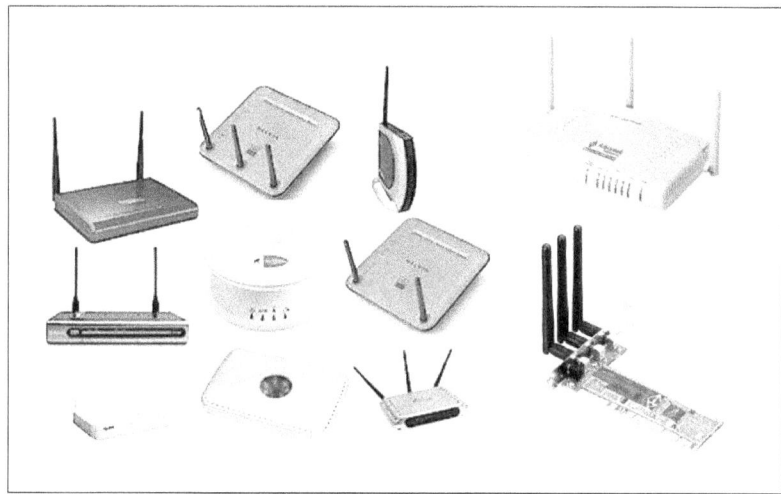

**Figure 15.a**

## 15.1 What came before MIMO?

The radio antennas have been used in different configurations since the advent of radio or wireless communications. Whereas the MIMO antenna configuration is currently finding applications in all wireless broadband technologies, such as LTE, WiMax and WiFi, there have been simpler configurations in use prior to MIMO. We will now consider each of these configurations, that is, SISO, SIMO, MISO and finally MIMO.

The *single input single output* (SISO) configuration has been in use since the beginning of wireless communications. **Figure 15.1.a** shows an illustration of SISO. It involves a single transmitting antenna and a single receiving antenna. In fact, SISO is the most commonly used configuration even today. Please note that the terms *input* and *output* are used with reference to the wireless or radio channel and not the wireless devices. The term *input* specifies the number of signal inputs into the radio channel by the transmitting antenna, and *output* specifies the number of signal outputs taken out from the radio channel by the receiving antenna.

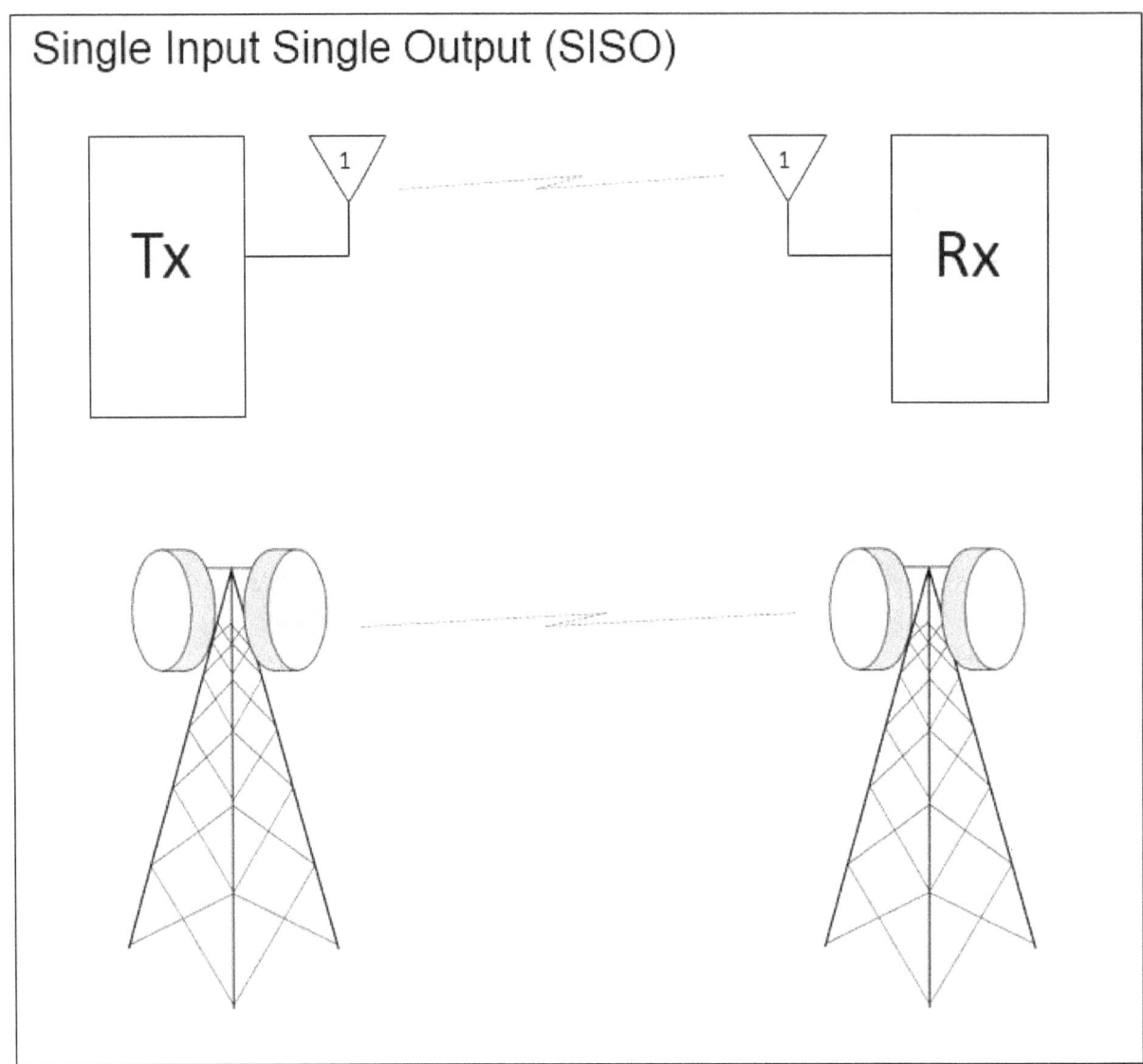

**Figure 15.1.a**

The *single input multiple output* (SIMO) configuration has single transmitting antenna, but multiple receiving antennas as shown in **Figure 15.1.b**. This is similar to *space* or *receive diversity* technique that we considered in **Section 12.3**. The multiple antennas enable the receiver to exploit multi path propagation of radio signals. The diversity combining techniques can be used to combine copies of radio signal received from multiple antennas and, thereby reduce the probability of signal fading. This is illustrated in **Figure 15.1.c**.

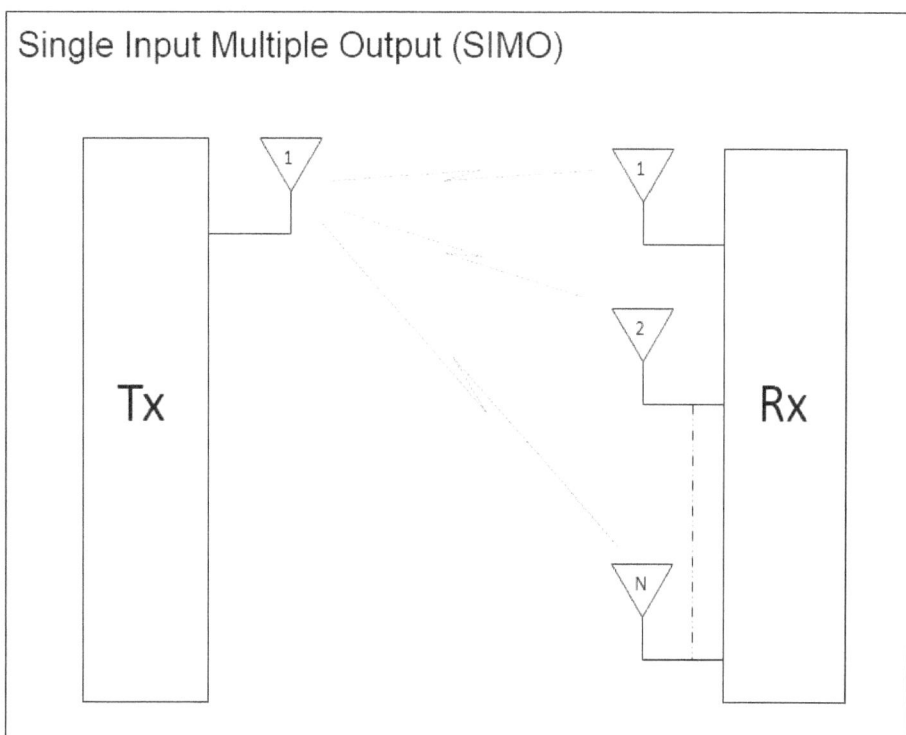

**Figure 15.1.b**

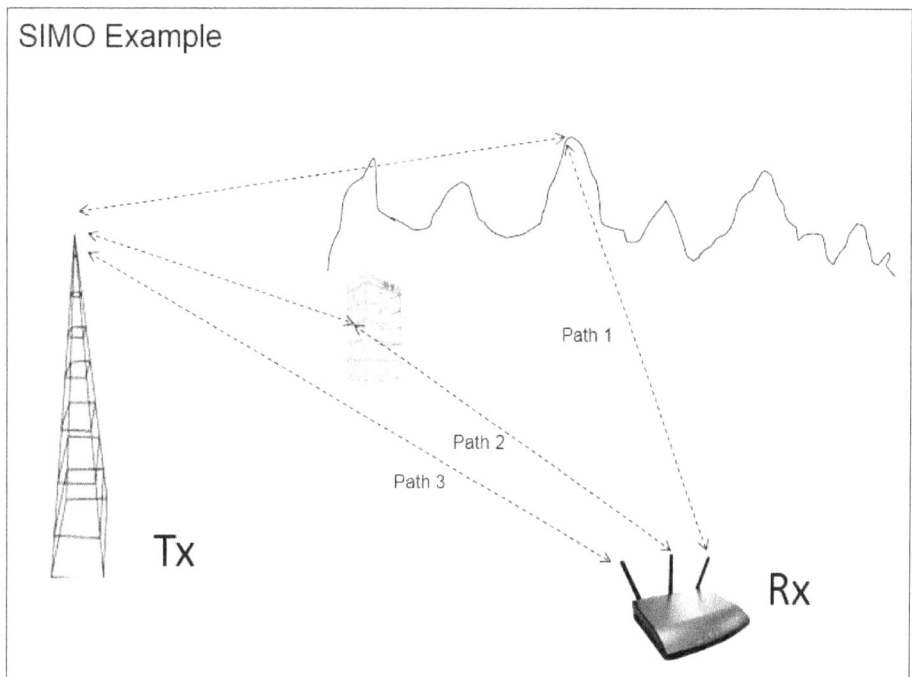

**Figure 15.1.c**

The *multiple input single output* (MISO) configuration has multiple transmitting antennas, but a single receiving antenna, as shown in **Figure 15.1.d**. This configuration can be used in wireless communications, where the base station can have multiple antenna elements transmitting redundant copies of the signal from multiple antenna elements, and the handset has the luxury of receiving the best signal out of the three transmitted, as illustrated in **Figure 15.1.e**.

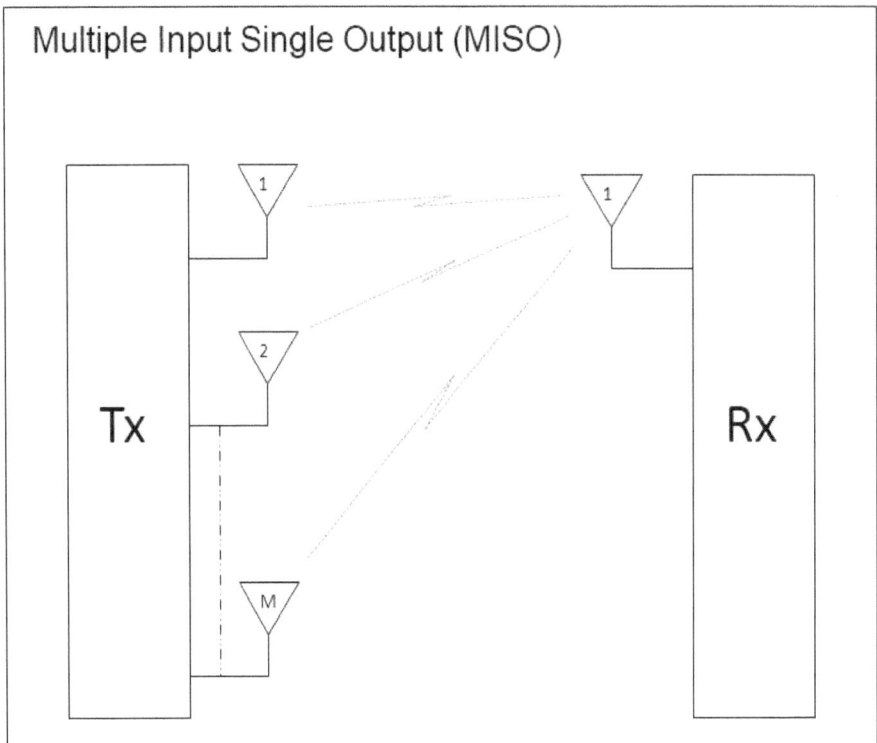

**Figure 15.1.d**

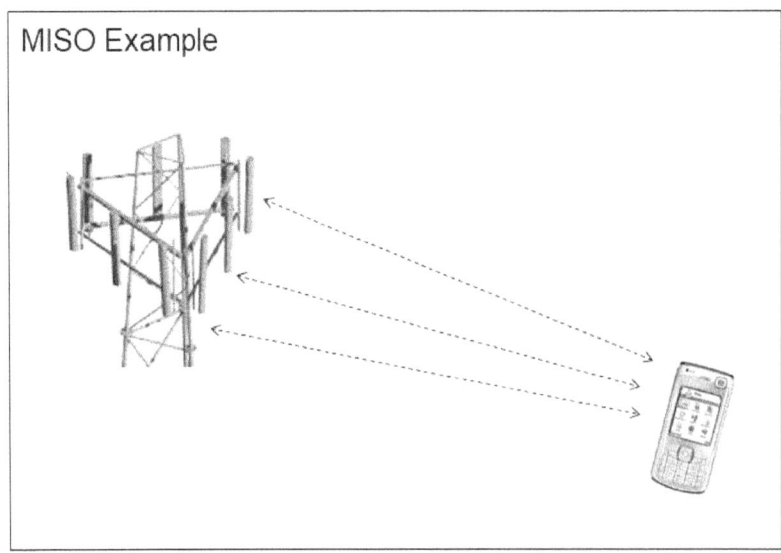

**Figure 15.1.e**

Finally, we have the *multiple input multiple output* (MIMO) configuration, as shown in **Figure 15.1.f**, where we have multiple transmitting as well as receiving antennas. With MIMO, we have the luxury of a robust wireless link as well as added system capacity in terms of bandwidth. This is illustrated in **Figure 15.1.g**.

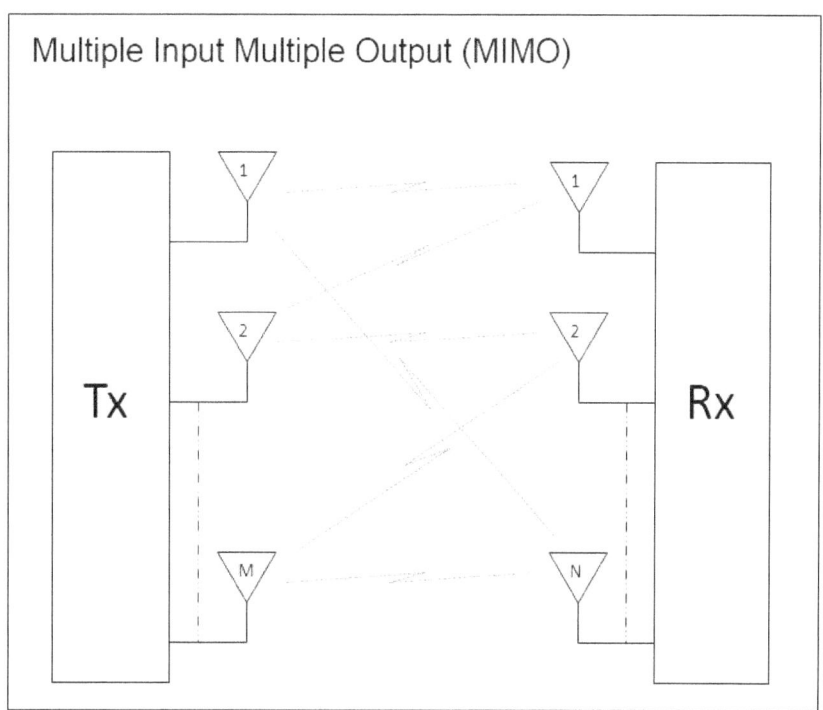

**Figure 15.1.f**

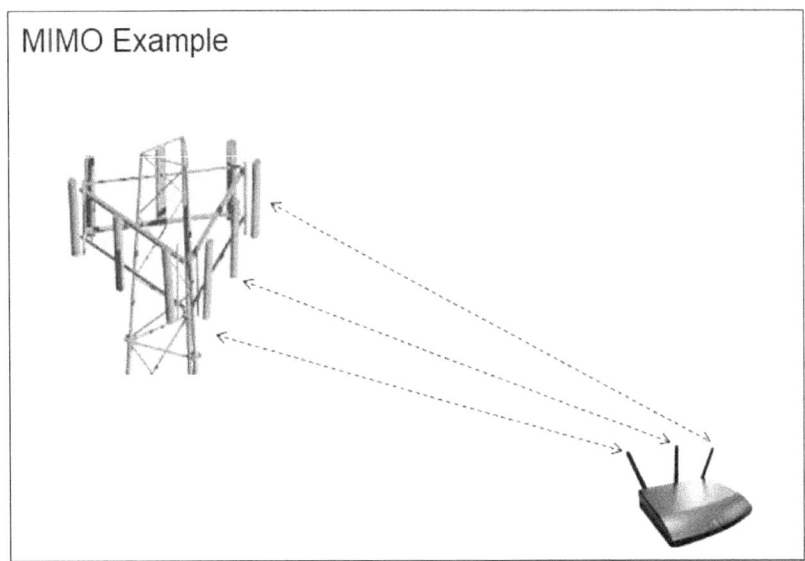

MIMO Example

**Figure 15.1.g**

In addition to mitigating the effects of fading due to multipath propagation, the MIMO antenna configuration also increases available throughput of the radio channel, which means higher data transfer rates in terms of Mbps (Megabits per sec). MIMO also enables several 4G wireless broadband applications for some of the leading wireless communication standards, such as, IEEE 802.11 n/ad (WLAN), IEEE 802.16m (WiMax) and 3GPP Rel 10 (LTE-A).

In next few sections, we will study different techniques used to realize the benefits of MIMO, which include *antenna diversity,* both transmit as well as receive, *beamforming* and *space division multiplexing* (SDM).

For *antenna diversity* techniques, please refer **Chapter 12** for detailed information.

## 15.2 Beamforming

The *beamforming* technique is used to control the shape and direction of radio beams transmitted from a MIMO antenna system. The multiple antennas on the transmitter of a base station can be treated as an array of antennas, which can be used to electrically steer the radio beam in the direction of a mobile hand set as shown in **Figure 15.2.a**. The signals fed to transmitting antenna elements can be pre-processed in such a way, that constructive interference happens in the preferred direction and destructive interference takes place in the undesired direction. In this way, in principle it is possible to target the radio beam in the direction of high density of mobile users, such as on a highway or at a public marketplace, such as shopping malls. For detailed information on beamforming, please refer **Chapter 8**.

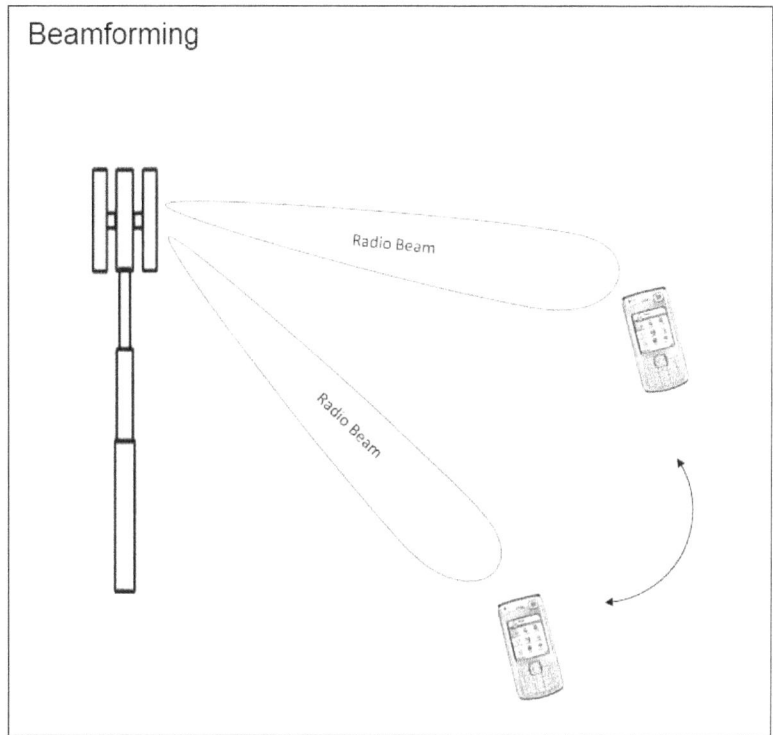

Beamforming

**Figure 15.2.a**

# 15.3 Space Division Multiplexing (SDM)

The technique of *space division multiplexing* (SDM) is used in MIMO configuration to enhance throughput of radio link, when *signal to noise ratio* (SNR) of the radio channel is high. The higher value of SNR implies, multiple spatially separated radio signals can be sent through the radio channel simultaneously, each carrying a separate data stream. This is akin to sending data over multiple physical network links in a wired network, except for the fact, that here we are using radio links and space is used as a resource. Also, the carrier frequency can be reused, because the radio signals are spatially separated. In this way, SDM in MIMO configuration, makes it possible to enhance the bandwidth of radio link without exhausting existing frequency spectrum.

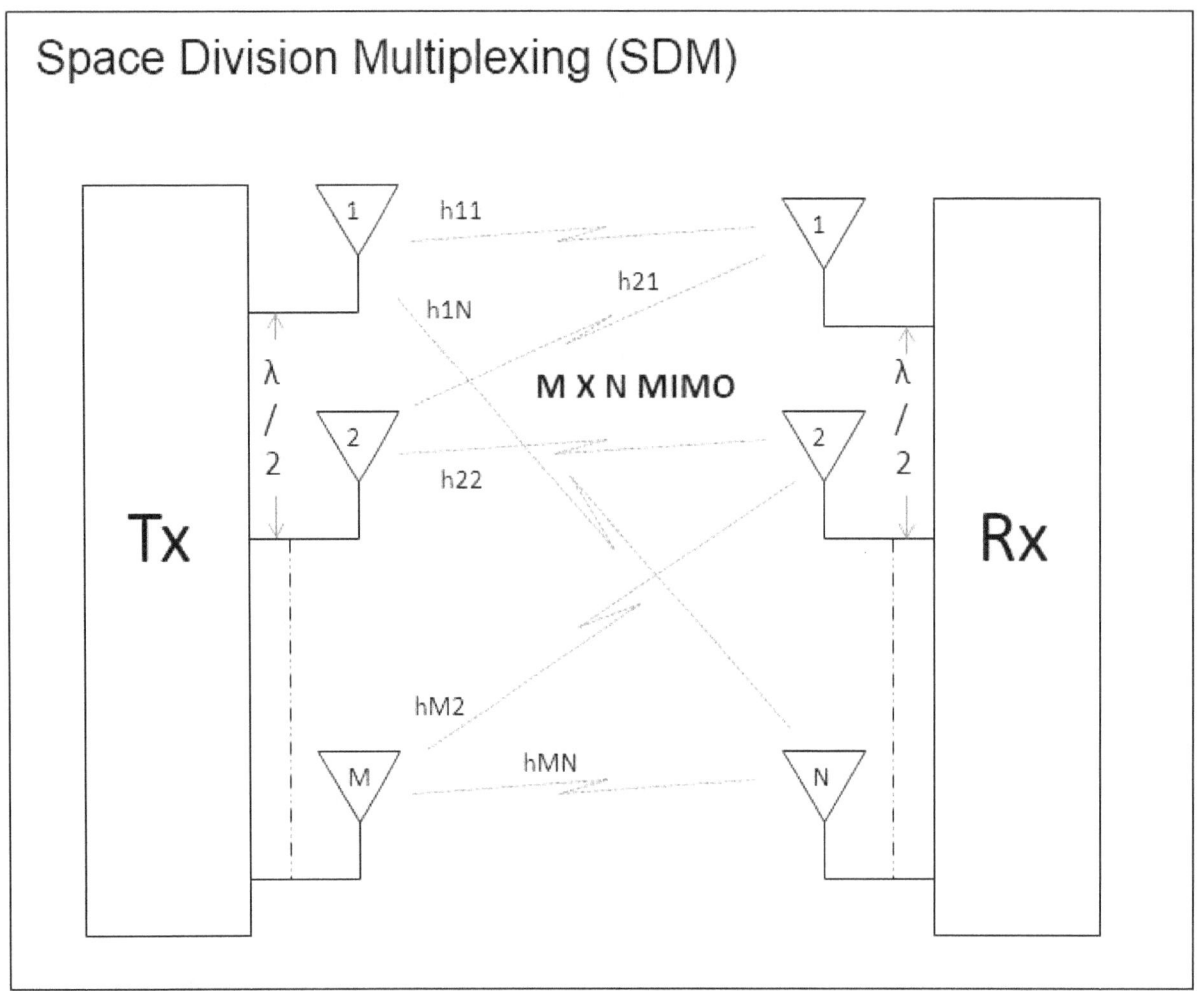

**Figure 15.3.a**

**Figure 15.3.a** shows a MIMO system with 1,2,…..,M transmitting antennas and 1,2,….., N receiving antennas. This is called M x N MIMO system. The minimum separation between antenna elements is kept at half of the wavelength to avoid mutual coupling between antenna elements. It is obvious that there are M x N different signal paths between the transmitter and the receiver. The *signal path coefficients* h11, h12, ……., hMN represent the signal amplitude and phase response for each path. These signal coefficients for each available path are pre-determined during the training sequence between transmitter and receiver. The transmitter, Tx generates a known training signal, which is processed by the receiver, Rx to determine the path responses.

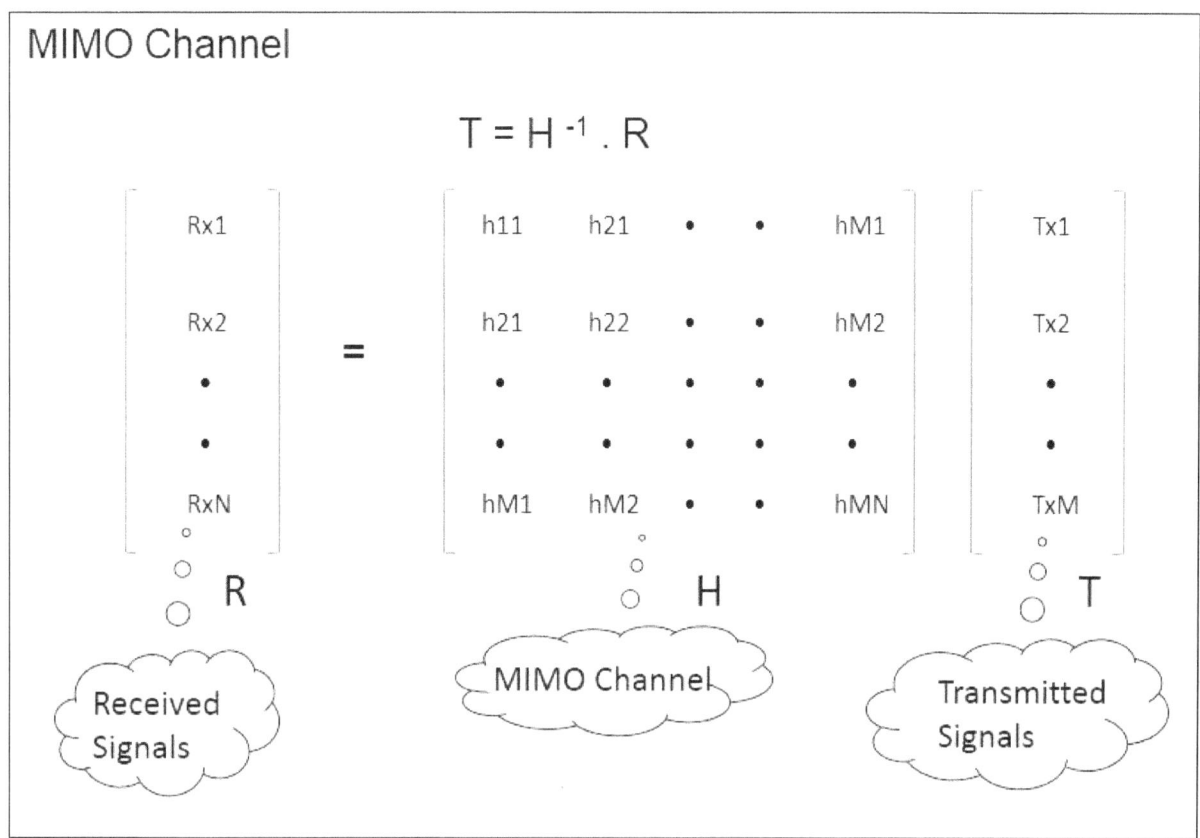

**Figure 15.3.b**

**Figure 15.3.b** shows the mathematical relationship between the transmitted signals, T, the MIMO channel, H, and the received signals, R. Notice that there are M transmitted signals, and N received signals for a M x N MIMO system, and that the MIMO channel is now represented by a matrix consisting of all possible signal path coefficients. The matrix, H, is determined during the training sequence between the transmitter and the receiver. The singular values of H provide a measure of the strength and separation of individual MIMO data streams. The *digital signal processor*, (DSP) at the receiver uses the inverse of matrix H and the matrix of received signals, R to compute the transmitted signal components, represented by the matrix, T.

## 15.4 Multiplexing Rate in MIMO

The *multiplexing rate* in MIMO is the number of distinctive data streams that can be received correctly and simultaneously. For M x N MIMO system, it is the minimum of M and N, expressed as min (M,N). The *multiplexing rate* is a measure of the throughput of a MIMO system. A higher *multiplexing rate* implies a higher *throughput* in Mbps.

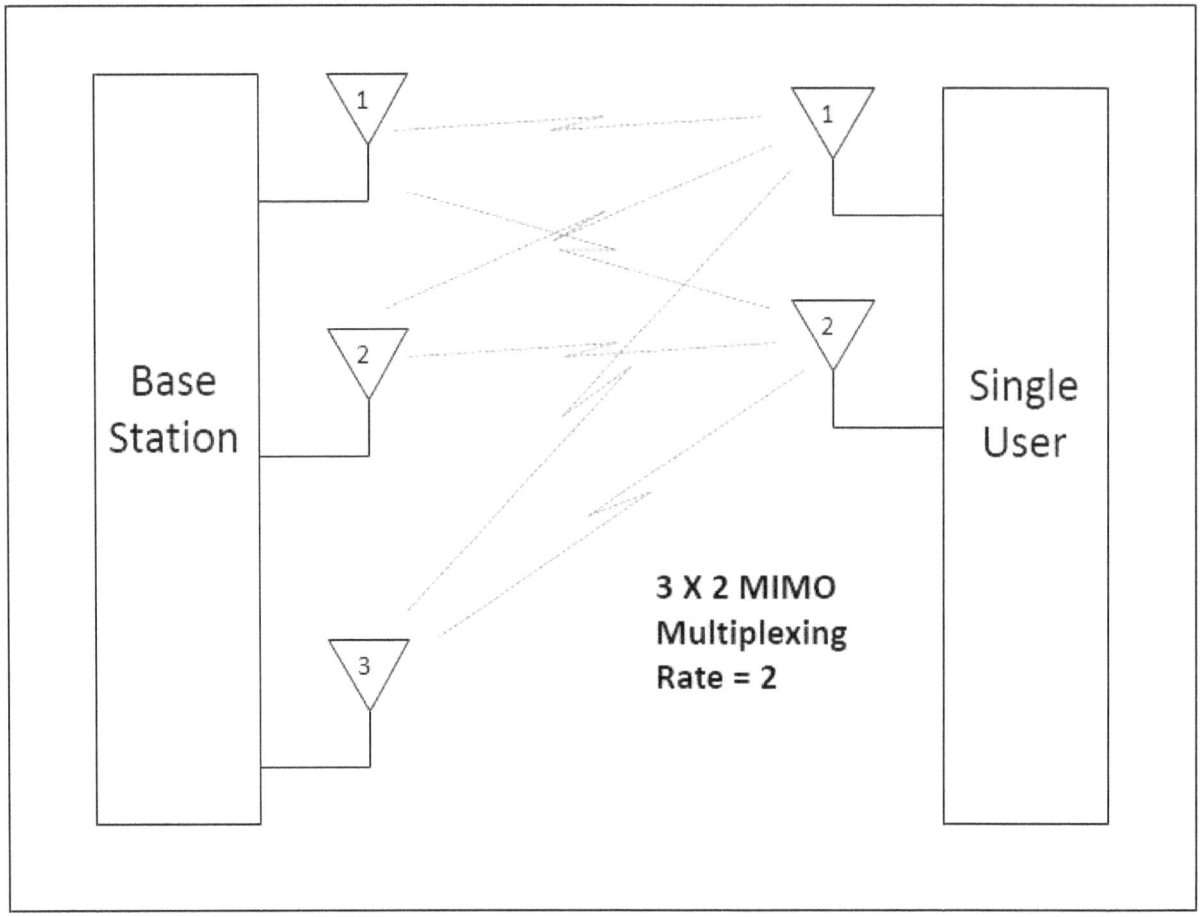

**Figure 15.4.a**

**Figure 15.4.a** shows a 3 x 2 MIMO system. So, in this case, min (3, 2) = 2. Therefore multiplexing rate for this example is 2.

## 15.5 Diversity Gain in MIMO

The *diversity gain* in M x N MIMO system, with the assumption that the radio channel is narrow band and slow fading, is the product of M and N and is expressed as M . N.

**Figure 15.5.a** shows a 3 x 2 MIMO system. In this case, 3.2 = 6. Therefore the diversity gain for this example is 6. This implies that the signal to noise ratio (SNR) is improved by 6 times compared to a SISO system that has no diversity, with the assumption that other factors are the same as in a MIMO system.

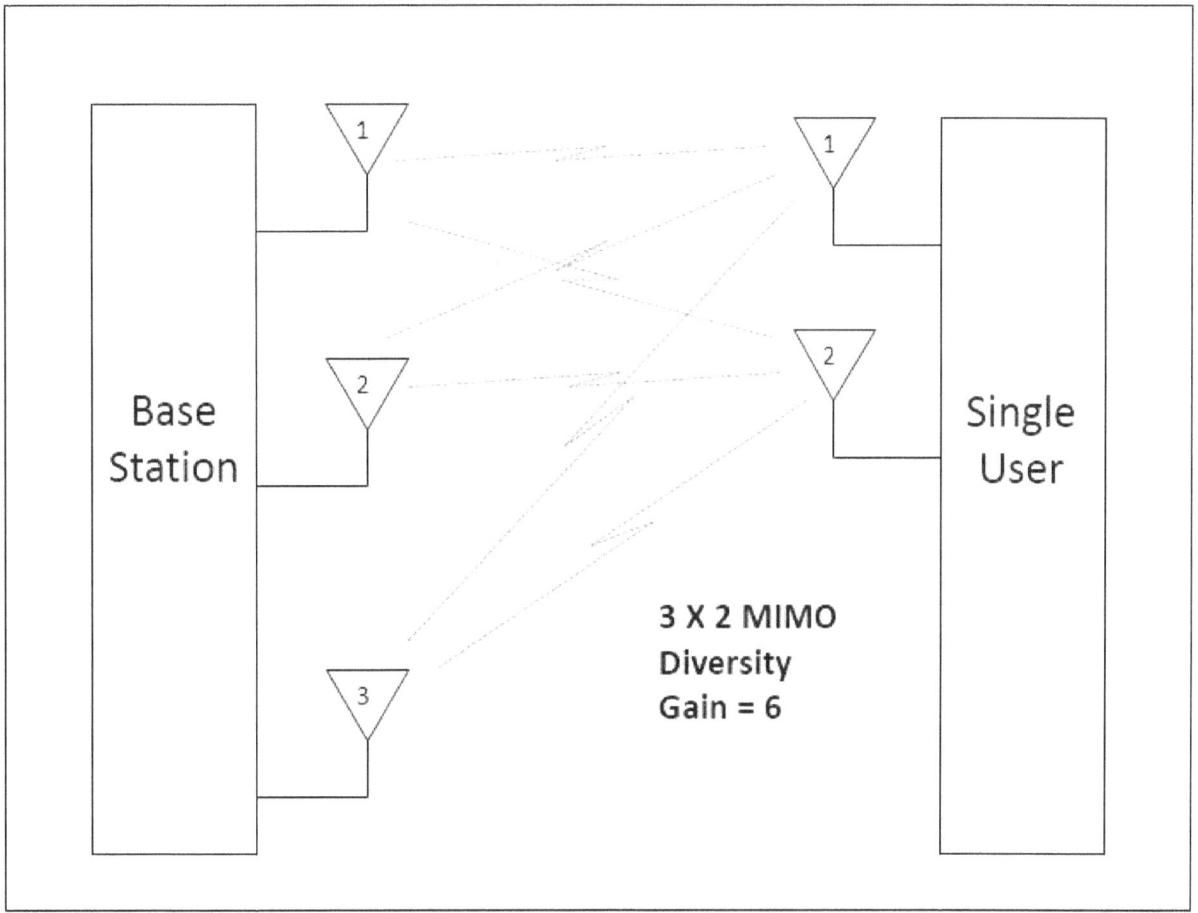

**Figure 15.5.a**

# 15.6 Multiplexing and Diversity Tradeoff

A MIMO system typically has competing demands on its performance for reliability of the received signal as well as throughput of the radio channel in Mbps.

On one hand there is a need to maximize throughput by using all transmit and receive antennas available at our disposal under the given channel conditions. Note that this can be achieved by having multiple data streams between transmitter and receiver, at the cost of reducing the diversity gain.

Also, there is a need to maximize the diversity gain so that effects of multipath propagation can be mitigated. Note that this can be achieved by having multiple antennas transmit and receive redundant copies of the same data stream, at the cost of reducing the overall throughput.

So, tradeoff between reliability of the signal and achievable throughput under given channel conditions is necessary. **Figure 15.6.a** shows an example of the tradeoff between reliability and throughput for a 5 x 4 MIMO system.

In Case 1, the objective is high reliability of the signal, but at the cost of lower throughput. A multiplexing rate of 2 is selected for this case, which means that there will be only 2 independent data streams that can be correctly and simultaneously received. So, at any instant 2 transmitter antennas and 2 receiver antennas must be available for data transmission. The remaining 3 antennas on the transmitter and 2 antennas on the receiver can be used to provide diversity gain, which is calculated to be 3.2 = 6.

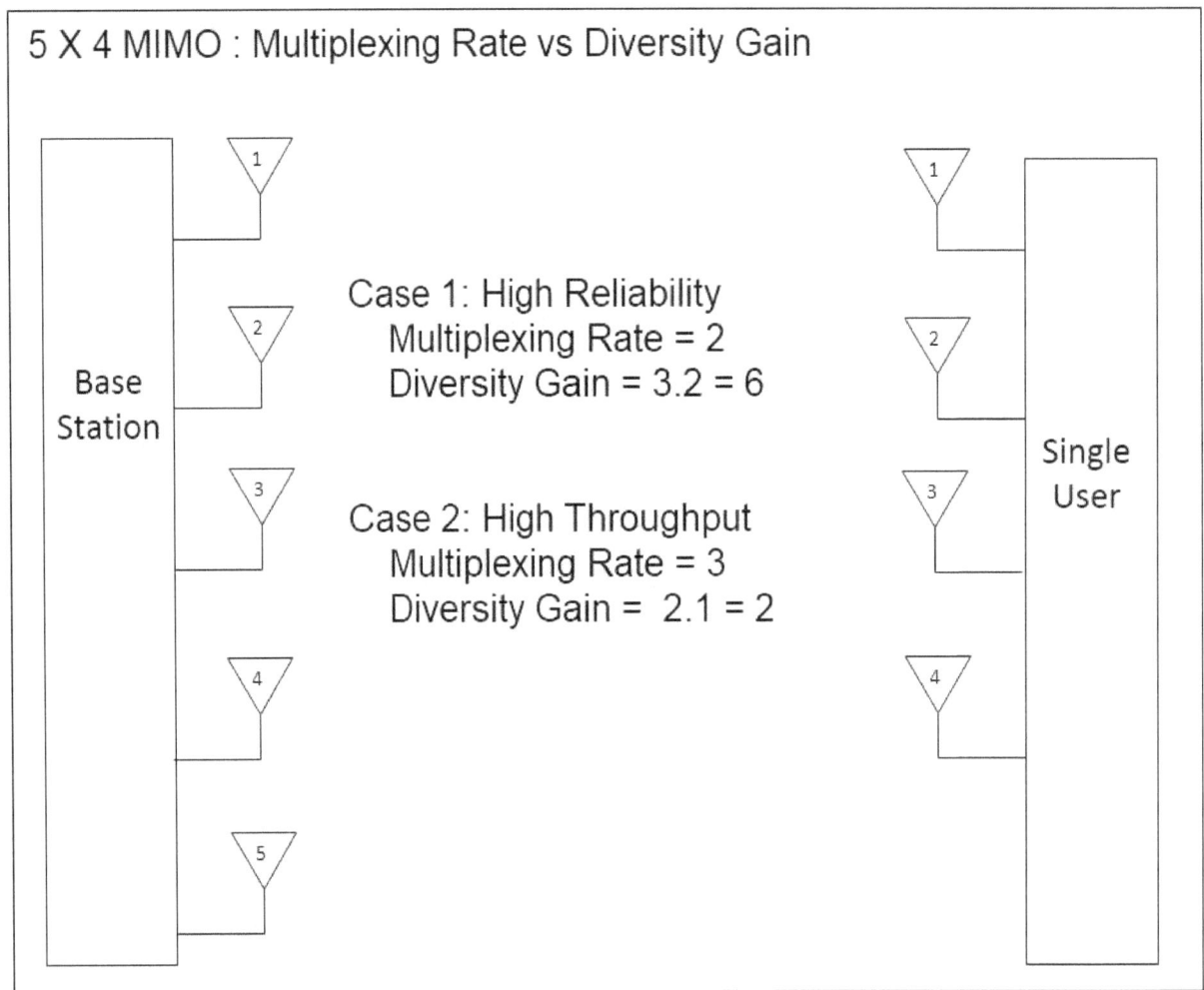

**Figure 15.6.a**

In Case 2, the objective is higher throughput and we select a multiplexing rate of 3. In this case there will now be 3 data streams that can be received correctly and simultaneously, which means 3 antennas on the transmitter and 3 antennas on the receiver must be available for data transmission. The remaining 2 antennas on the transmitter and 1 antenna on the receiver can be used to provide diversity gain, which is calculated to be 2.1 = 2.

# 15.7 MIMO Antennas in WLAN

**Figure 15.7.a** shows MIMO antennas for a WLAN *access point* (AP). This is a narrow band implementation with three monopole antennas each of quarter wavelength height, and a separation of a half wavelength between them. The separation reduces mutual coupling between the radiating antenna elements.

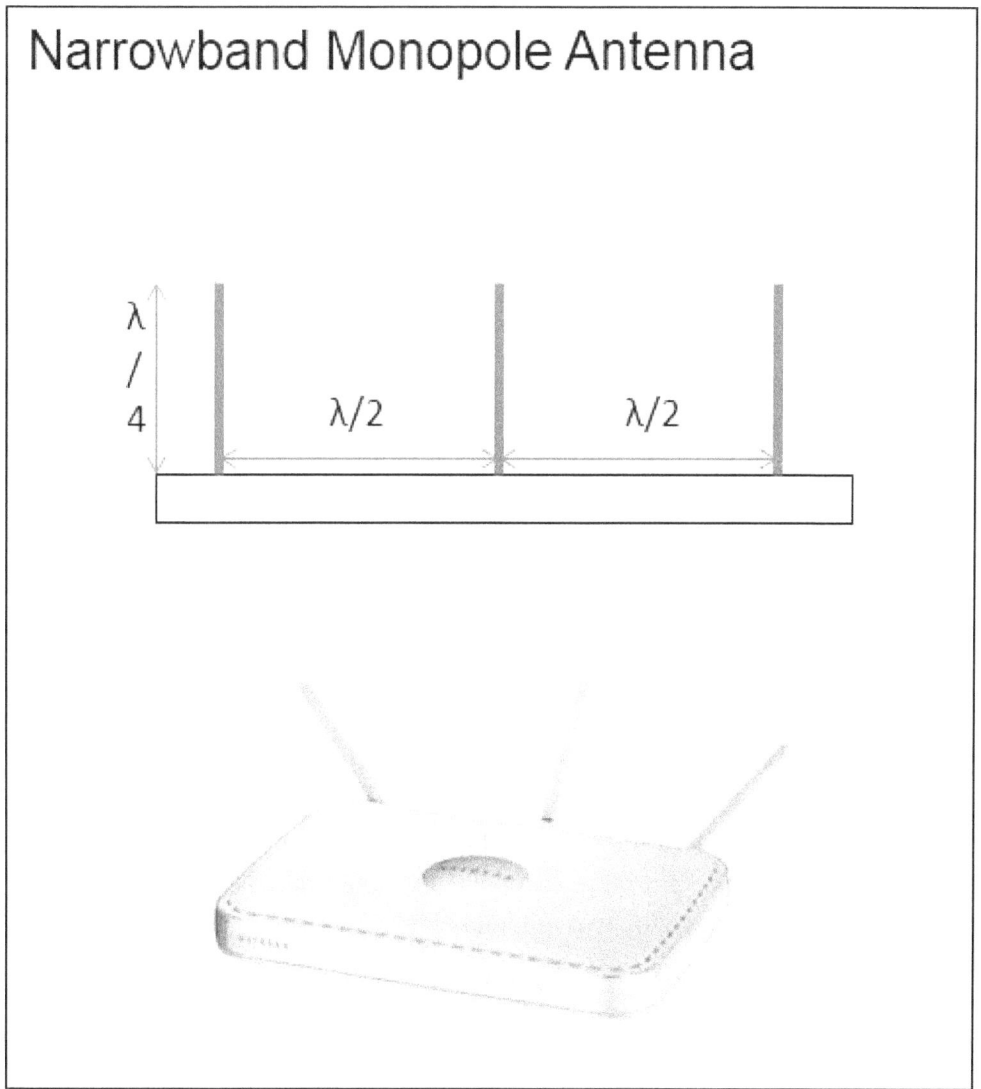

**Figure 15.7.a**

If the separation between the antennas is reduced to less than a half wavelength, the mutual coupling between elements induces large impedance at the feed, which ultimately deteriorates impedance matching, radiation efficiency, radiation patterns and diversity performance. With the monopole antenna design, it is also difficult to ensure omnidirectional coverage in both azimuth and elevation planes.

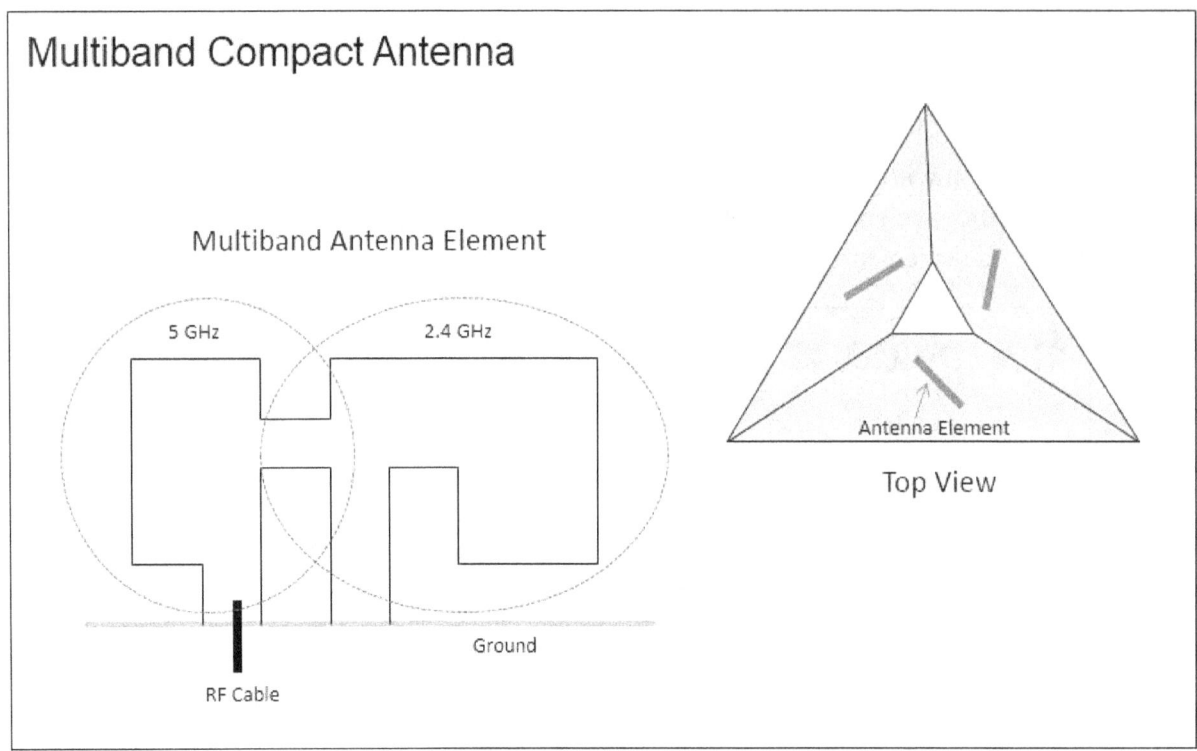

**Figure 15.7.b**

**Figure 15.7.b** shows a multiband compact MIMO antenna design for WLAN *access point* (AP). The antenna element has inverted F shape and is designed for 2.4 and 5 GHz. The top view shows three such slanted antenna elements spaced 120 deg apart, that provide omni directional coverage. The antennas are fabricated on a printed circuit board (PCB) and therefore are low profile.

**Figure 15.7.c** shows two cases of MIMO antenna configuration for portable devices, such as lap tops and smartphones. In Case 1, diversity is achieved by placing a second multiband antenna element along the length of the ground plane. In Case 2, the second antenna element is placed beside the first one to make it more compact. This placement limits diversity, however, the feed points are much closer for the two antenna elements.

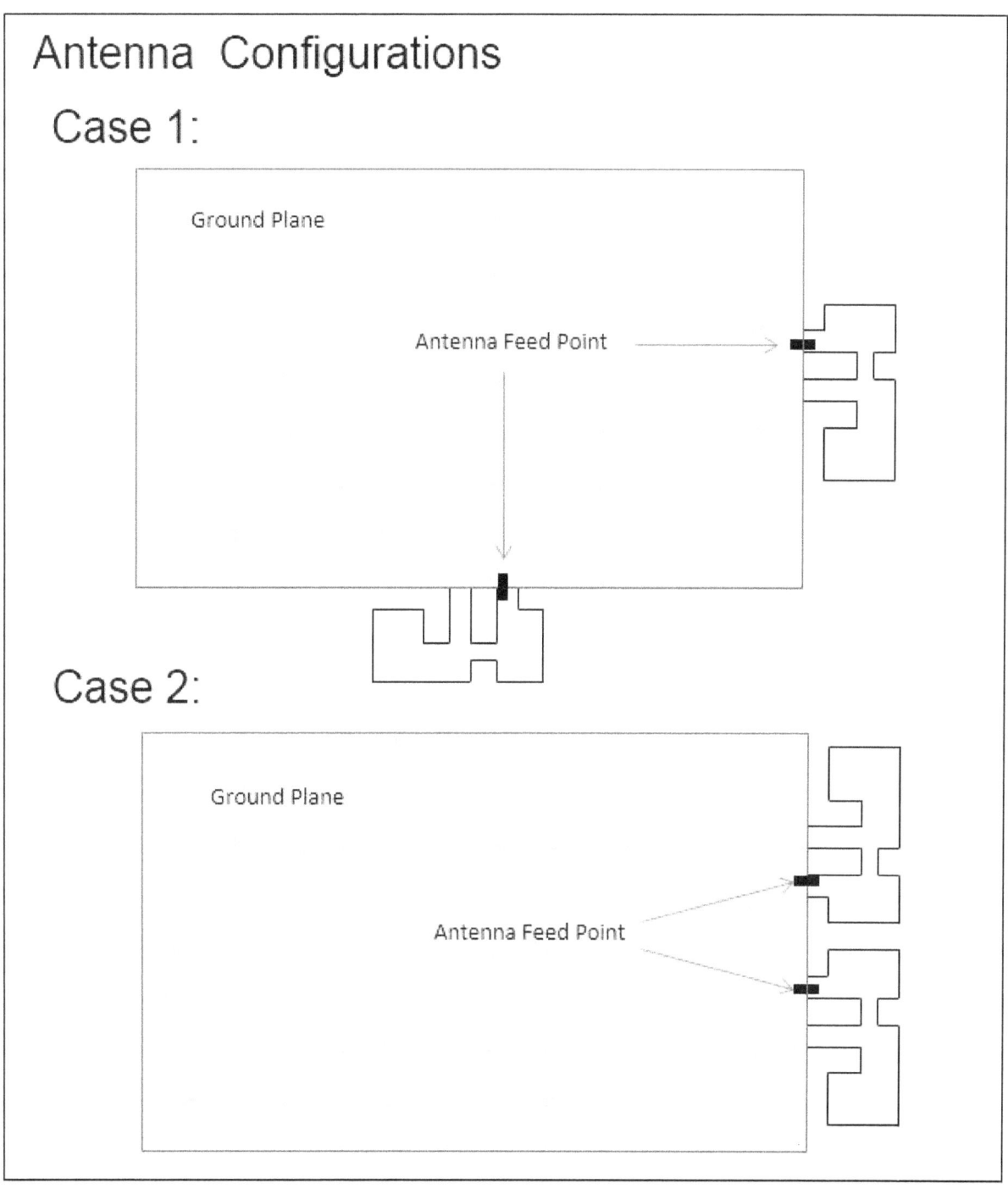

**Figure 15.7.c**

## 15.8 Single User MIMO (SU-MIMO)

In a *single user* MIMO (SU-MIMO) configuration, there is a single dedicated user at the receiving end. This configuration can be expensive to set up and maintain and is usually limited to high bandwidth or mission critical communication applications. A good application is for reliable high bandwidth backhaul wireless communication link for point to point connection, or for connection to enterprise networks. **Figure 15.8.a** shows the illustration.

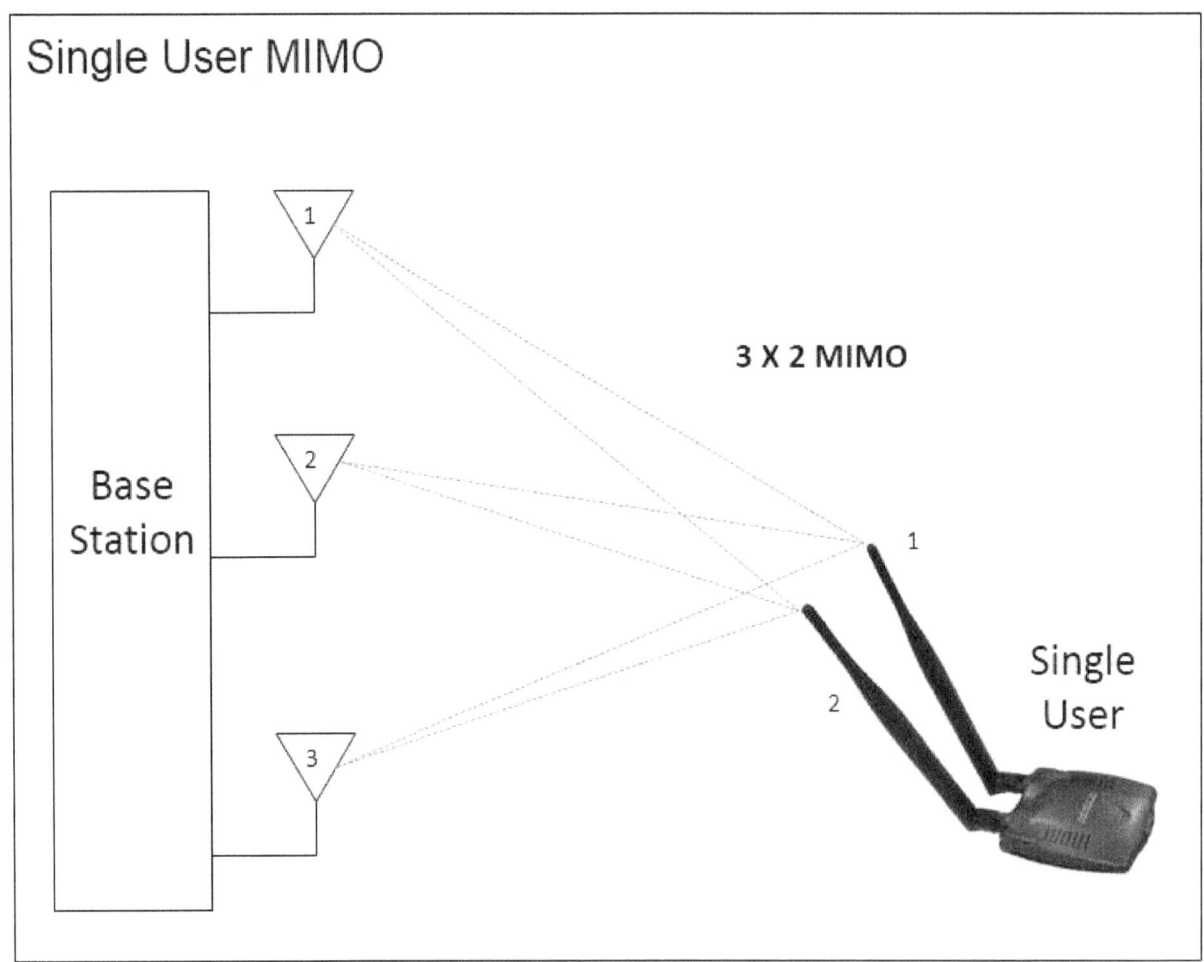

**Figure 15.8.a**

## 15.9 Multi User MIMO (MU-MIMO)

In multi user MIMO (MU-MIMO) configuration, the available time and frequency resource of a MIMO system is shared by multiple users. This is most commonly used in commercial wireless broadband applications. The set up and operating cost is shared by a large number of users. This ultimately improves the quality of service (QoS) and meets the bandwidth requirement for each individual user. **Figure 15.9.a**. shows the illustration.

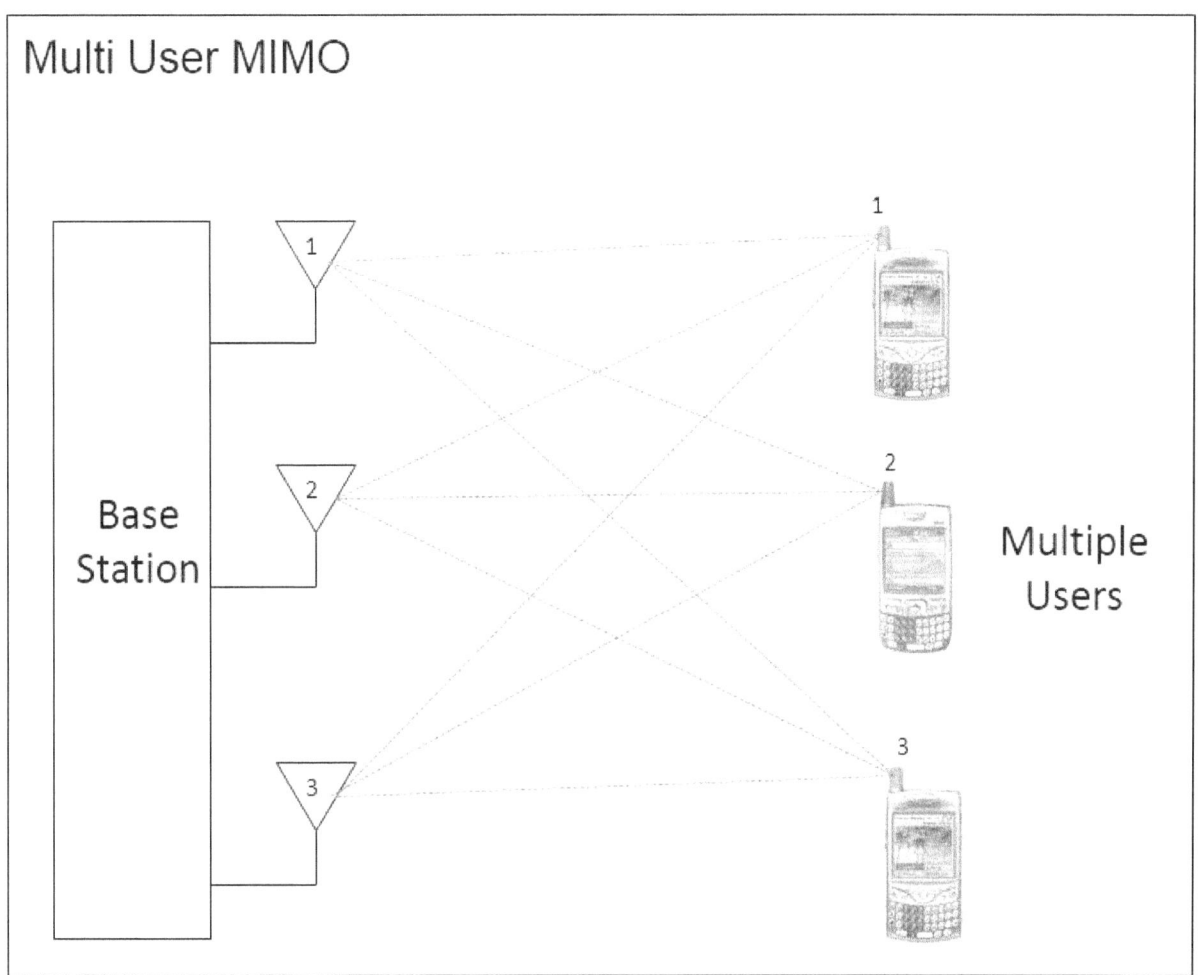

**Figure 15.9.a**

## 15.10    Massive MIMO

There is a growing demand for bandwidth and quality of service (QoS) in 4G wireless broadband applications, and the demand is only going to increase further in the future. This is evidenced by a surge in video traffic on wired as well as wireless networks. MIMO technology has the promise of enabling higher bandwidth. However, it is a prerequisite to have multiple antennas on the device before we can begin to reap its benefits. The size of antenna element depends on frequency, and therefore wavelength used for communication. It is relatively easy to have multiple antennas on the base station or the cell tower, because *size, weight* and *power* (SWAP) are not critical factors. However, it is much more challenging to implement multiple antenna elements in a mobile hand set. One possible solution to this problem is to reduce the size of antenna element by increasing the frequency of wireless communications to the range of 10s of GHz.

IEEE 802.11ad (WiGig) standard aims to exploit the 60 GHz carrier frequency. It recommends 4 wireless channels each of 2 GHz bandwidth at this frequency, and a peak throughput of 7 Gbps. At 60 Ghz, the wavelength, λ, turns out to be only 5 mm. With these dimensions, it is possible to implement, not just one or two, but an array of small antenna elements in a *massive MIMO* configuration.

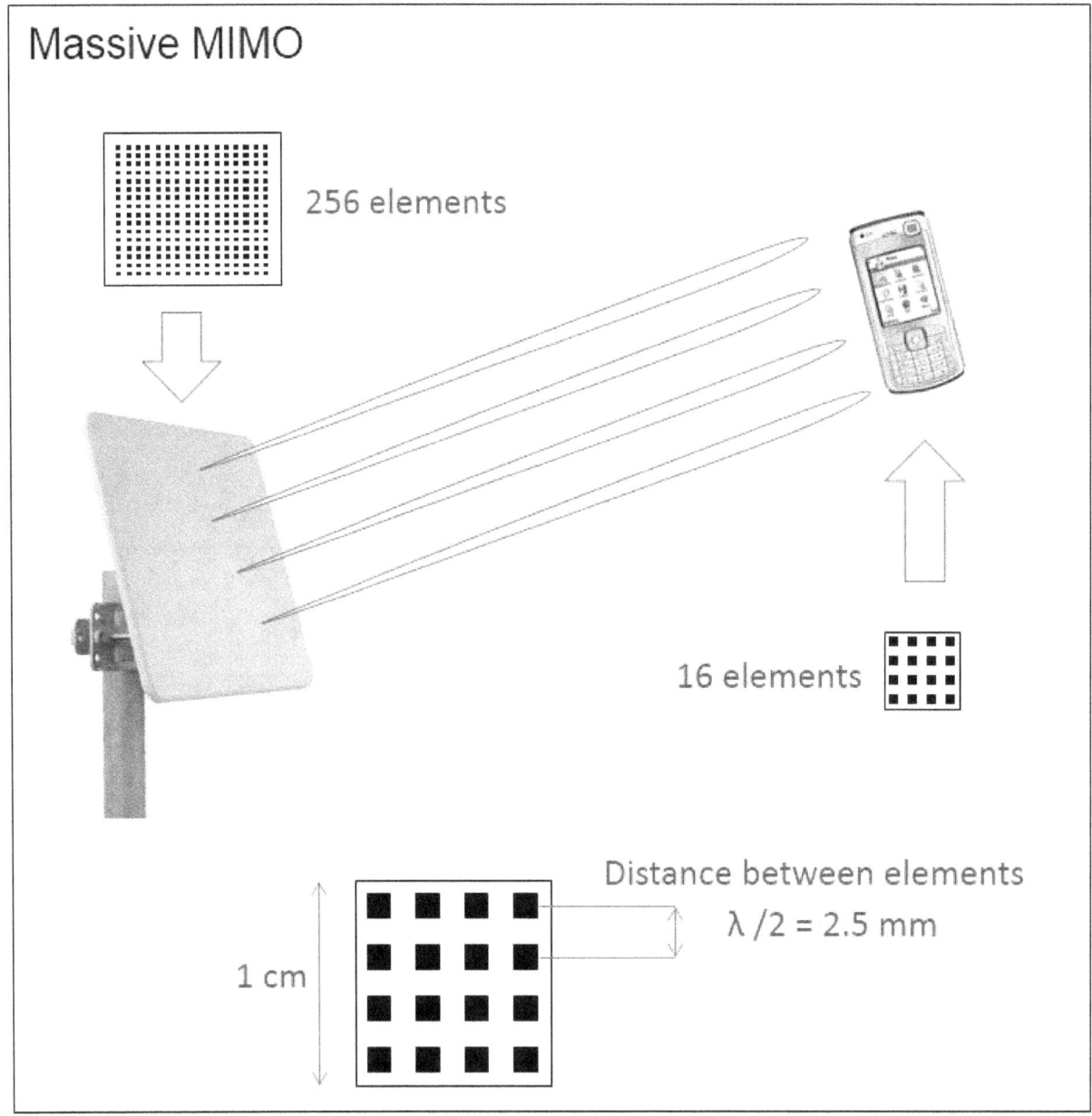

**Figure 15.10.a**

**Figure 15.10.a** shows an illustration of *massive MIMO*. It shows the antenna elements fabricated in the shape of little squares on a surface, such as a *printed circuit board* (PCB). An array of 16 such elements can be arranged in a square of only 1 cm size, which is a good size to fit inside the enclosure of a mobile hand set. Notice that the required minimum separation between adjacent elements is only 2.5 mm. This is to avoid mutual coupling between the antenna elements.

Similarly, an array of 256 such elements will occupy a square of the size of a pizza box approximately, which is a good enough size for a base station or an access point that serves a big enterprise network. With this implementation, it is possible to deploy a 256 x 16 *massive MIMO* antenna configuration.

# 16.  Statistical Fading Models

In this chapter, we will learn about statistical fading models, which are used to characterize a wireless communications channel. These models have practical use in planning a radio network with respect to correct estimation of path loss and signal fading characteristics in a given user environment.

## 16.1 Types of Fading

The phenomenon of radio signal *fading* can be classified on the basis of *radio channel*, or the *radio frequency* used or the *rate of change of amplitude* and *phase* of the radio signal.

The radio channels are of three types, the *Gaussian* channel, the *Rayleigh* channel and the *Rician* channel. Let us first study the characteristics of *fading* for each of these three types of radio channels.

A *Gaussian* channel is the ideal radio channel in which the signal quality is only impaired by the *additive white Gaussian noise* (AWGN) and the signal does not fade. The AWGN is the natural wideband noise generated by many sources, such as, thermal vibrations of atoms, black body radiation from earth, solar and galactic sources. It has constant spectral density (W/Hz) over the range of frequencies under consideration for a radio link. **Figure 16.1.a** shows an illustration. It is a challenge to achieve the ideal *Gaussian* channel in a real world mobile wireless environment.

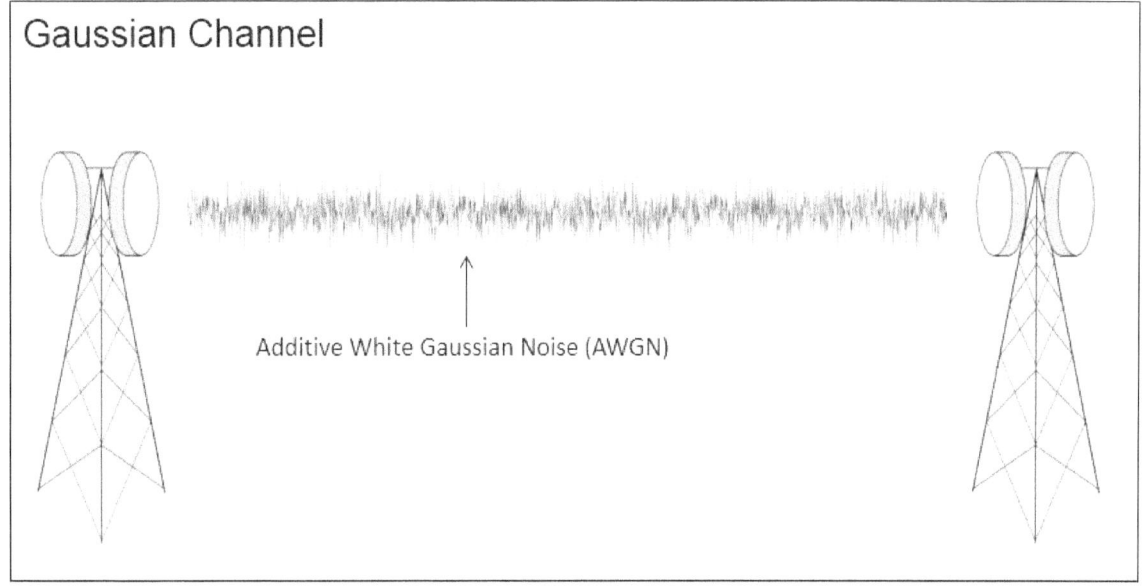

**Figure 16.1.a**

A *Rayleigh* channel represents the other extreme of signal fading. It is the worst case scenario of signal fading and is much closer to the real world scenario encountered in a dense urban environment, where it is normal for the radio signal to suffer multiple reflections, diffraction or scattering and follow multiple paths of propagation, with no *line of sight* (LOS) signal component. This means the handset cannot "see" the cell tower, and all received signals are indirect signals. **Figure 16.1.b** shows an illustration.

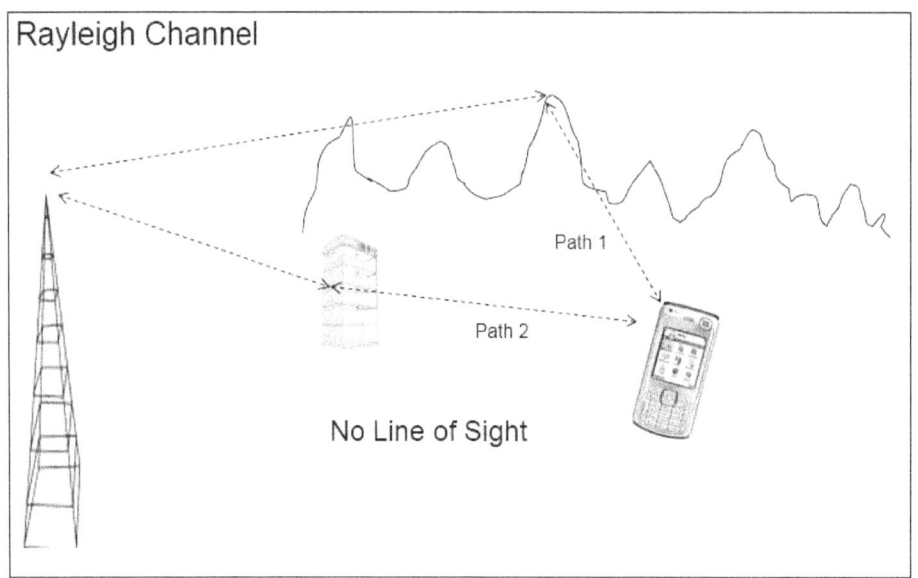

**Figure 16.1.b**

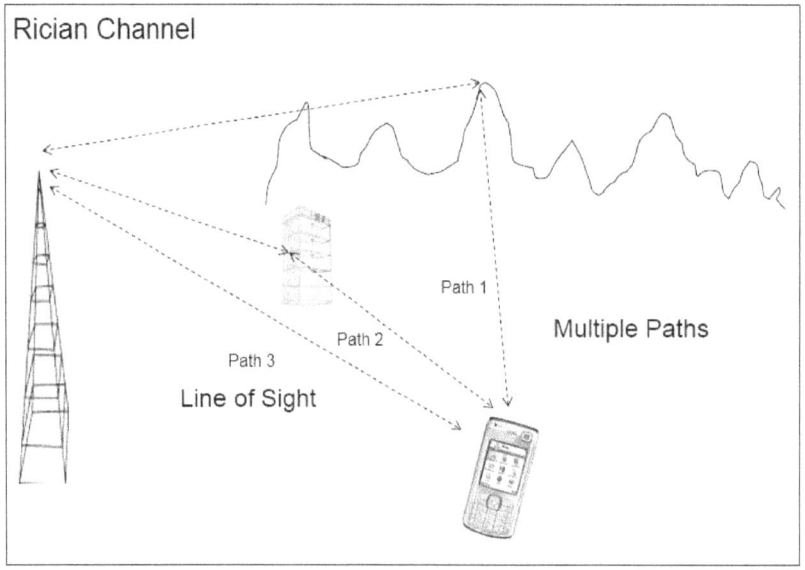

**Figure 16.1.c**

A *Rician* channel is the scenario in between the ideal case of *Gaussian* channel and the worst case of *Rayleigh* channel. In a *Rician* channel, the line of sight (LOS) signal component is present and is the dominant of all other signal components received by the hand set over multiple paths. This is a possible scenario in suburban areas which are interspersed with wide open spaces and often the hand set can "see" the cell tower. However, the landscape is dotted with buildings that cause multi path propagation due to reflection or knife edge diffraction. **Figure 16.1.c** shows an illustration.

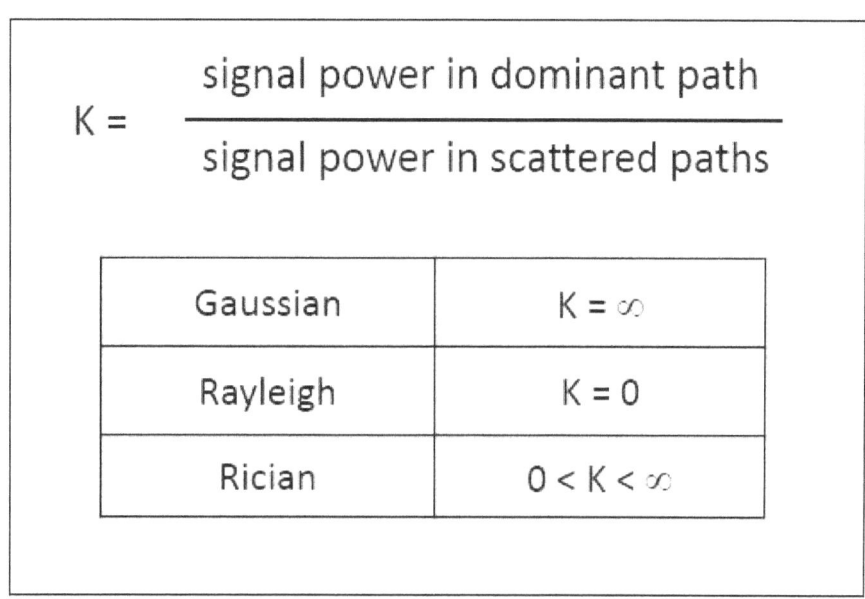

**Figure 16.1.d**

The ratio, *K*, is defined as the ratio of signal power in dominant path of propagation, such as LOS, to the ratio of signal power in scattered paths. K can be used to characterize the *Gaussian*, *Rayleigh* and *Rician* radio channels.

For a *Gaussian* channel, there are no scattered paths, so there is no signal power wasted in scattered paths, therefore K is infinity.

For a *Rayleigh* channel, the LOS dominant signal is missing, so there is no received signal power in the dominant path, therefore, K is 0.

For a *Rician* channel, K lies between 0 and infinity, as shown in **Figure 16.1.d**.

Let us now study the phenomenon of signal fading based on *radio frequency*. In this classification, fading can be either *flat* fading or *frequency selective* fading.

In *flat* fading, all frequencies in the radio channel fade equally to the same level. However, in *frequency selective* fading, different frequencies in the radio channel are affected to different levels in amplitude and phase, as illustrated in **Figure 16.1.e**. In this case, some form of frequency equalization is required at the receiver to mitigate the effects of fading. Some digital modulation techniques, such as OFDM, are able to spread data over a wide channel bandwidth, and only a portion of data is lost due to *frequency selective* fading. The lost data can be reconstituted using *forward error correction* (FEC) techniques.

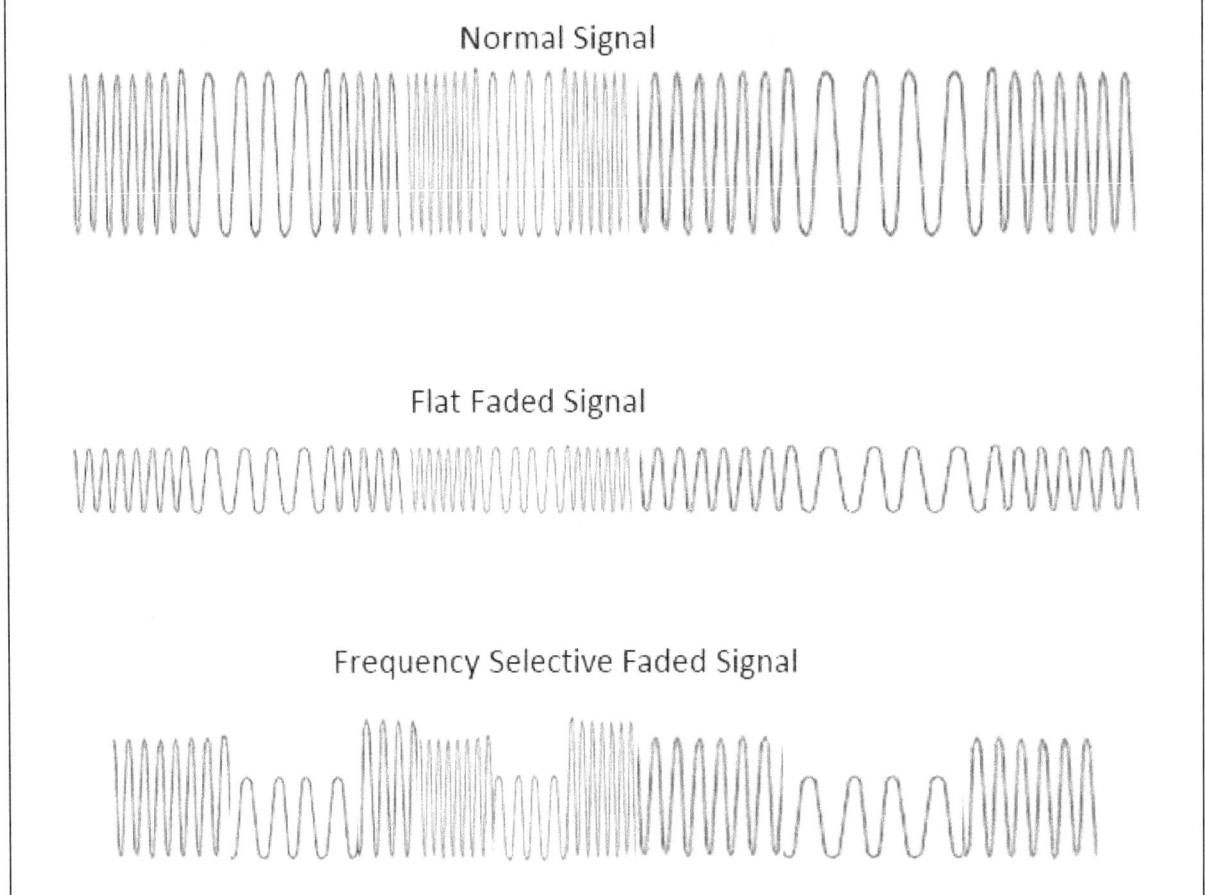

**Figure 16.1.e**

Another way to classify *fading* is based on the *rate of change of amplitude* and *phase* of the radio signal; *slow* or *fast* fading, as illustrated in **Figure 16.1.f**.

*Slow* fading can be caused by shadowing due to an obstruction such as a hill or a building. In slow fading, *time diversity* technique cannot be used, because the receiver sees only a single realization of the radio channel, If there are deep fades, they will last over the entire duration of transmission.

However, in *fast fading*, time diversity technique can be useful and can help mitigate the effects of fading. This is achieved by transmitting redundant copies of data over different time intervals, so that reliable data is still received, even though radio signal may have faded for some intervals. See **Section 12.4** for more details on time diversity.

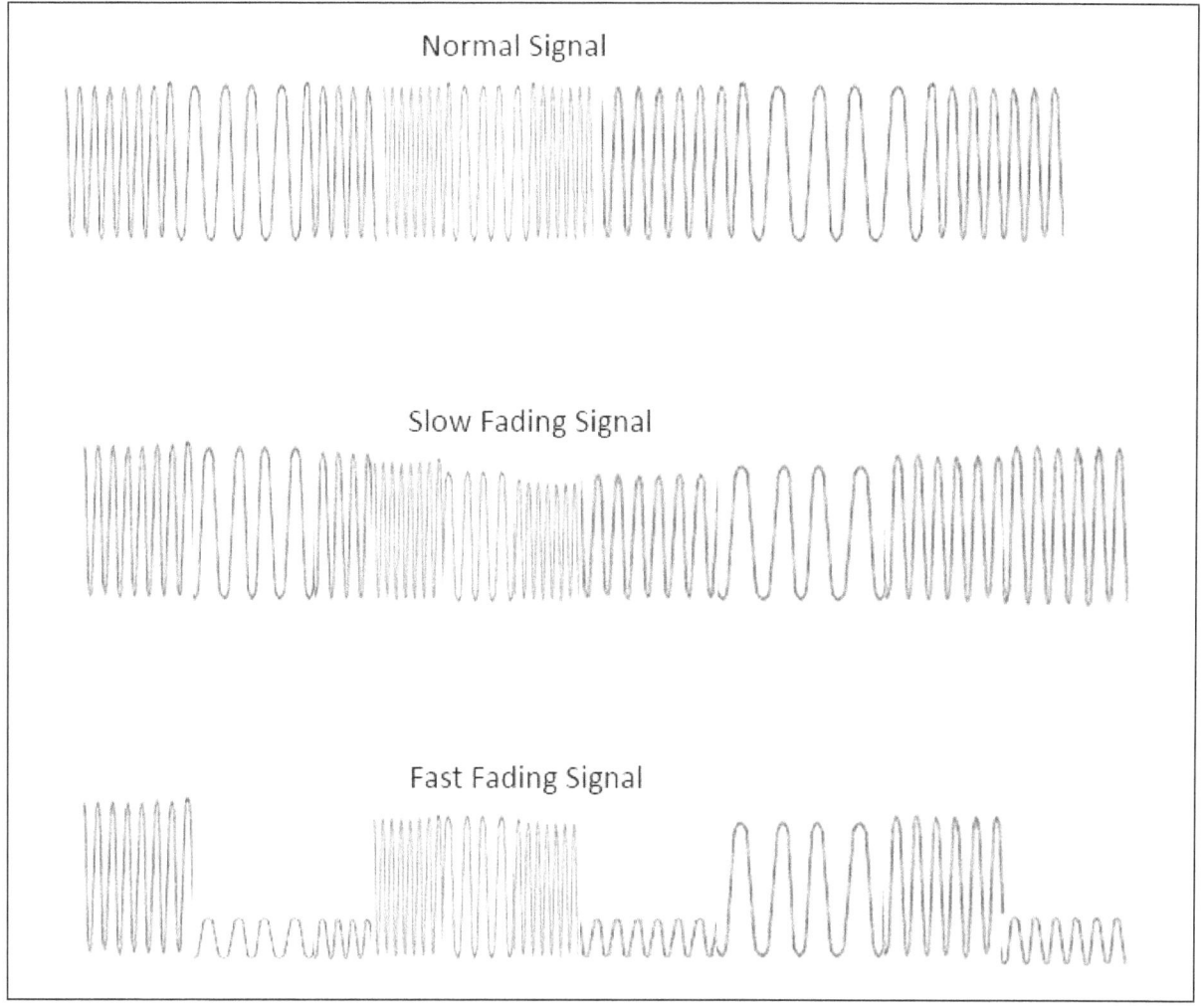

**Figure 16.1.f**

# 16.2 Types of Cells

In cellular wireless communications, the *cells* are classified into various types based on the radius of the cell. The position of base station antenna also depends on size of the cell. **Table 16.2.a** can be used as a reference.

There are several statistical models that we will study in the remaining sections of this chapter. Each model operates under specific assumptions, restrictions and limitations. It is important to understand, which model is suitable for a certain cell type. **Tables 16.2.b**, **16.2.c** and **16.2.d** can be used as a reference for this determination.

| Category | Cell Type | Cell Radius | Position of base station antenna |
|---|---|---|---|
| Large Cells | Macro Cell | 1-30 km | outdoor; mounted above medium roof-top level, heights of all surrounding buildings are below base station antenna height |
| | Small Macro Cell | 0.5 – 3 km | outdoor; mounted above medium roof-top level, heights of some surrounding buildings are above base station antenna height |
| Small Cells | Micro Cell | < 1 km | outdoor; mounted below medium roof top level |
| | Pico Cell | < 500m | indoor or outdoor (mounted below roof-top level) |
| | Femto Cell | < 10 m | Indoor, mounted below roof-top level or ground level |

**Table 16.2.a**

| Category | Cell Type | Cell Radius | Prediction Model |
|---|---|---|---|
| Large Cells | Macro Cell | 1-30 km | Egli, Okumara, Hata, COST-231 |
| | Small Macro Cell | 0.5 – 3 km | COST-231, COST-WI |
| Small Cells | Micro Cell | < 1 km | See Table 16.2.c and Table 16.2.d |
| | Pico Cell | < 500m | |
| | Femto Cell | < 10 m | |

**Table 16.2.b**

| Prediction Model | Method | Features | Terrain Data | Results |
|---|---|---|---|---|
| Uni-Lund | Empirical | BS below roof top | 2D building layout | Path Loss |
| CNET micro cell model | Analytical LOS + NLOS Model | 2D horizontal plane + 2D over-roof-top | 2D building layout | Path Loss |
| RT-Swiss Telecom PTT | Ray Tracing | 2D horizontal plane | 2D building layout | Path Loss and CIR (Channel Impulse Response) |
| Univ. Geneva/ Swiss Telecom PTT | TLM like (Transmission Line Method) | 2D plane | 2D building layout | Path Loss |
| 2D-URBAN-PICO Univ. Karlsruhe | Ray Launching | 2D horizontal plane | 2D building layout | Path Loss and CIR |
| Telekom | Analytical LOS + NLOS Model | 2D horizontal plane + 2D over-roof-top | 2D building layout | Path Loss |
| Ericsson | Ray Tracing + COST-WI | 2D horizontal plane + 2D over-roof-top | 2D building layout | Path Loss |
| COST-231 small-cell | Walfisch-Ikagami Model | 2D over-roof-top | Building classes | Path Loss |
| Univ.. Valencia | Walfisch-Bertoni Model | 2D vertical plane + 3D reflections at Rx | 2D building layout + building height | Path Loss, FS distribution |
| MCOR-Swiss Telecom PTT | Modified Deygout | 2D over-roof-top | 2D building layout + building height | Path Loss |
| CSELT | Deygout | 2D over-roof-top, BS above roof top | 3D raster data | Path Loss |

**Table 16.2.c**

| Prediction Model | Method | Features | Terrain Data | Results |
|---|---|---|---|---|
| CNET Ray Launching Model | Ray Launching | 3D no diffraction at vertical wedges | 3D building layout | Path Loss and CIR |
| ASCOM-ETH | Ray-tracing by image source | 3D, only reflections | 2D building layout + building height | Path Loss and CIR |
| Villa Griffone Lab, Bologna | Ray Tracing, Saunders-Bonar | Transverse plane+ground reflection, 2D over-roof-top | 2D building layout + building height | Path Loss and CIR |
| Uniiv. Stuttgart | Ray Launching+ W/I model for 2D case | 3D (2 diffraction + 6 refection processes), 2D vertical plane | 2D building layout + building height | Path Loss and CIR |
| 3D-URBAN-MICRO, Univ. Karlsruhe | Ray Tracing | 2D transverse plane, 3D surface scatter | 2D building layout + building height or raster data | Path Loss and CIR |

**Table 16.2.d**

In cellular communications, it is useful to know, which RF propagation mechanisms come into play for each type of cell coverage area, i.e., *macro cells*, *micro cells* and *indoor areas*. For example, in *macro cells*, the propagation mechanisms involved are *LOS, reflection, diffraction, multiple diffraction, reflection* and *scattering*. In *micro cells*, these are *LOS, reflection* and *diffraction*. For *indoor areas*, these are *LOS, reflection, diffraction, absorption* and *guided wave*. Each of these propagation mechanisms have been determined to cause the path loss to a different extent, that is proportional to the distance involved, d, and an empirically determined exponent. The details and illustrations for each of these mechanisms are shown in **Figures 16.2.e, 16.2.f** and **16.2.g**.

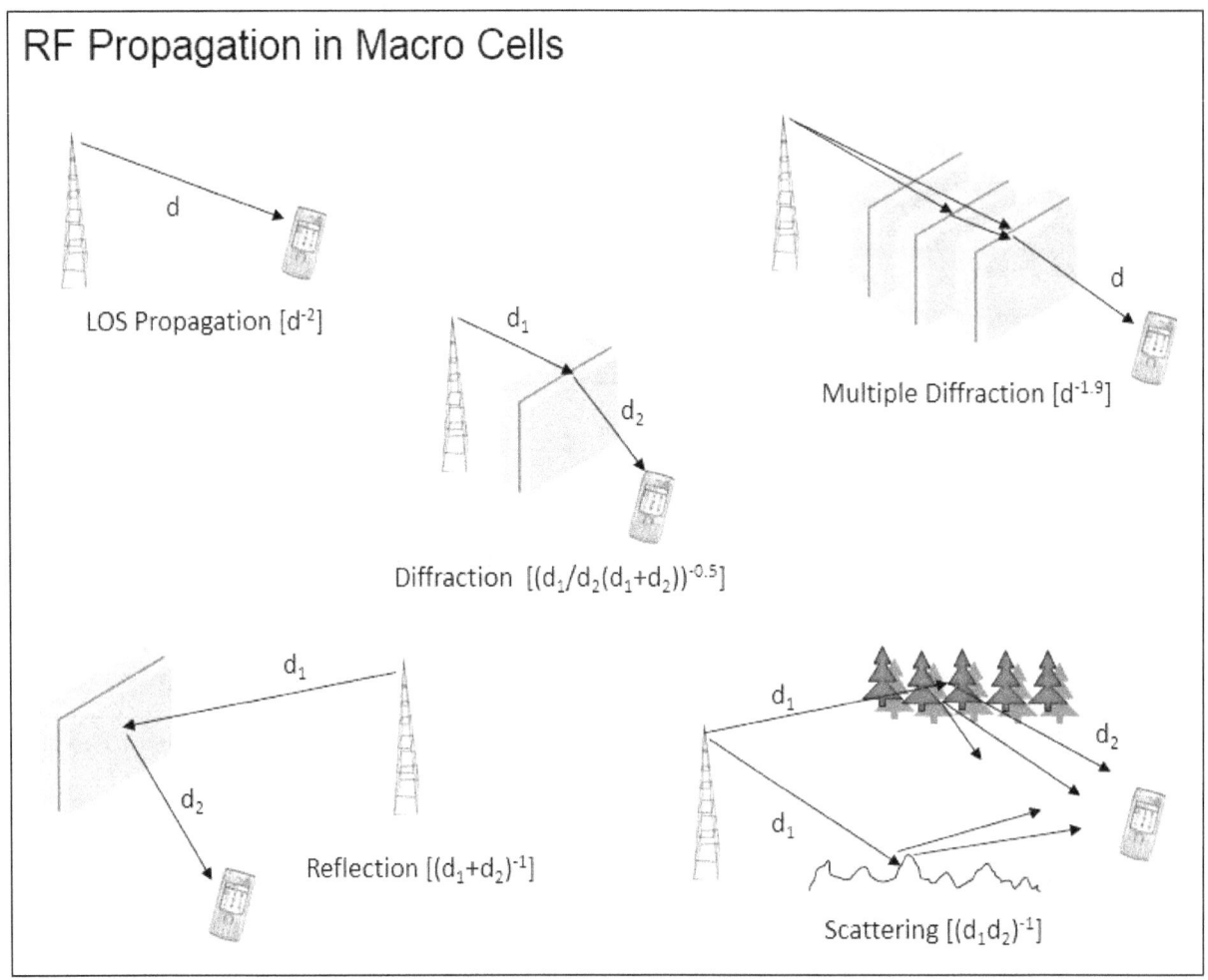

**Figure 16.2.e**

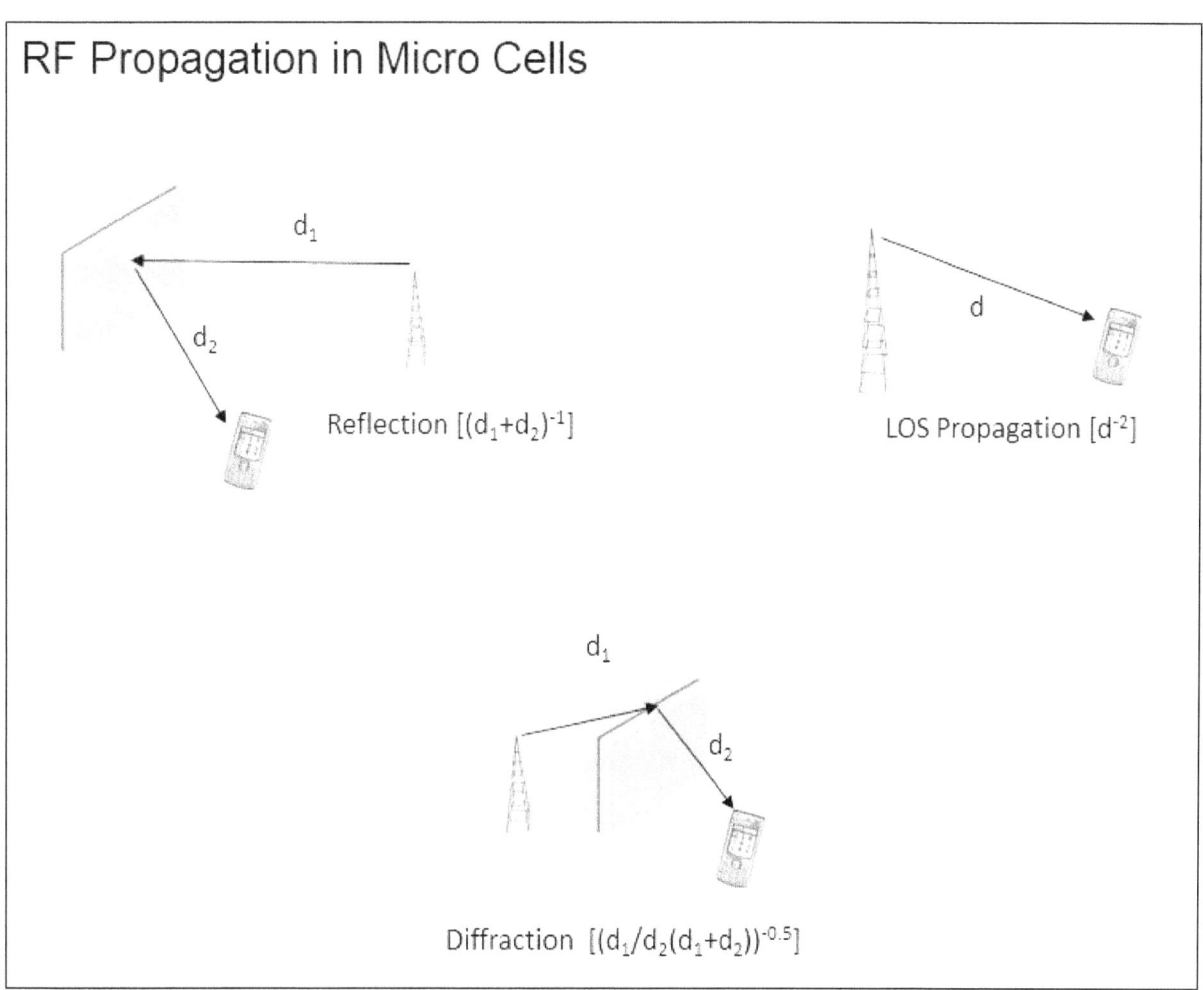

**Figure 16.2.f**

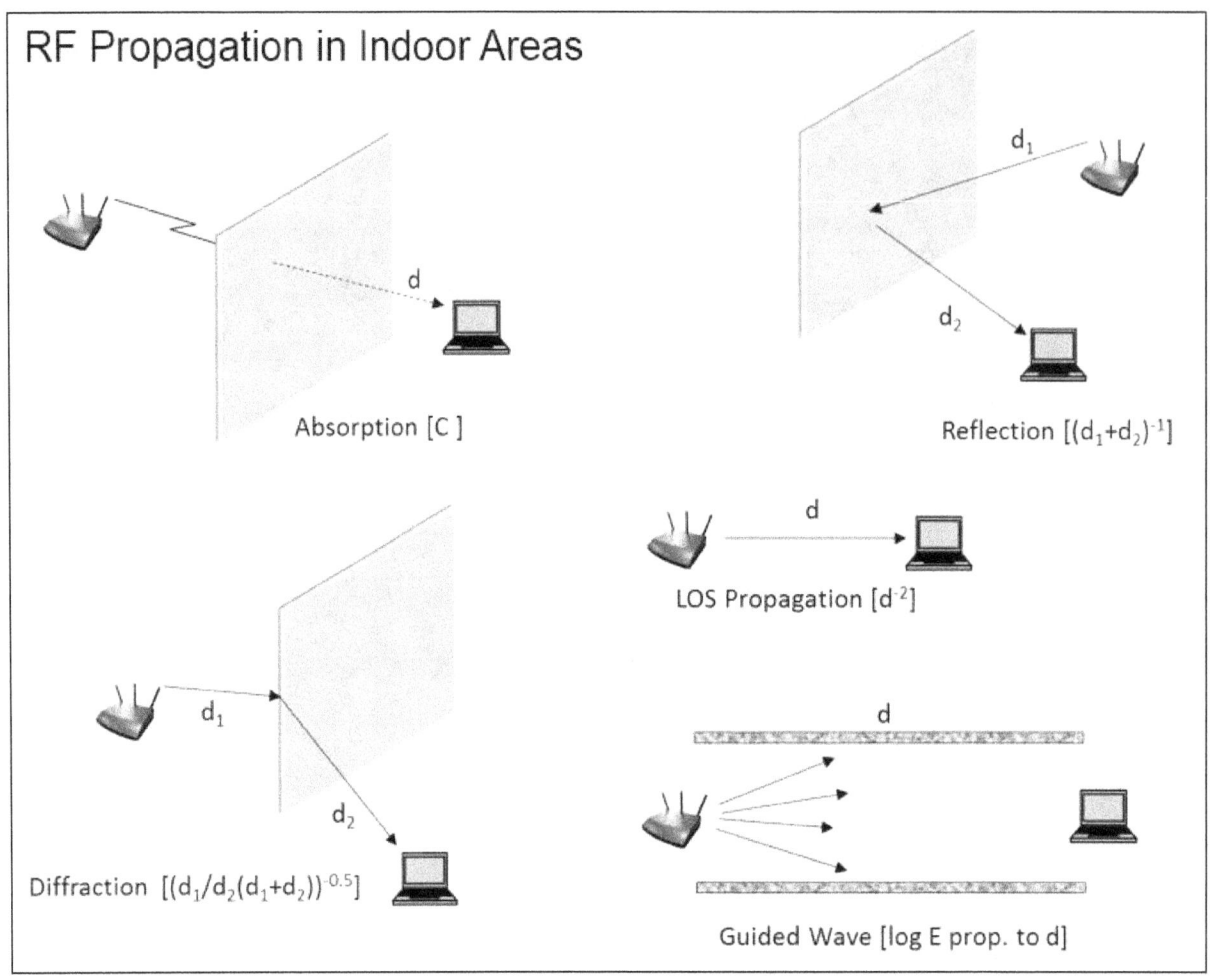

**Figure 16.2.g**

# 16.3 Log Distance Path Loss (PL) Model

The log distance *path loss* (PL) model is illustrated in **Figure 16.3.a**. It can be applied to a mobile communications environment for the calculation of *path loss* (PL) and accommodates *large scale* as well as *small scale* signal variations (See **Chapter 3**). Notice that the PL exponent changes from n = 2 for free space to n = 6 for dense urban areas. This implies that much higher path loss is experienced in dense urban areas compared to open or free space.

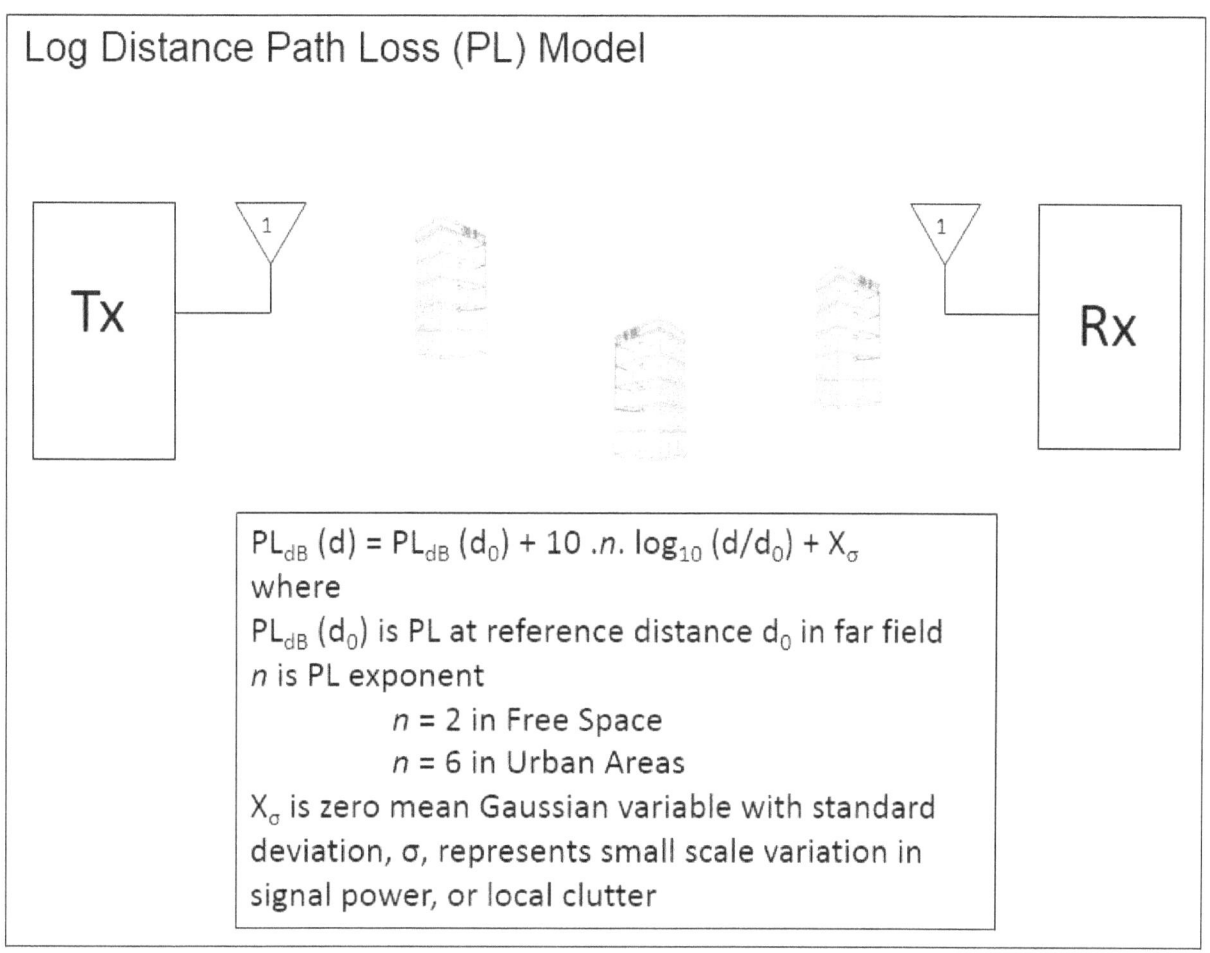

**Figure 16.3.a**

## 16.4 Egli Model

The *Egli Model* was first introduced by John Egli in his 1957 paper and it was derived from real world data on VHF and UHF television transmissions, and can be easily applied to mobile communications. The assumption for terrain in this model is, gently rolling hills with average hill height of 50 ft. Because of this assumption, no terrain data is needed. Instead the free space propagation loss is adjusted for the height of base station antenna and mobile hand set antenna. **Figure 16.4.a** illustrates the model along with restrictions for distance and frequency range.

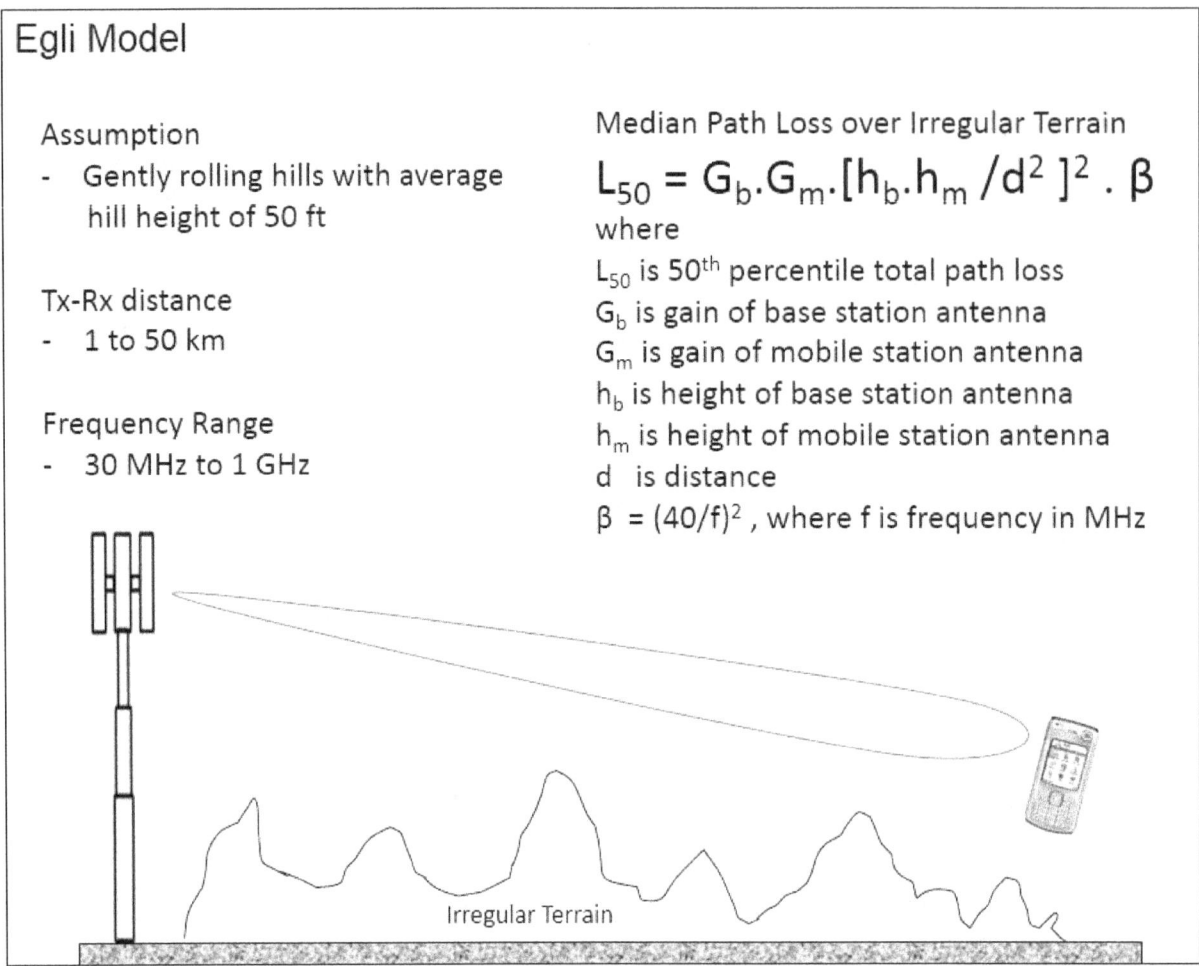

## Egli Model

**Assumption**
- Gently rolling hills with average hill height of 50 ft

**Tx-Rx distance**
- 1 to 50 km

**Frequency Range**
- 30 MHz to 1 GHz

**Median Path Loss over Irregular Terrain**

$$L_{50} = G_b . G_m . [h_b . h_m /d^2 ]^2 . \beta$$

where

$L_{50}$ is 50th percentile total path loss

$G_b$ is gain of base station antenna

$G_m$ is gain of mobile station antenna

$h_b$ is height of base station antenna

$h_m$ is height of mobile station antenna

d   is distance

$\beta = (40/f)^2$ , where f is frequency in MHz

Irregular Terrain

**Figure 16.4.a**

By assuming a log normal distribution of terrain height, Egli generated a family of curves showing the *terrain factor* or adjustment to the *median path loss* for the desired fade probability. This way the mean or median signal level at a given percentage of locations for a circle of radius d, can be determined. In other words, the *Egli model* provides the median path loss due to terrain loss. If a terrain loss point other than the median (50th percentile) is desired, the adjustment factor in dB can be inferred from the table in **Figure 16.4.b**, that is derived from *Egli's* family of curves.

For example, if $L_{90dB}$ , that is *path loss* at 90th percentile is to be determined, which also means that level of *path loss* will be exceeded 10% of times, then from *Egli's* family of curves (read from table in **Figure 16.4.b**), 10 dB correction can be added for the *terrain factor*.

Egli Model

$$L_{ndB} = L_{50dB} + \text{Terrain Factor (dB)}$$

Example

For f = 1 GHz and n= 99 percentile

Terrain Factor = −29 dB

$$L_{90dB} = L_{50dB} - 29$$

| Percentile (n) | f (MHz) | Terrain Factor (dB) |
|---|---|---|
| 10 | 40 | +8 |
| 10 | 1000 | +16 |
| 20 | 40 | +5 |
| 20 | 1000 | +11 |
| 60 | 40 | -1 |
| 60 | 1000 | -3 |
| 80 | 40 | -5 |
| 80 | 1000 | -11 |
| 99 | 40 | -14 |
| 99 | 1000 | -29 |

**Figure 16.4.b**

# 16.5 Okumara Model

The *Okumara Model* is based on measurements taken in Tokyo in 1960 and was developed for urban and suburban areas in large cells. **Figure 16.5.a** and **16.5.b** illustrate the model along with restrictions for distance, frequency range and base station antenna height. The urban areas are defined as heavily built areas with tall buildings and apartments. The suburban areas are residential with 1-2 storey houses with some obstacles near the mobile hand set but are not very congested. The open areas are open fields and farms with no tall trees or buildings.

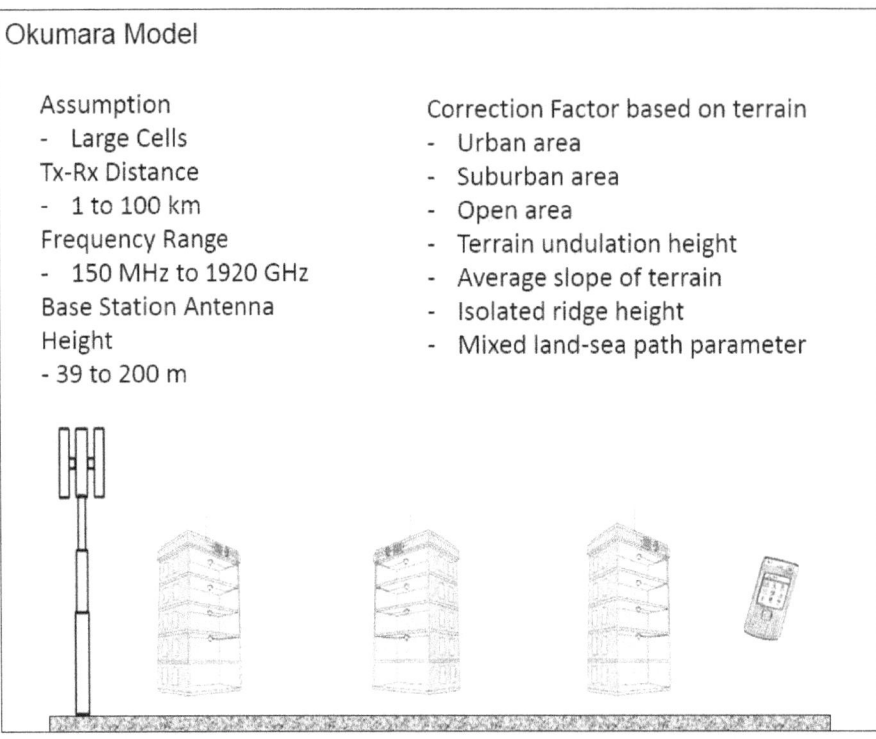

Okumara Model

Assumption
- Large Cells
Tx-Rx Distance
- 1 to 100 km
Frequency Range
- 150 MHz to 1920 GHz
Base Station Antenna
Height
- 39 to 200 m

Correction Factor based on terrain
- Urban area
- Suburban area
- Open area
- Terrain undulation height
- Average slope of terrain
- Isolated ridge height
- Mixed land-sea path parameter

**Figure 16.5.a**

# Okumara Model

## Median Attenuation Equation

$$L_{50dB} = L_{free} + A_{mu}\ (f,d) - G(h_{te}) - G(h_{re}) - G_{area}$$

where

$L_{50dB}$ is 50th percentile or median value of propagation path loss

$L_{free} = 32.45 + 20 \log f + 20 \log d$

f is frequency in MHz

d is distance in km

$A_{mu}\ (f,d)$ is median attenuation relative to free space loss in urban area, with
- Base station antenna height, $h_{te}$ = 200 m
- Mobile antenna height, $h_{re}$ = 3 m
- function of frequency, f and distance, d

$G(h_{te})$ is Base station antenna gain factor due to height

$G(h_{re})$ is Mobile antenna gain factor due to height

$G_{area}$ is gain factor due to terrain

Corrections to antenna gain factors if base and mobile antenna heights are different than 200 m and 3 m

$G(h_{te}) = 20 \log_{10} (h_{te} /200)$    $30\ m < h_{te} < 100\ m$
$G(h_{re}) = 10 \log_{10} (h_{re} /3)$        $h_{re} < 3\ m$
$G(h_{re}) = 20 \log_{10} (h_{re} /3)$      $3\ m < h_{re} < 10\ m$

**Figure 16.5.b**

# 16.6 Hata Model

The *Hata Model* also applies to large cells, however, it is a simplified version of Okumara model. **Figures 16.6.a**, **16.6.b** and **16.6.c** illustrate this model with all its assumptions and restrictions.

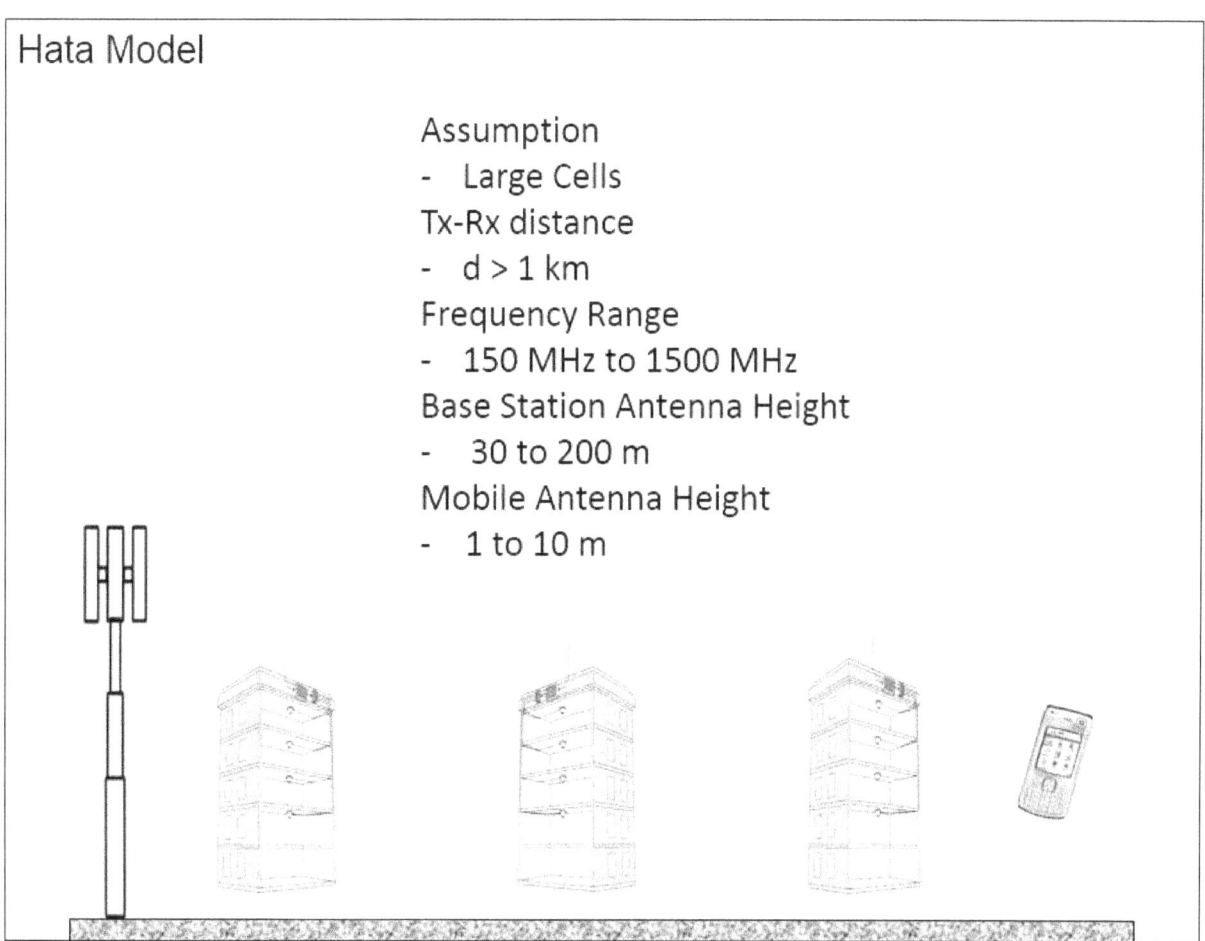

## Hata Model

Assumption
- Large Cells
Tx-Rx distance
- d > 1 km
Frequency Range
- 150 MHz to 1500 MHz
Base Station Antenna Height
- 30 to 200 m
Mobile Antenna Height
- 1 to 10 m

**Figure 16.6.a**

Hata Model
Median Attenuation Equation for Urban Areas

$$L_{50dB,urban} = 69.55 + 26.16\log_{10}f_c - 13.82\log_{10}f_c$$
$$- 13.82\log_{10}h_{te} - a(h_{re}) + 44.9$$
$$- 6.55\log_{10}h_{te} + 10\log_{10}d$$

where

$L_{50dB, urban}$ is 50th percentile of propagation path loss in urban areas
$h_{te}$ is Effective Base station antenna height (30-200 m)
$h_{re}$ is Effective Mobile antenna height (1-10 m)
$f_c$ is frequency (150-1500 MHz)
d is Tx-Rx distance or size of large cell (d > 1 km)
$a(h_{re})$ is correction factor for effective mobile antenna height

 - Function of size of coverage area
 - For Small and Medium Cities:
   $$a(h_{re}) = (1.11\log_{10}f_c - 0.7) - (1.56\log_{10}f_c - 0.8)$$
 - For Large Cities, $f_c \leq 300$ MHz
   $$a(h_{re}) = 8.29(\log_{10}(1.54h_{re}))^2 - 1.1$$
 - For Large Cities, $f_c > 300$ MHz
   $$a(h_{re}) = 3.2(\log_{10}(11.75h_{re}))^2 - 4.98$$

**Figure 16.6.b**

Hata Model
Median Attenuation Equation for Suburban Areas

$$L_{50dB,suburban} = L_{50dB,urban} - 2[\log_{10}(f_c/28)]^2 - 5.4$$

Median Attenuation Equation for Open Areas

$$L_{50dB,open} = L_{50dB,urban} - 4.78[\log_{10}(f_c)]^2$$
$$+ 18.33\log_{10}f_c - 40.94$$

**Figure 16.6.c**

## 16.7 COST-231 Model

The *COST-231 Model* is developed for higher frequency range in large cells and is developed by modifying the Okumara Model. **Figures 16.7.a** and **16.7.b** illustrate this model.

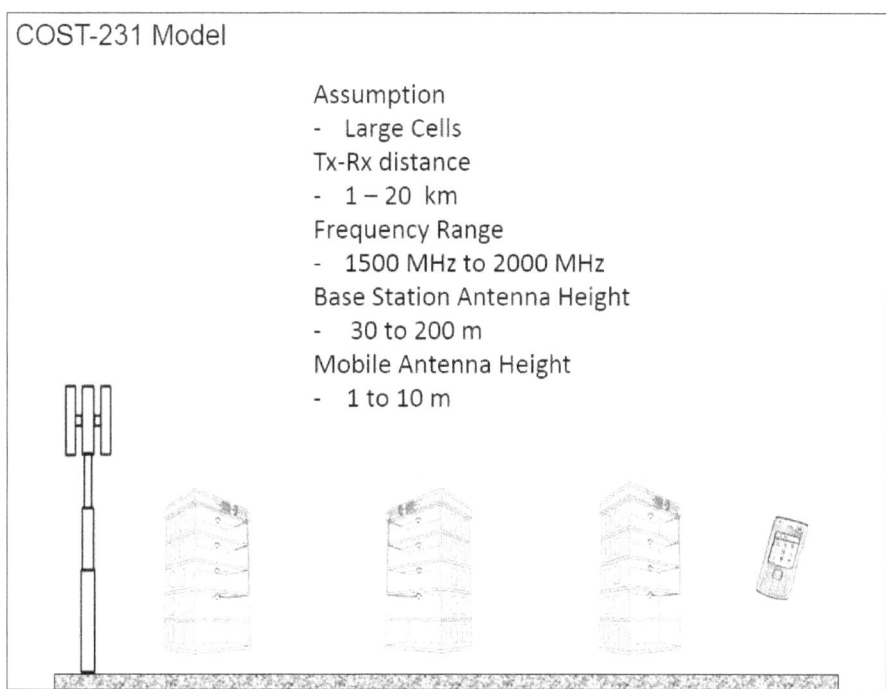

COST-231 Model

Assumption
- Large Cells

Tx-Rx distance
- 1 – 20 km

Frequency Range
- 1500 MHz to 2000 MHz

Base Station Antenna Height
- 30 to 200 m

Mobile Antenna Height
- 1 to 10 m

**Figure 16.7.a**

COST-231 Model
Median Attenuation Equation for Urban Areas

$$L_{50dB,urban} = 46.3 + 33.9 \log_{10} f_c - 13.82 \log_{10} h_{te} - a(h_{re})$$
$$+ [44.9 - 6.55 \log_{10} h_{te}] \log_{10} d + C$$

where

$L_{50dB, urban}$ is 50th percentile of propagation path loss in urban areas
$h_{te}$ is Effective Base station antenna height (30-200 m)
$h_{re}$ is Effective Mobile antenna height (1-10 m)
$f_c$ is frequency (1500-2000 MHz)
d is Tx-Rx distance or size of large cell (1-20 km)
C is 0 dB for Medium Cities and Suburban Areas,
    3 dB for Metropolitan Areas
$a(h_{re})$ is correction factor for effective mobile antenna height
        - For Small and Medium Cities:
            $a(h_{re}) = (1.11 \log_{10} f_c - 0.7) - (1.56 \log_{10} f_c - 0.8)$
        - For Large Cities:
            $a(h_{re}) = 3.2 (\log_{10} (11.75 h_{re}))^2 - 4.98$

**Figure 16.7.b**

# 16.8 COST-WI (Walfisch Ikigami) Model

The *COST-WI* Model is suitable for small macro cells in urban and sub urban areas, and takes into account the corrections for path loss due to heights of buildings, width of roads, separation between buildings, orientation of the road with respect to the direct radio path and multiple screen diffraction loss. **Figure 16.8.a** illustrates the restrictions and limitations of COST-WI model.

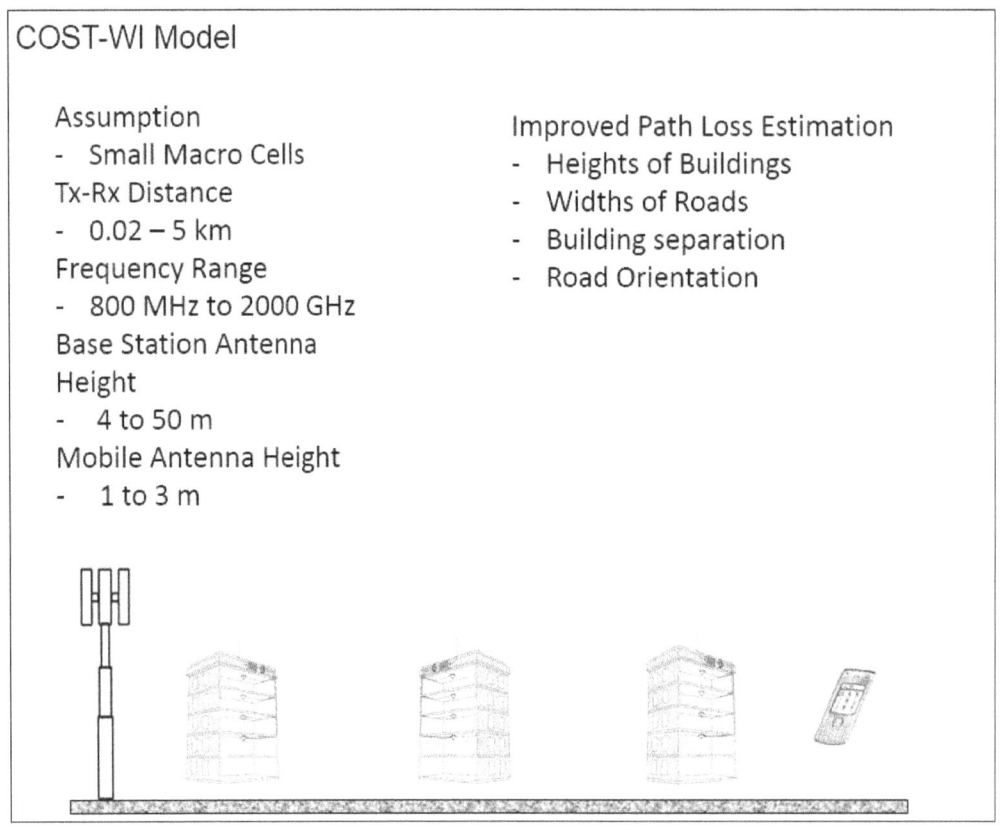

**Figure 16.8.a**

**Figure 16.8.b** and **16.8.c** define various parameters, such as height of building, width of road, building separation, heights of base station and mobile antennas, and street orientation.

**Figure 16.8.d** illustrates the propagation loss model for *line of sight* (LOS) case.

**Figure 16.8.e** to **16.8.l** illustrates the model for *near line of sight* (NLOS) case. The directly applicable equations for free space loss, roof top to street diffraction loss, street orientation loss and multiple screen diffraction loss are also illustrated.

Finally, **Figure 16.8.m** shows the recommended default values for roof height, road width, building separation and street orientation, which can be used for the NLOS case, if their actual measured values are not known.

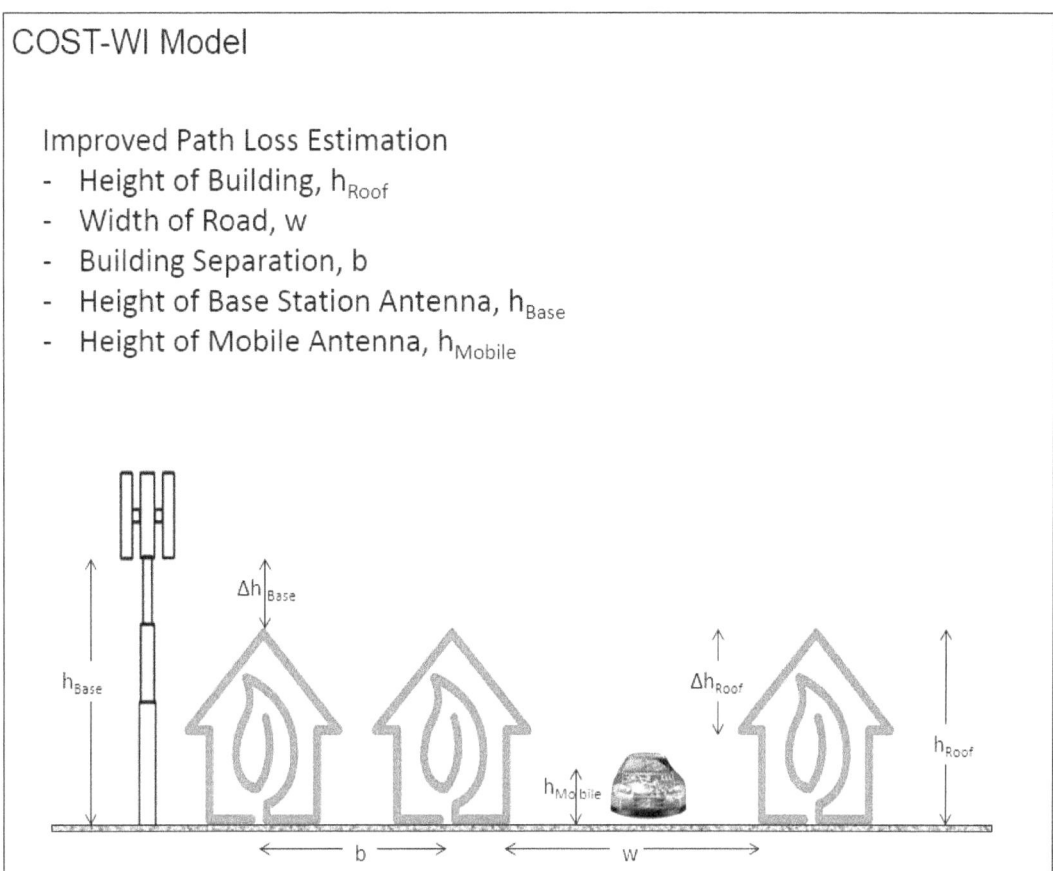

**Figure 16.8.b**

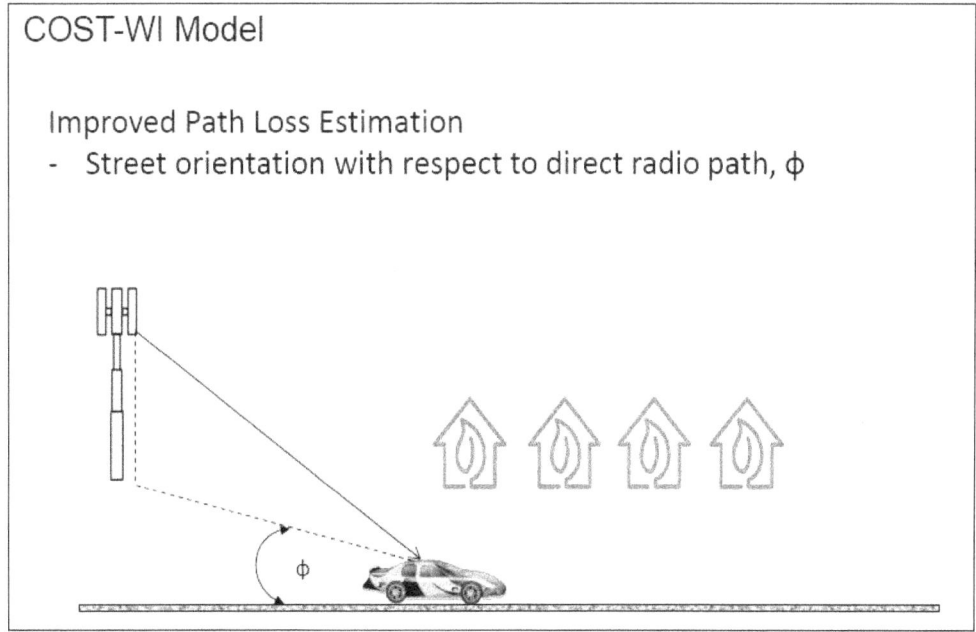

**Figure 16.8.c**

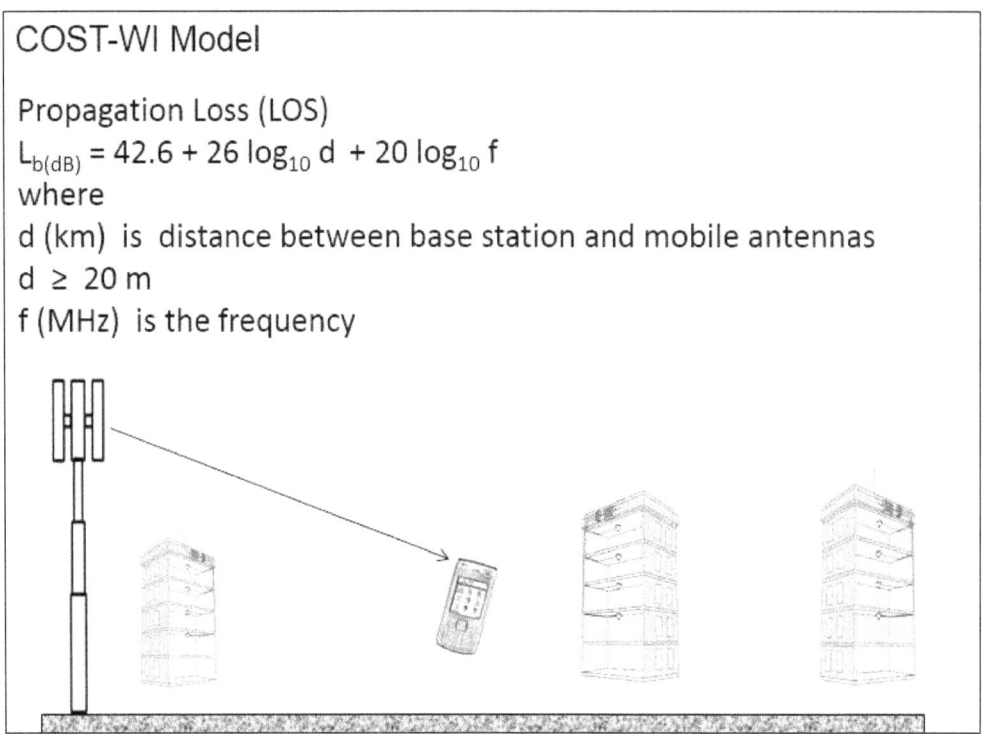

**Figure 16.8.d**

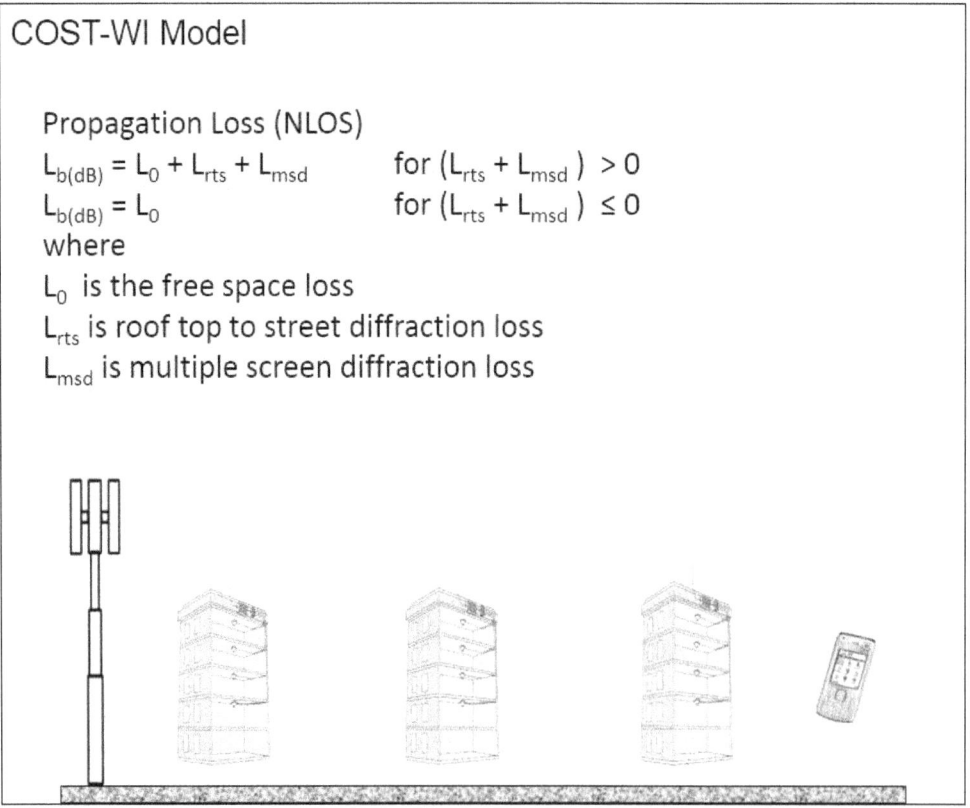

**Figure 16.8.e**

COST-WI Model

Free Space Loss (NLOS), $L_0$
$L_{0(dB)}$ = 32.4 + 20 log d + 20 log f
where
d (km)  is  distance between base station and mobile antenna
f (MHz)  is the frequency

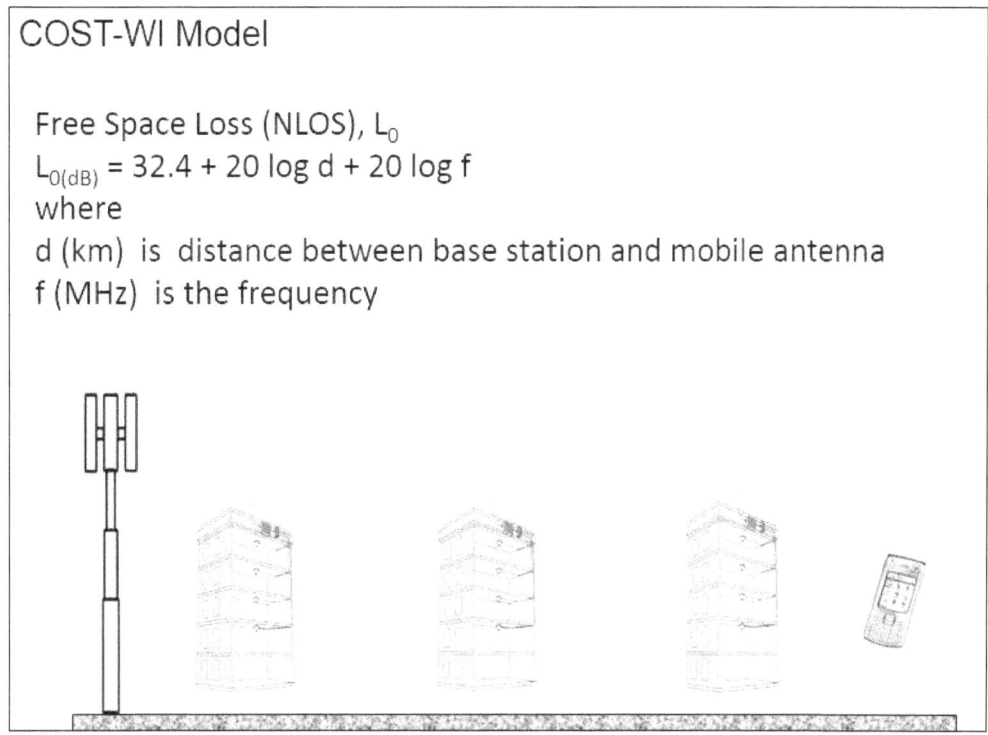

**Figure 16.8.f**

COST-WI Model

Roof Top to Street Diffraction Loss (NLOS), $L_{rts}$
$L_{rts(dB)}$ = $-$ 16.9 $-$ 10 log w + 10 log f + 20 log $\Delta h_{Mobile}$ + $L_{Ori}$
where
w (m)  is  street width
f (MHz)  is the frequency
$\Delta h_{Mobile} = h_{Roof} - h_{Mobile}$
$L_{Ori}$ is the Street Orientation Loss

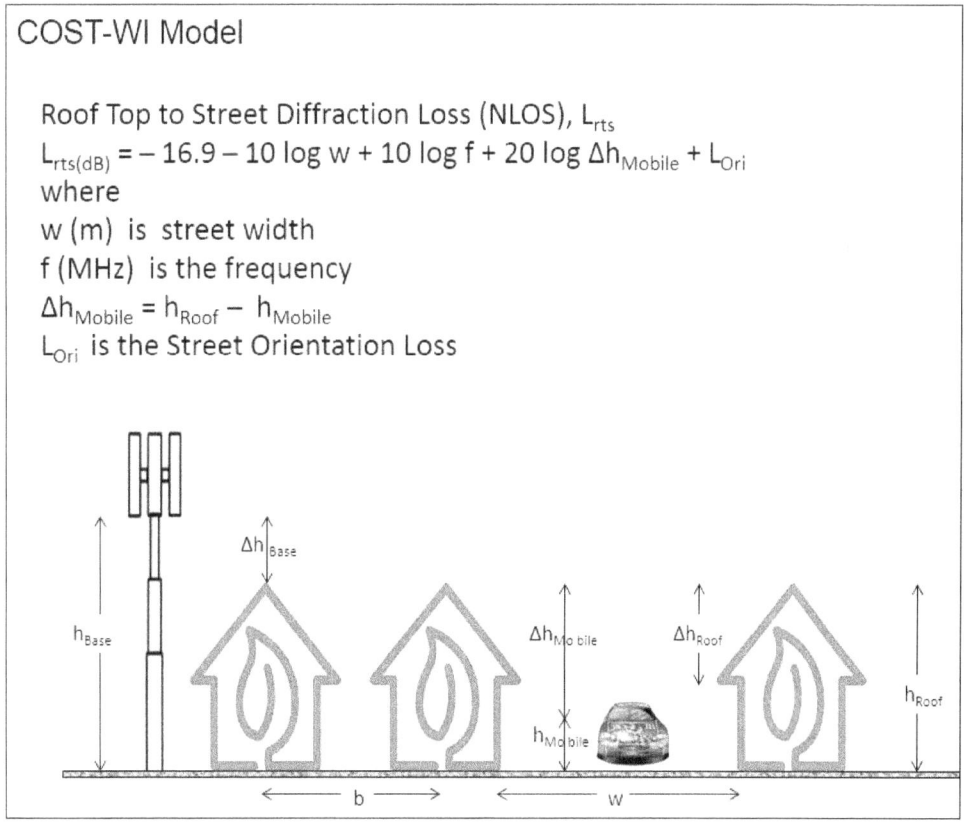

**Figure 16.8.g**

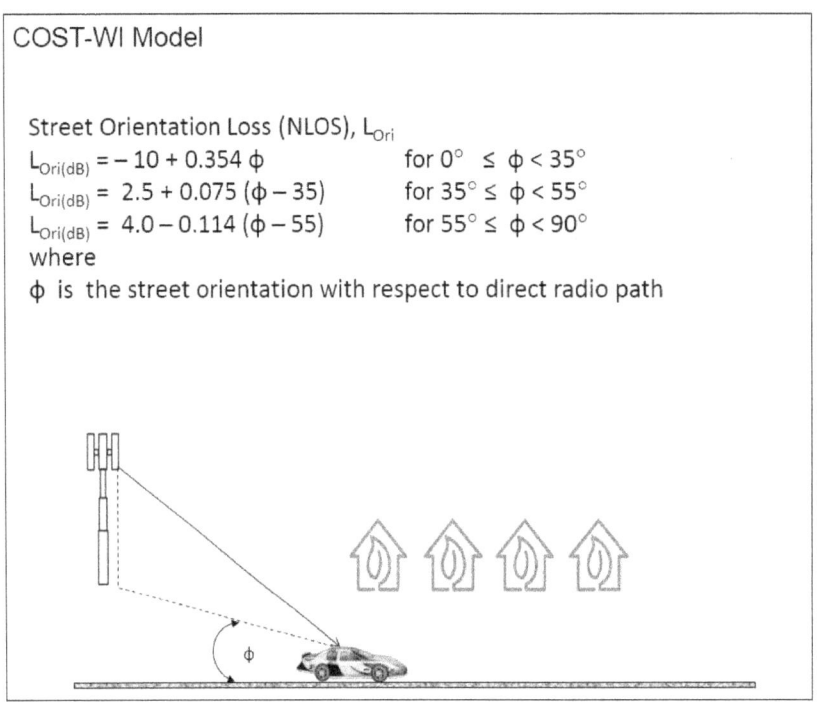

**Figure 16.8.h**

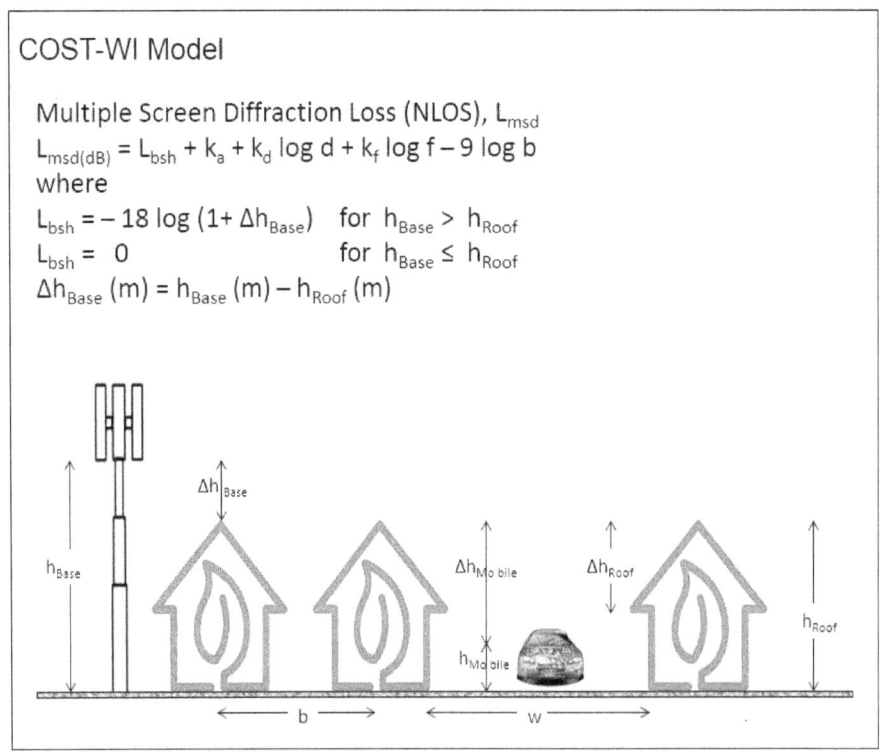

**Figure 16.8.i**

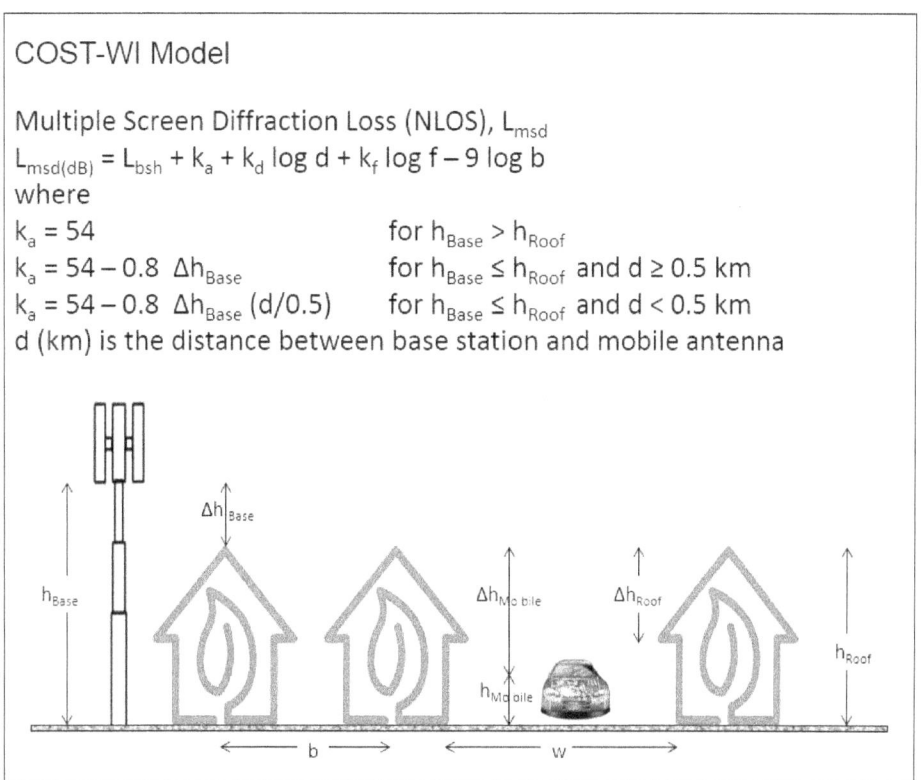

**Figure 16.8.j**

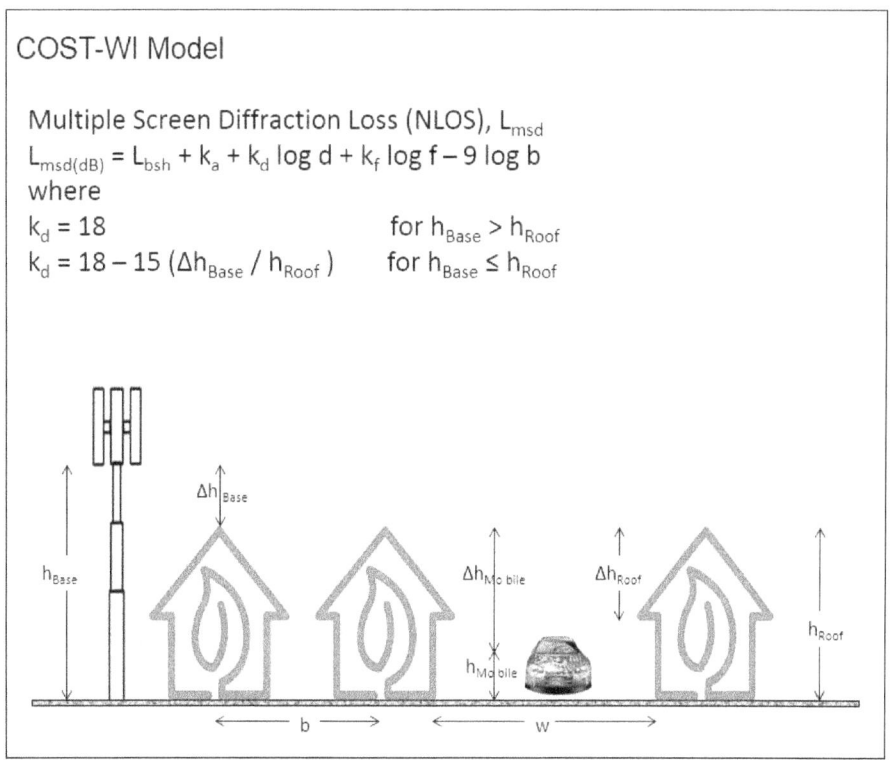

**Figure 16.8.k**

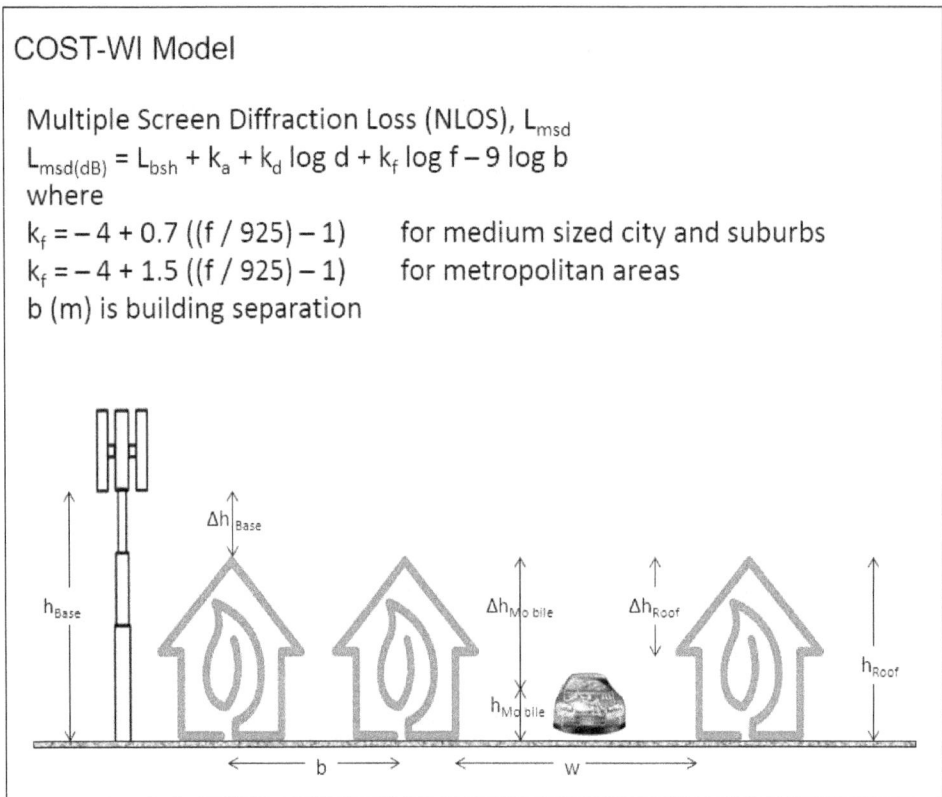

**Figure 16.8.l**

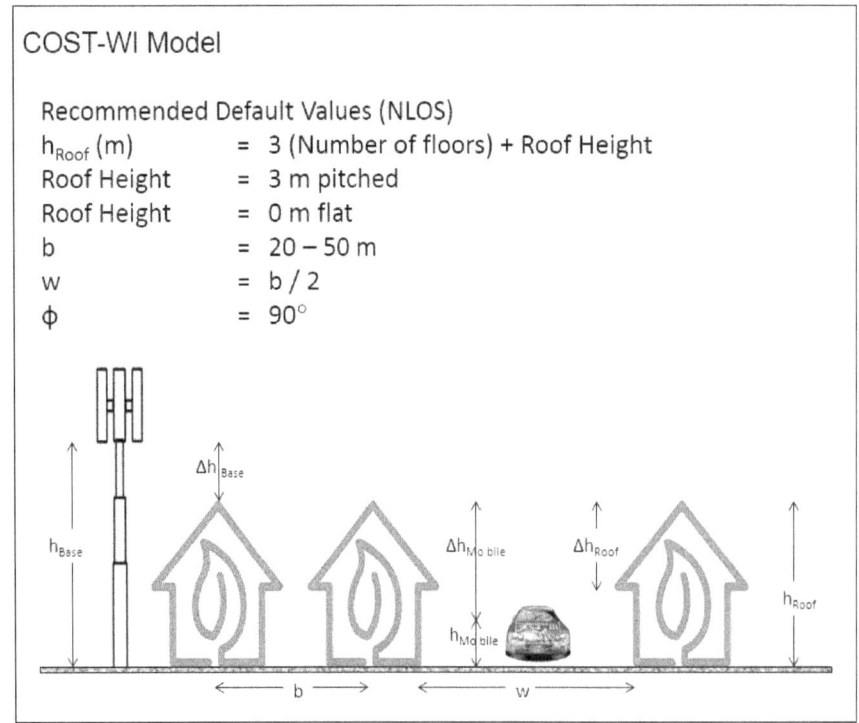

**Figure 16.8.m**

# 16.9 Foliage Loss

For areas with thick coverage of leafy trees, the *Enhanced Fitted ITU-R Model*, can be used as illustrated in **Figure 16.9.a**.The path loss can be calculated from the radio frequency and the average depth of trees. There is a path loss equation for out-of-leaf case, i.e. fall season, and a separate equation for in-leaf case, i.e. spring season.

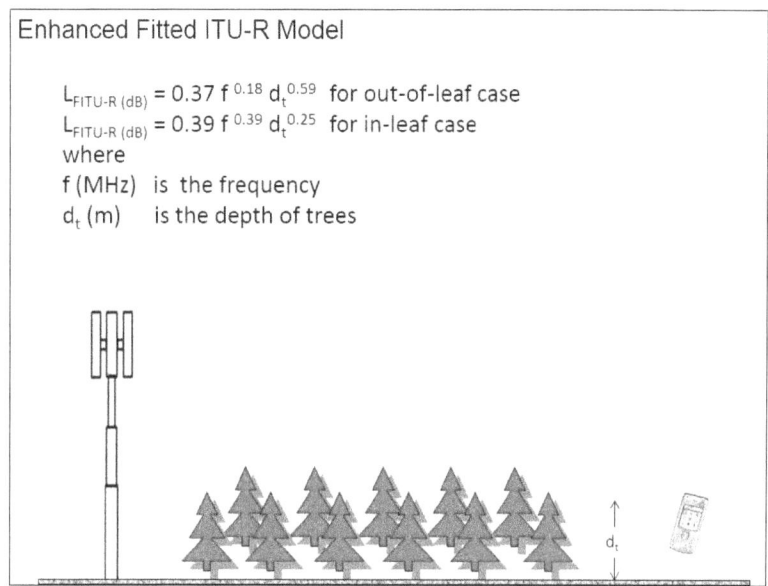

Enhanced Fitted ITU-R Model

$L_{FITU\text{-}R\,(dB)} = 0.37\,f^{\,0.18}\,d_t^{\,0.59}$ for out-of-leaf case
$L_{FITU\text{-}R\,(dB)} = 0.39\,f^{\,0.39}\,d_t^{\,0.25}$ for in-leaf case
where
$f$ (MHz)  is the frequency
$d_t$ (m)   is the depth of trees

**Figure 16.9.a**

# 16.10    Building Penetration Loss

In this section we will consider models for the losses incurred when the radio signal penetrates through the walls and floors of a building. This is a typical scenario for a user inside a building trying to receive the radio signal from a cell tower that is outside the building.

The *wall loss* is the penetration loss through the wall as shown in **Figure 16.10.a**. It depends on the angle at which the radio signal is incident on the wall. The true wall loss is difficult to determine due to multiple reflections from the walls inside the building and from the furniture and other obstacles inside the building. In general, a ray of radio signal incident perpendicular to the wall from outside suffers less penetration loss than a ray that is incident at a grazing angle.

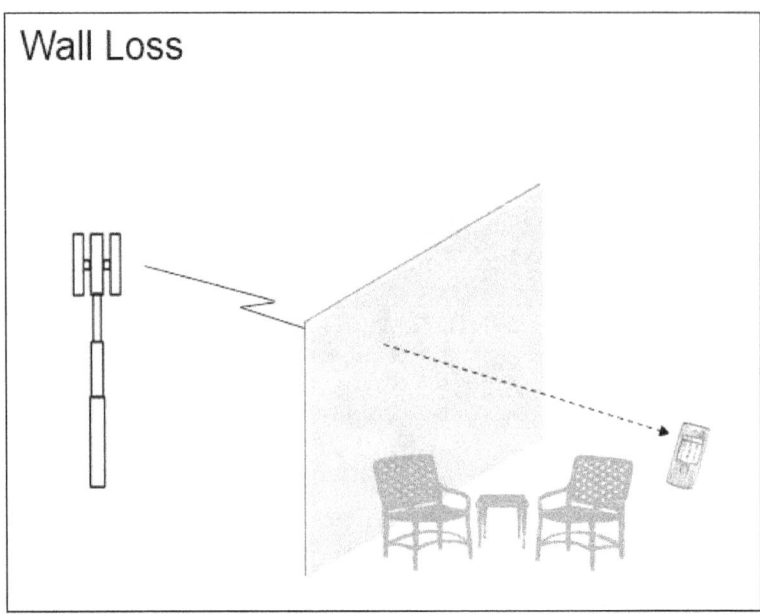

**Figure 16.10.a**

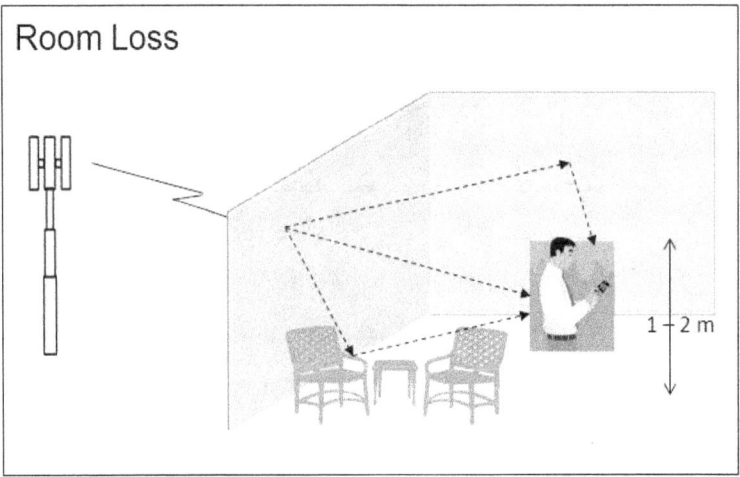

**Figure 16.10.b**

The *room loss* is the median loss determined from measurements taken in the whole room at a height of about 1 to 2 m above the floor, as illustrated in **Figure 16.10.b**.

The *floor loss* is the median loss in all rooms of same floor in the building as illustrated in **Figure 16.10.c**.

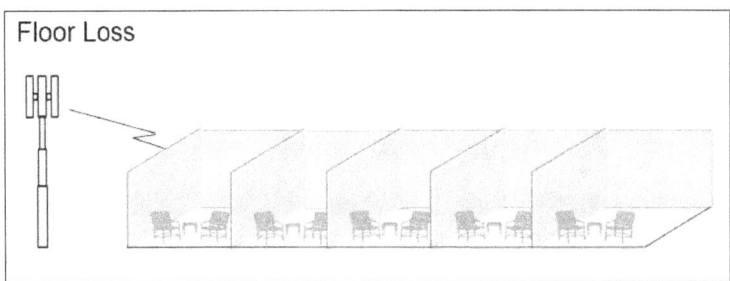

**Figure 16.10.c**

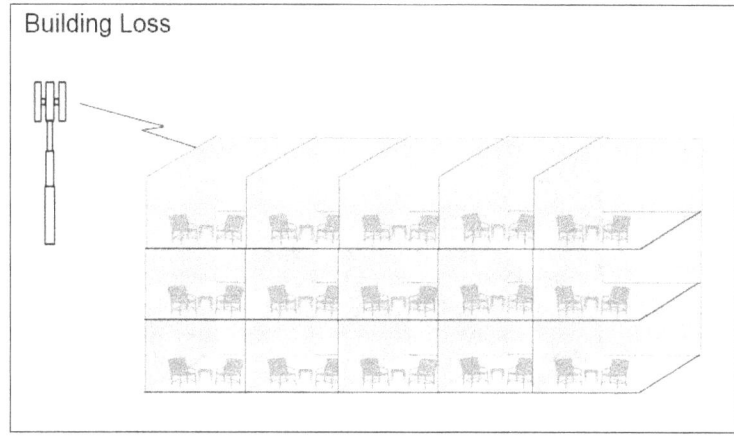

**Figure 16.10.d**

The *building loss* is the same as *floor loss*, except that it is taken over all rooms on all floors of the building. With this type of information, it is important to indicate if the basement is included or not in the loss calculations.

The *building penetration loss* is modelled in **Figure 16.10.e** and **16.10.g** for *line of sight* (LOS) conditions. This implies that the radio signal from cell tower is directly incident at an angle to the external wall. Notice the recommended default values for walls of different types, such as concrete or wood.

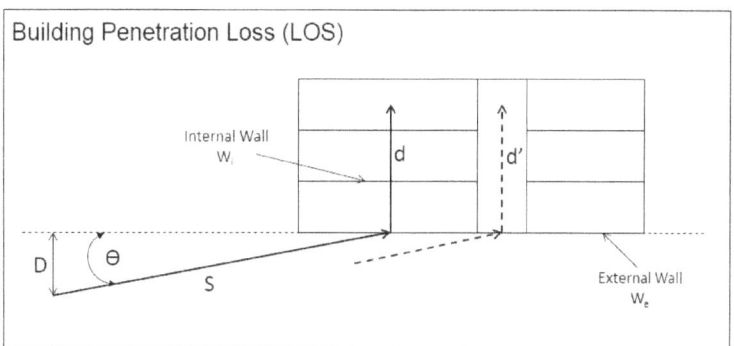

**Figure 16.10.e**

Building Penetration Loss (LOS)

$L_{dB} = 32.4 + 20 \log f + 20 \log (S+d) + W_e + WG_e (1 - (D/S))^2 + \max (\Gamma_1, \Gamma_2)$

where

f (GHz) is the frequency

S (m) is the distance between external antenna and the wall

d(m) is the distance between external wall and mobile antenna

D (m) is the perpendicular distance between external antenna and wall plane

$\Theta$ (°) is the grazing angle

$\Gamma_1$ (dB) = $W_i$ . p

$\Gamma_2$ (dB) = $\alpha$. (d – 2).$(1 - (D/S))^2$

p is  the number of penetrated internal walls (p = 0,1,2,...)

$\alpha$ (dB/m)  is the rate of loss when there are no internal walls

$W_i$ (dB) is the loss in internal walls

$W_e$ (dB) is the loss in external wall when grazing angle $\Theta = 90°$

$WG_e$ (dB) is the loss in external wall when grazing angle $\Theta = 0°$

Recommended Default Values

$W_e$ = 4 - 10 dB

$W_e$ = 7 dB for concrete walls with normal window size

$W_e$ = 4 dB  for wood

$W_i$ = 4 – 10 dB

$W_i$  = 7 dB for concrete walls

$W_i$  = 4 dB for wood and plaster

$WG_e$ = 20 dB

$\alpha$  = 0.6 dB/m

**Figure 16.10.f**

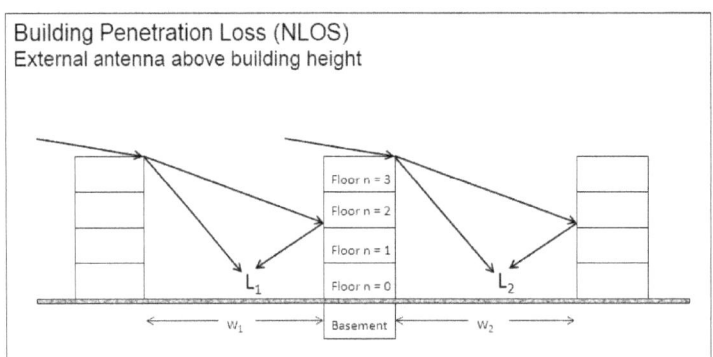

**Figure 16.10.g**

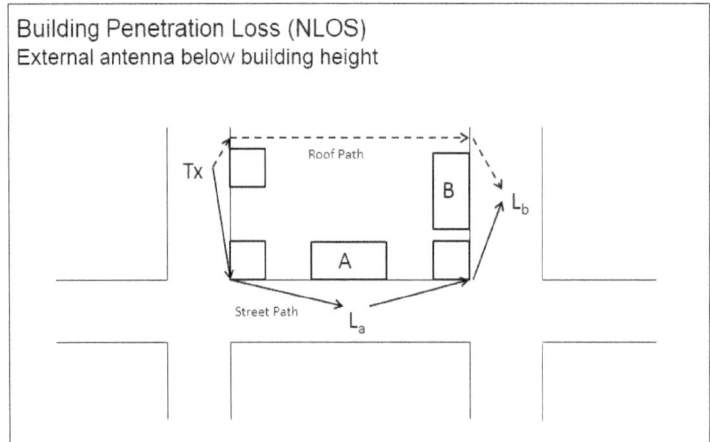

**Figure 16.10.h**

The *building penetration loss* is modelled in **Figure 16.10.g** to **16.10.j** for *near line of sight* (NLOS) conditions. There are two scenarios, one with external antenna above building height, and another with antenna below the building height. Notice that in NLOS conditions, the radio signal from cell tower is not directly incident on the external wall. It suffers reflections and diffractions before it impinges on the external wall, and then part of its energy penetrates to reach inside of the building. In order to arrive at a reliable measure of penetration loss, it is required to have a measure of the signal power lost because of diffraction and reflection at certain locations outside the building, as shown in **Figures 16.10.g** and **16.10.h**. These locations are marked as $L_1$, $L_2$, $L_a$, $L_b$.

**Figure 16.10.j** shows the recommended default values that can be used in the model.

Building Penetration Loss (NLOS)

$L_{dB} = L_{outside} + W_e + Wg_e + \max(\Gamma_1, \Gamma_3) - G_{FH}$

where

$L_{outside}$ (dB) = $L_1$, $L_2$, $L_a$, $L_b$

$W_e$ (dB) is the loss in external wall when grazing angle, $\Theta = 90°$

$W_{ge}$ (dB) is the loss in external wall when grazing angle $\Theta = 0°$

$\Gamma_1$ (dB) = $W_i \cdot p$

$W_i$ (dB) is the loss in internal walls

p is the number of penetrated internal walls (p = 0,1,2,...)

$\Gamma_3$ (dB) = $\alpha \cdot d$

$\alpha$ (dB/m) is the rate of loss when there are no internal walls

d(m) is the distance between external wall and mobile antenna

$G_{FH}$ (dB) = $n \cdot G_n$

$G_{FH}$ (dB) = $h \cdot G_h$

n is the floor number (n=0,1,2...)

$G_n$ (dB/floor) is the floor height gain

h (m) is the height above the outdoor reference path level

$G_h$ (dB/m) is the height gain

**Figure 16.10.i**

```
Building Penetration Loss (NLOS)
L_dB = L_outside + W_e + Wg_e + max (Γ_1 , Γ_3) − G_FH

Recommended Default Values
W_e = 4 - 10 dB
W_e = 7 dB for concrete walls with normal
window size
W_e = 4 dB  for wood
W_i  = 4 − 10 dB
W_i  = 7 dB for concrete walls
W_i  = 4 dB for wood and plaster
Wg_e (900 MHz) = 3 - 4 dB
Wg_e (1800 MHz) = Wg_e (900 MHz)  + 2 dB
α  = 0.6 dB/m
```

**Figure 16.10.j**

# 16.11    Indoor Propagation Models

There are three indoor propagation *path loss* models illustrated in this section, the *one slope model* (1SM), *multi wall model* (MWM) and the *linear attenuation model* (LAM), in **Figures 16.11.a** to **16.11.c**. The recommended optimized values of coefficients used in these three models are tabulated in **Table 16.11.d**.

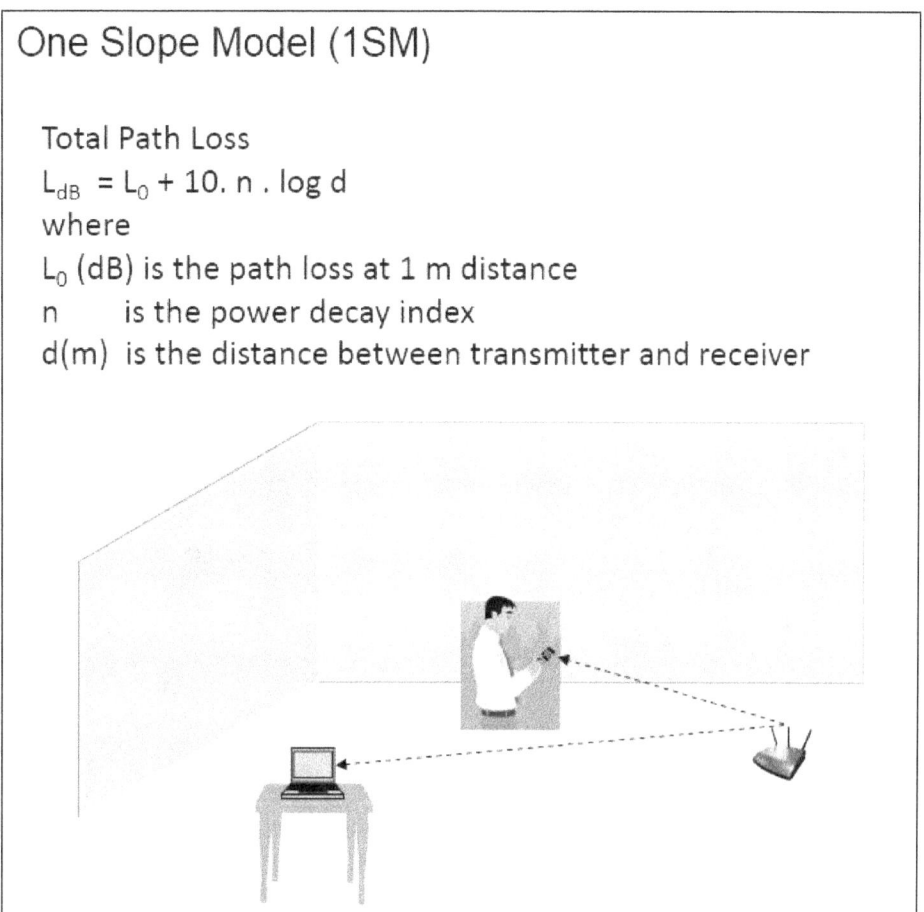

**Figure 16.11.a**

The *one slope model* (1SM) and the linear attenuation model (LAM) are simplest to apply for a single room situation, where the access point or eNodeB and the handset or PC are located in the same room within line of sight.

The *multi wall model* (MWM) is appropriate for situations where the access point or the eNodeB is located in a building with multiple rooms or office spaces on multiple floors. Notice in **Figure 16.11.b**, that the model also includes losses based on the number of walls and floors and the types of walls.

## Multi Wall Model (MWM)
Total Path Loss

$$L_{dB} = L_{FS} + L_c + \sum_{i=1}^{I} k_{wi} L_{wi} + k_f^{[(kf + 2)/(kf + 1) - b]} \cdot L_f$$

where

$L_{FS}$ (dB) is the free space loss between transmitter and receiver

$L_c$ (dB) is the negligible constant loss from determination of measurement results using multiple linear regression, $L_c \sim 0$

$k_{wi}$ is the number of penetrated walls of type i

$k_f$ is the number of penetrated floors

$L_{wi}$ (dB) is the loss of wall type i

$L_f$ (dB) is the loss between adjacent floors

b is the empirical parameter

I is the number of wall types

$L_{w1}$      Light Wall, a wall that is not bearing load, e.g. plasterboard, particle board (< 10 cm), light concrete wall

$L_{w2}$      Heavy Wall, a load bearing wall or other thick wall (> 10 cm) made of concrete or brick

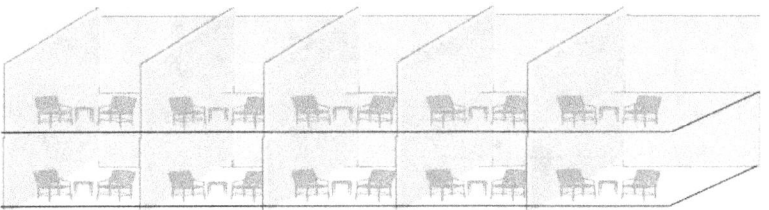

**Figure 16.11.b**

## Linear Attenuation Model (LAM)

Total Path Loss
$$L_{dB} = L_{FS} + \alpha \cdot d$$
where
$L_{FS}$ (dB) is the free space loss between transmitter and receiver
$\alpha$ (dB/m) is the rate of loss when there are no internal walls
d(m) is the distance between transmitter and receiver

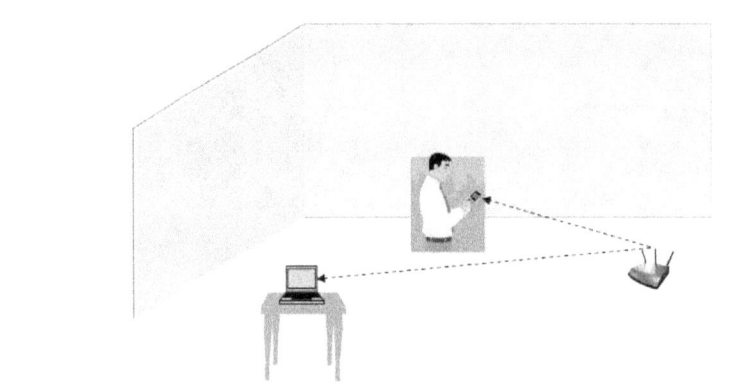

**Figure 16.11.c**

| Optimized Model Coefficients for f = 1800 MHz | | | | | | | |
|---|---|---|---|---|---|---|---|
| Environment | | One Slope Model (ISM) | | Multi Wall Model (MWM) | | | | Linear Attenuation Model (LAM) |
| | | $L_0$ (dB) | n | $L_{w1}$ (dB) | $L_{w2}$ (dB) | $L_f$ (dB) | b | $\alpha$ (dB/m) |
| Dense | one floor | 33.3 | 4.0 | 3.4 | 6.9 | 18.3 | 0.46 | 0.62 |
| | two floors | 21.9 | 5.2 | - | - | - | - | - |
| | Multi floors | 44.9 | 5.4 | - | - | - | - | 2.8 |
| Open | - | 42.7 | 1.9 | 3.4 | 6.9 | 18.3 | 0.46 | 0.22 |
| Large | - | 37.5 | 2.0 | 3.4 | 6.9 | 18.3 | 0.46 | - |
| Corridor | - | 39.2 | 1.4 | 3.4 | 6.9 | 18.3 | 0.46 | - |

**Table 16.11.d**

# 17.  Link Budget Analysis Examples

In this chapter, we will apply some of the statistical fading models that we studied in **Chapter 16** to some practical examples of *macro*, *micro* and *femto* cells.

## 17.1 LTE Macro Cell Example

The user equipment (UE) in a typical LTE macro cell is served by eNodeB. In this example, the area covered by eNodeB is assumed to be dense urban with tall buildings and a cell radius of 8 km. The uplink and downlink frequencies for FDD LTE and the antenna heights are given in **Figure 17.1.a**.

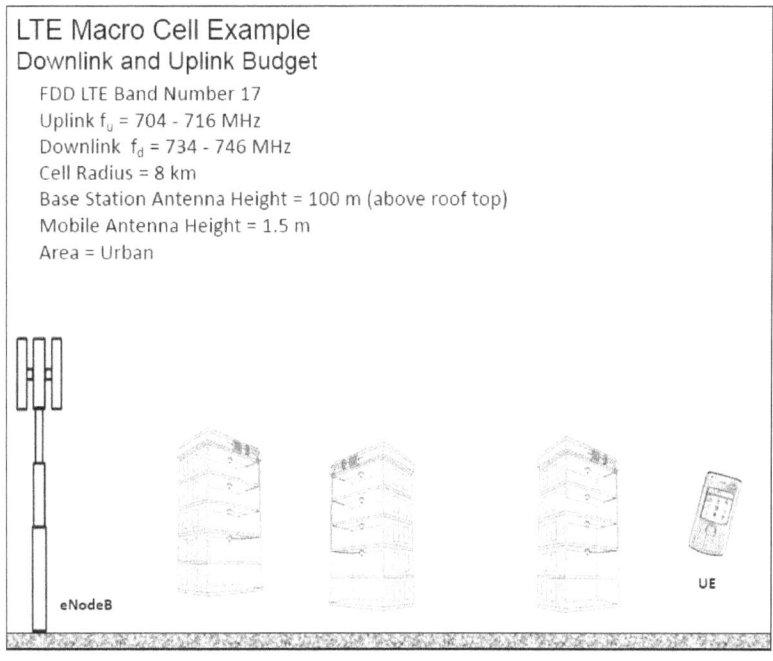

**Figure 17.1.a**

*Hata Model* (See **Section 16.6**) is used to calculate the $50^{th}$ percentile of path loss in urban area for uplink path as well as downlink path. Please note that that the *uplink path* refers to transmission of radio signal from UE to eNodeB at the *uplink frequency* band of 704 – 716 MHz, and the *downlink path* refers to transmission of radio signal from eNodeB to UE at the *downlink frequency* band of 734 – 746 MHz for FDD LTE Band Number 17. The path loss calculations are shown in **Figure 17.1.b** and **17.1.d**. The downlink and uplink budgets are illustrated in **Figure 17.1.c** and **Figure 17.1.e**. Please review **Chapter 10** to understand the link budget calculations for this example.

Notice that in the *downlink budget* (**Figure 17.1.c**) the link margin is calculated to be 32 dB and in the *uplink budget* (**Figure 17.1.e**), the link margin is 30 dB, and that provides a good resilience against signal fading effects.

---

**LTE Macro Cell Example**

Use Hata Model for Downlink Path Loss:
$$L_{50dB,urban} = 69.55 + 26.16\log_{10}f_c - 13.82\log_{10}f_c - 13.82\log_{10}h_{te} - a\,(h_{re}) + 44.9 - 6.55\log_{10}h_{te} + 10\log_{10} d$$

Substitute values:
$h_{te} = 100$ m,  $h_{re} = 1.5$ m , d = 8 km
$f_c$ = centre $f_d$ = 740 MHz
$a(h_{re}) = 3.2\,(\log_{10}(11.75\,h_{re}))^2 - 4.98$

Calculate:
$a(h_{re}) = -0.011$
$$L_{50dB,urban} = 69.55 + 26.16\log_{10}740 - 13.82\log_{10}740 - 13.82\log_{10}100 + 0.011 + 44.9 - 6.55\log_{10}100 + 10\log_{10} 8$$

$L_{50dB,urban} = 118.16$

---

**Figure 17.1.b**

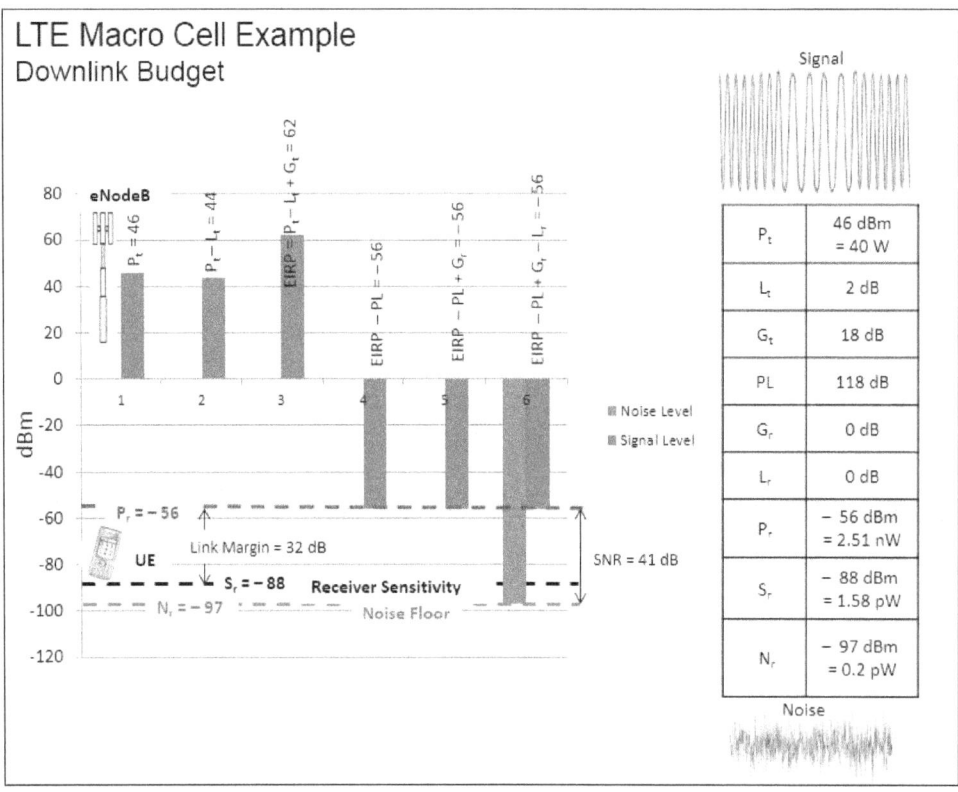

**Figure 17.1.c**

## LTE Macro Cell Example

Use Hata Model for Uplink Path Loss:

$L_{50dB, urban} = 69.55 + 26.16\log_{10}f_c - 13.82\log_{10}f_c - 13.82\log_{10}h_{te} - a\,(h_{re}) + 44.9$
$\qquad - 6.55\log_{10}h_{te} + 10\log_{10}d$

Substitute values:

$h_{te} = 100$ m, $h_{re} = 1.5$ m , $d = 8$ km
$f_c$ = centre $f_u = 710$ MHz
$a(h_{re}) = 3.2\,(\log_{10}(11.75\,h_{re}))^2 - 4.98$

Calculate:

$a(h_{re}) = -0.011$
$L_{50dB, urban} = 69.55 + 26.16\log_{10}710 - 13.82\log_{10}710 - 13.82\log_{10}100 + 0.011 + 44.9$
$\qquad - 6.55\log_{10}100 + 10\log_{10}8$

$L_{50dB, urban} = 117.94$

**Figure 17.1.d**

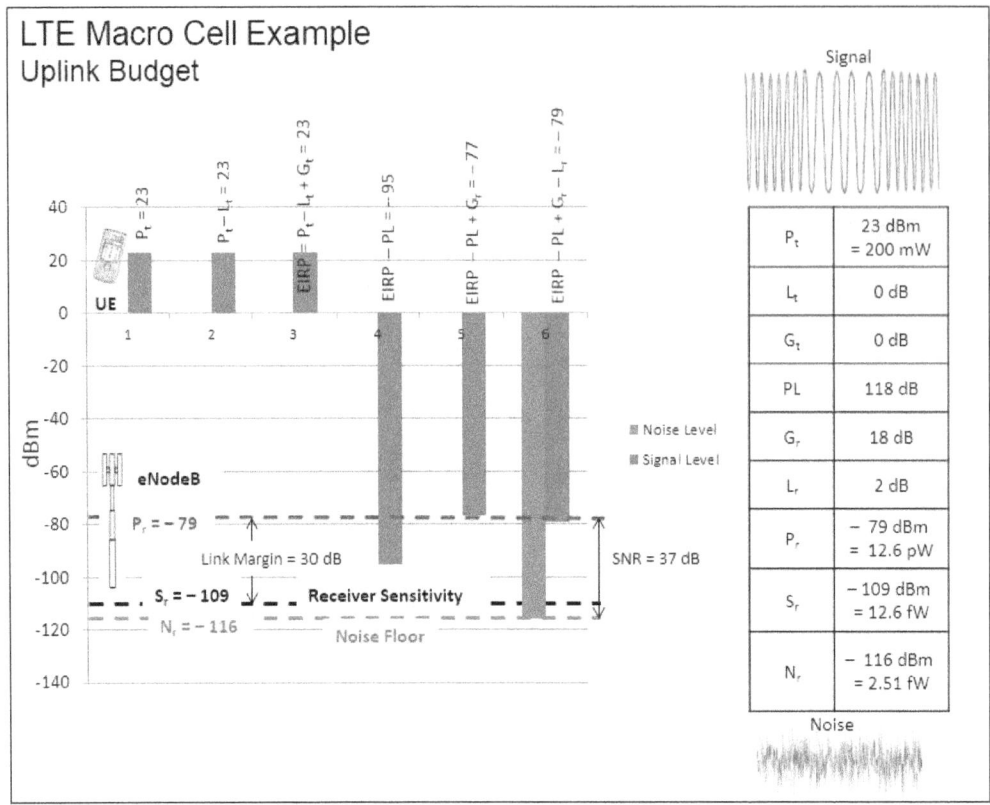

**Figure 17.1.e**

## 17.2 UMTS Macro Cell Example

The user equipment (UE) in a UMTS macro cell is served by a NodeB. In this example, the area covered by NodeB is assumed to be dense urban with tall buildings and a cell radius of 8 km. The uplink and downlink frequencies for PCS FDD Band II (W-CDMA 1900) and the antenna heights are given in **Figure 17.2.a**.

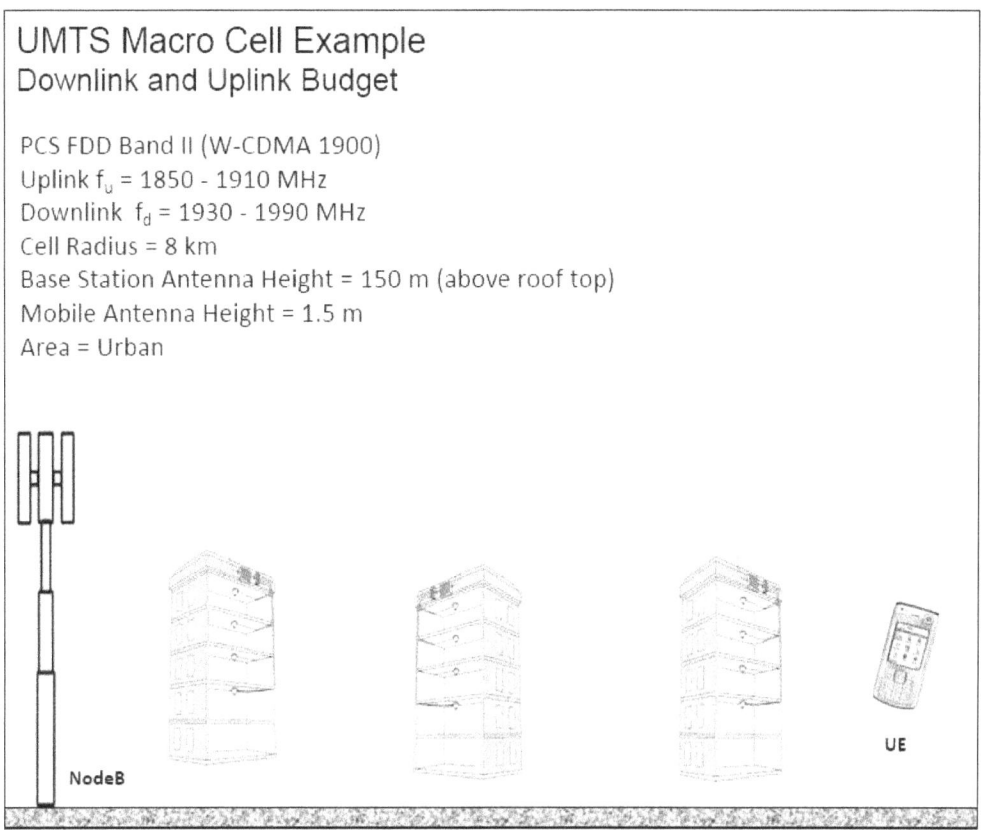

UMTS Macro Cell Example
Downlink and Uplink Budget

PCS FDD Band II (W-CDMA 1900)
Uplink $f_u$ = 1850 - 1910 MHz
Downlink $f_d$ = 1930 - 1990 MHz
Cell Radius = 8 km
Base Station Antenna Height = 150 m (above roof top)
Mobile Antenna Height = 1.5 m
Area = Urban

NodeB

UE

**Figure 17.2.a**

*COST 231 Model* (See **Section 16.7**) is used to calculate the 50th percentile of path loss in urban area for uplink path as well as downlink path. Please note that that the *uplink path* refers to transmission of radio signal from UE to NodeB at the *uplink frequency* band of 1850 – 1950 MHz, and the *downlink path* refers to transmission of radio signal from NodeB to UE at the *downlink frequency* band of 1930 – 1990 MHz. The path loss calculations are shown in **Figure 17.2.b** and **17.2.d**. The downlink and uplink budgets are illustrated in **Figure 17.2.c** and **Figure 17.2.e**. Please review **Chapter 10** to understand the link budget calculations for this example.

Notice that in the *downlink budget* (**Figure 17.2.c**), the *link margin* is calculated to be only 6 dB. Also, in *uplink budget* (**Figure 17.2.e**), there is no *link margin*. In order to improve *link margin* and for better resilience against signal fading effects, either the cell radius must be reduced, or the NodeB power must be increased to recalculate the path loss and received signal power level.

UMTS Macro Cell Example

Use COST 231 Model for Downlink Path Loss:
$L_{50dB,urban}$ = 46.3 + 33.9 $\log_{10} f_c$ − 13.82 $\log_{10} h_{te}$ − a($h_{re}$)+[44.9 − 6.55 $\log_{10} h_{te}$ ] $\log_{10} d$ + C

Substitute values:
$f_c$ = centre $f_d$ = 1960 MHz
$h_{te}$ = 150 m
$h_{re}$ = 1.5 m
d = 8 km
C = 3 dB for metropolitan areas
a($h_{re}$) for large cities = 3.2 ($\log_{10}$ (11.75 $h_{re}$))$^2$ − 4.98 = − 0.01

Calculate:
$L_{50dB,urban}$ = 46.3 + 33.9 $\log_{10}$ 1960 − 13.82 $\log_{10}$ 150 − (− 0.01) + [44.9 − 6.55 $\log_{10}$ 150] $\log_{10}$ 8 + 3

$L_{50dB,urban}$ = 158.52 dB

**Figure 17.2.b**

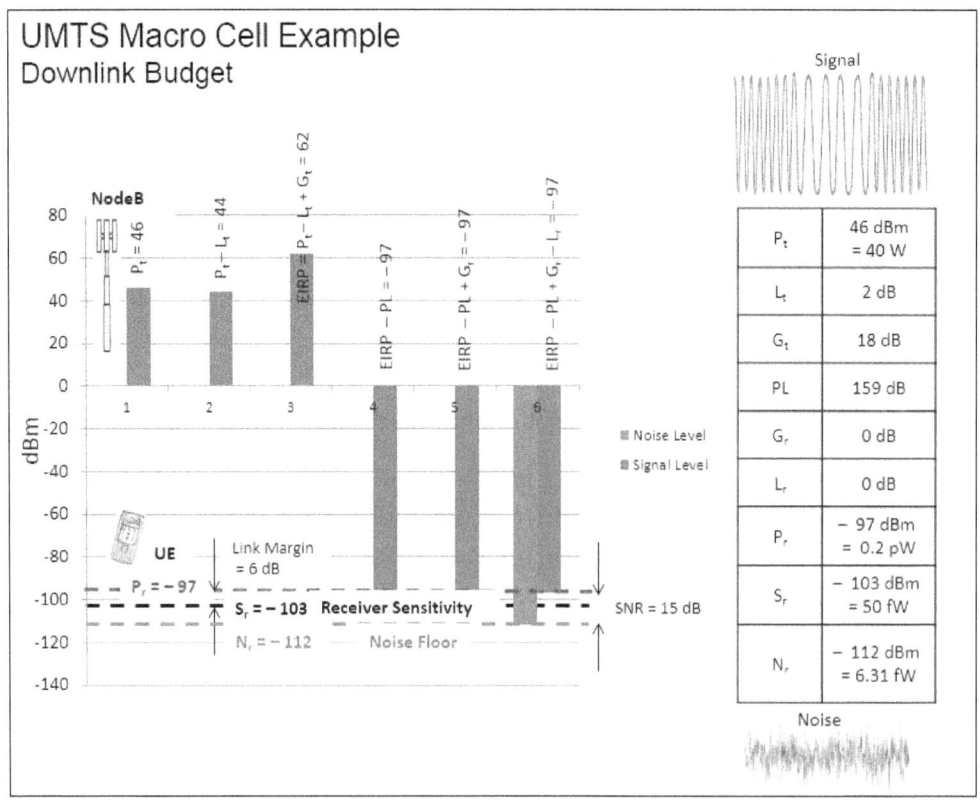

**Figure 17.2.c**

## UMTS Macro Cell Example

Use COST 231 Model for Uplink Path Loss:

$L_{50dB,urban}$ = 46.3 + 33.9 $\log_{10} f_c$ – 13.82 $\log_{10} h_{te}$ – $a(h_{re})$+[44.9 – 6.55 $\log_{10} h_{te}$ ] $\log_{10} d$ + C

Substitute values:
$f_c$ = centre $f_u$ = 1880 MHz
$h_{te}$ = 150 m
$h_{re}$ = 1.5 m
d = 8 km
C = 3 dB for metropolitan areas
$a(h_{re})$ for large cities = 3.2 $(\log_{10} (11.75\ h_{re}))^2$ – 4.98 = – 0.01

Calculate:
$L_{50dB,urban}$ = 46.3 + 33.9 $\log_{10}$ 1880 – 13.82 $\log_{10}$ 150 – (– 0.01) + [44.9 – 6.55 $\log_{10}$ 150] $\log_{10}$ 8 + 3

$L_{50dB,urban}$ = 158 dB

**Figure 17.2.d**

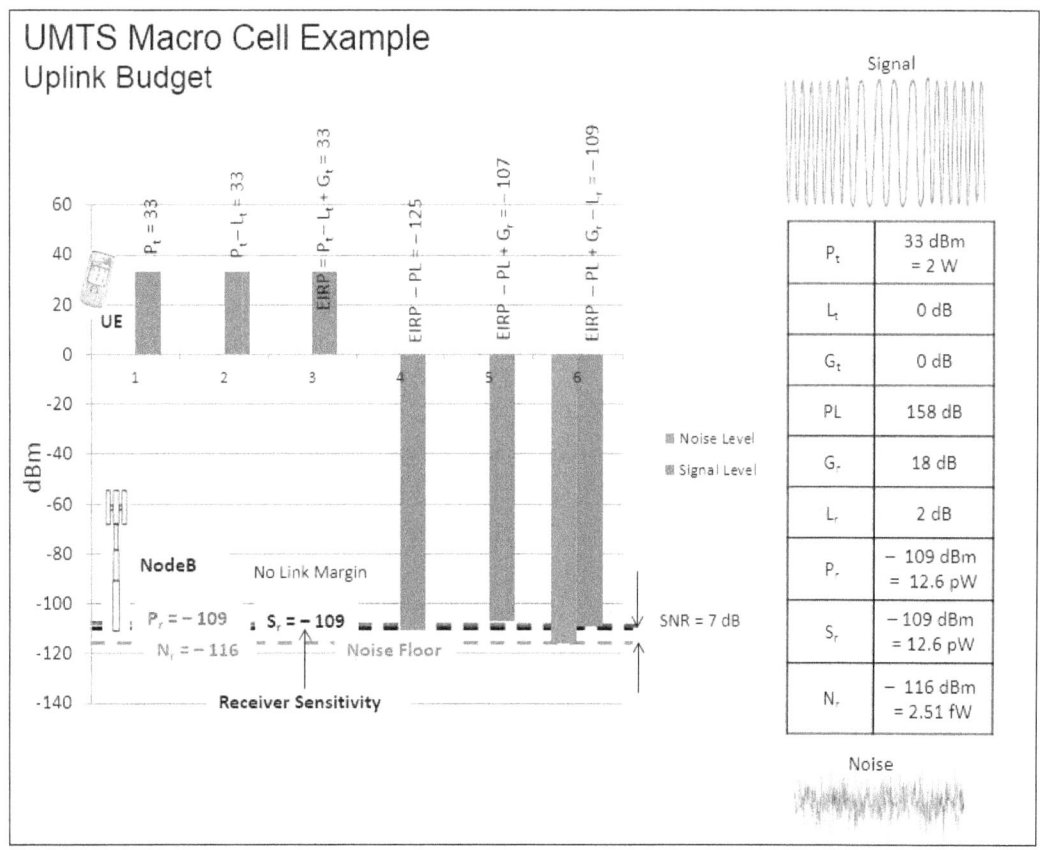

**Figure 17.2.e**

# 17.3 GSM Macro Cell Example

The user equipment (UE) in a GSM macro cell is served by a BTS. In this example, the area covered by BTS is assumed to be dense urban with tall buildings and a cell radius of 8 km. The uplink and downlink frequencies for P-GSM-900 Band 900 Channel 1-124 and the antenna heights are given in **Figure 17.3.a**.

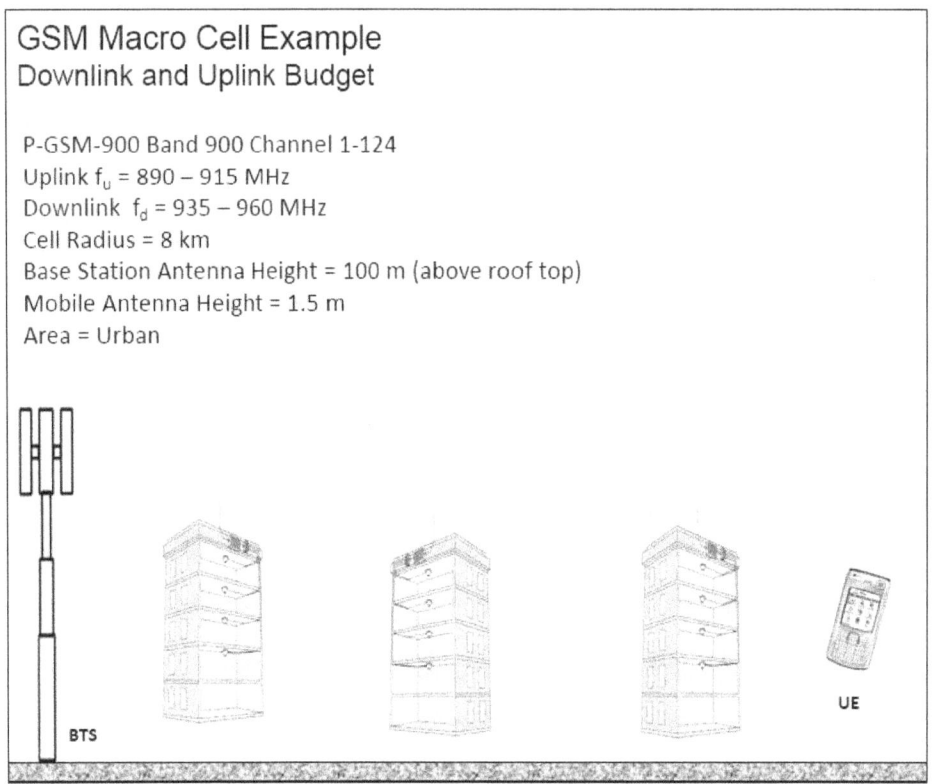

**GSM Macro Cell Example**
**Downlink and Uplink Budget**

P-GSM-900 Band 900 Channel 1-124
Uplink $f_u$ = 890 – 915 MHz
Downlink $f_d$ = 935 – 960 MHz
Cell Radius = 8 km
Base Station Antenna Height = 100 m (above roof top)
Mobile Antenna Height = 1.5 m
Area = Urban

BTS

UE

**Figure 17.3.a**

*Hata Model* (See **Section 16.6**) is used to calculate the 50th percentile of path loss in urban area for uplink path as well as downlink path. Please note that that the *uplink path* refers to transmission of radio signal from UE to BTS at the *uplink frequency* band of 890 – 915 MHz, and the *downlink path* refers to transmission of radio signal from BTS to UE at the *downlink frequency* band of 935 – 960 MHz. The path loss calculations are shown in **Figure 17.3.b** and **17.3.d**. The downlink and uplink budgets are illustrated in **Figure 17.3.c** and **Figure 17.3.e**. Please review **Chapter 10** to understand the link budget calculations for this example.

Notice that in the *downlink budget* (**Figure 17.3.c**), the *link margin* is calculated to be 31 dB. Also, in *uplink budget* (**Figure 17.3.e**), the *link margin* is 26 dB. This amount of link margin provides a good cushion against signal fading.

## GSM Macro Cell Example

Use Hata Model for Downlink Path Loss:

$L_{50dB,urban} = 69.55 + 26.16\log_{10}f_c - 13.82\log_{10}f_c - 13.82\log_{10}h_{te} - a(h_{re}) + 44.9$
$\qquad\qquad - 6.55\log_{10}h_{te} + 10\log_{10} d$

Substitute values:
$h_{te} = 100$ m, $h_{re} = 1.5$ m, $d = 8$ km
$f_c = $ centre $f_d = 947$ MHz
$a(h_{re}) = 3.2 (\log_{10}(11.75\, h_{re}))^2 - 4.98$

Calculate:
$a(h_{re}) = -0.011$
$L_{50dB,urban} = 69.55 + 26.16\log_{10}947 - 13.82\log_{10}947 - 13.82\log_{10}100 + 0.011 + 44.9$
$\qquad\qquad - 6.55\log_{10}100 + 10\log_{10} 8$

$L_{50dB,urban} = 119.48$

**Figure 17.3.b**

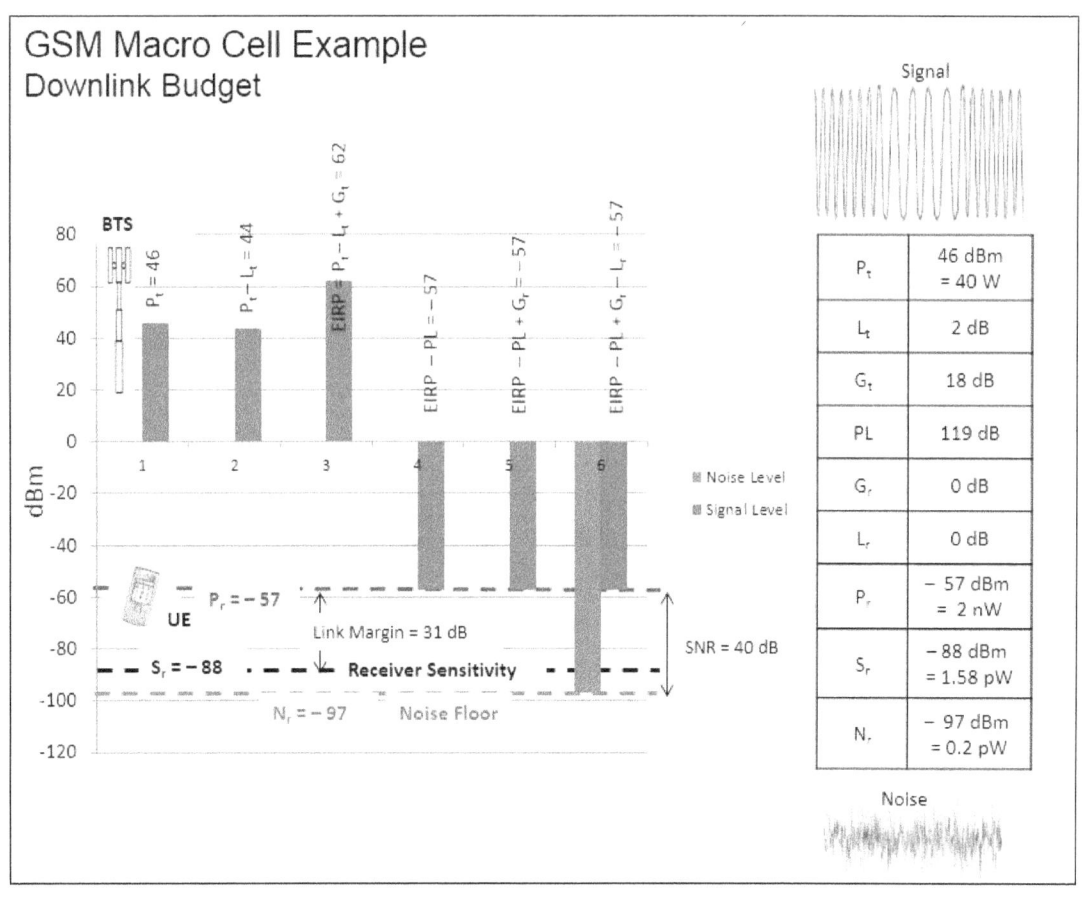

**Figure 17.3.c**

## GSM Macro Cell Example

Use Hata Model for Uplink Path Loss:

$L_{50dB,urban} = 69.55 + 26.16\log_{10}f_c - 13.82\log_{10}f_c - 13.82\log_{10}h_{te} - a(h_{re}) + 44.9$
$\qquad - 6.55\log_{10}h_{te} + 10\log_{10}d$

Substitute values:

$h_{te} = 100$ m, $h_{re} = 1.5$ m, $d = 8$ km
$f_c = $ centre $f_u = 902$ MHz
$a(h_{re}) = 3.2(\log_{10}(11.75\, h_{re}))^2 - 4.98$

Calculate:

$a(h_{re}) = -0.011$
$L_{50dB,urban} = 69.55 + 26.16\log_{10}902 - 13.82\log_{10}902 - 13.82\log_{10}100 + 0.011 + 44.9$
$\qquad - 6.55\log_{10}100 + 10\log_{10}8$

$L_{50dB,urban} = 119.22$

**Figure 17.3.d**

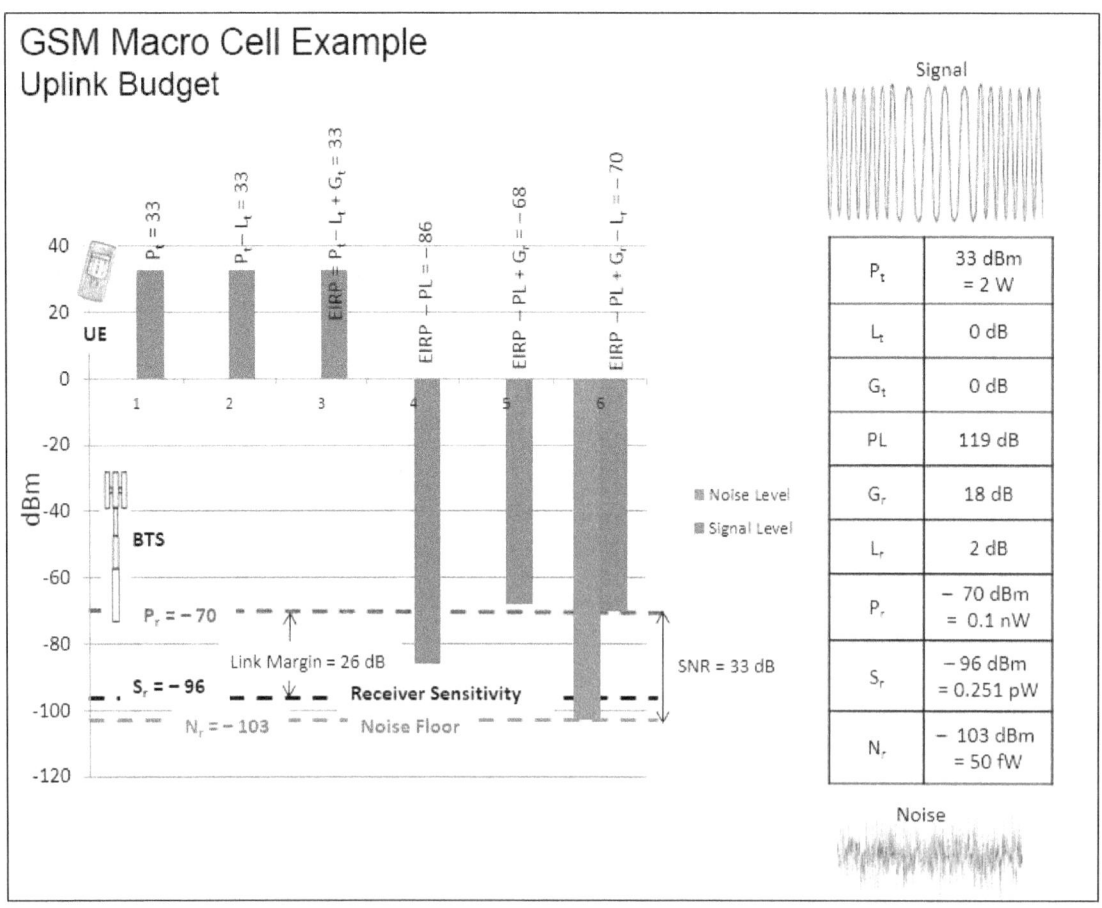

**Figure 17.3.e**

# 17.4 LTE Micro Cell Example

The user equipment (UE) in a LTE micro cell is served by eNodeB. In this example, the area covered by eNodeB is assumed to be dense urban with tall buildings and a cell radius of 1 km. The uplink and downlink frequencies for TDD LTE Band Number 33, the antenna heights, the street width and orientation, roof height and building separation for a *near line of sight* (NLOS) case are given in **Figure 17.4.a**.

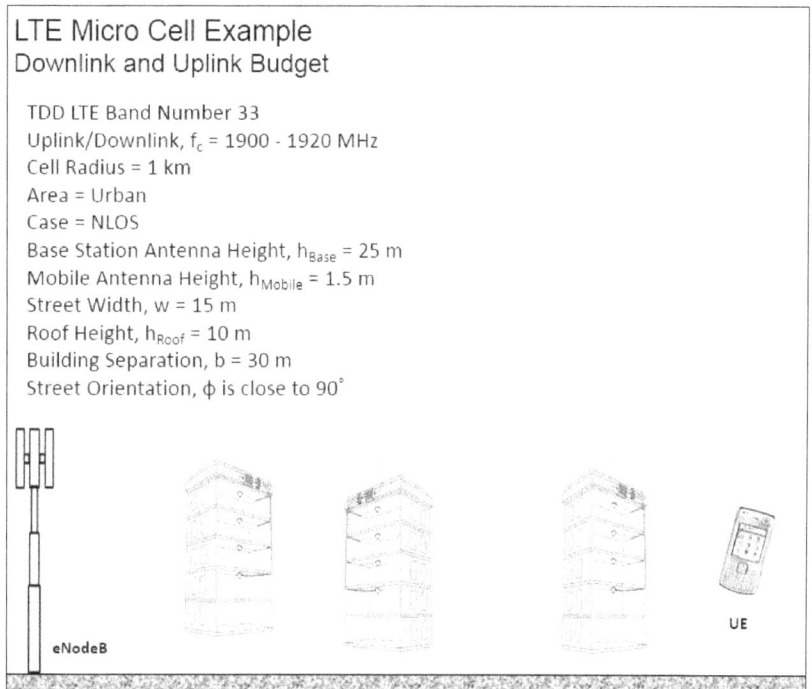

**Figure 17.4.a**

*COST-WI Model* (See **Section 16.8**) is used to calculate the path loss in urban area for uplink path as well as downlink path. Please note that the *uplink path* and downlink path are in the same frequency band of 1900-1920 MHz, because *time division duplexing* (TDD) is used in this band for transmission of data in both directions. The path loss calculations are shown in **Figure 17.4.b**. The downlink and uplink budgets are illustrated in **Figure 17.4.c** and **Figure 17.4.d**. Please review **Chapter 10** to understand the link budget calculations for this example.

Notice that in the *downlink budget* (**Figure 17.4.c**), the *link margin* is calculated to be 18 dB. Also, in *uplink budget* (**Figure 17.4.d**), the *link margin* is 16 dB. This amount of link margin provides a good cushion against signal fading. However, it could be further improved by reducing the size of micro cell or by increasing the transmitted power.

## LTE Micro Cell Example

### Use COST-WI Model NLOS Case for Path Loss:

$L_{b(dB)} = L_0 + L_{rts} + L_{msd}$
$L_{0(dB)} = 32.4 + 20 \log d + 20 \log f$
$L_{rts(dB)} = -16.9 - 10 \log w + 10 \log f + 20 \log \Delta h_{Mobile} + L_{Ori}$
$\Delta h_{Mobile} = h_{Roof} - h_{Mobile}$
$L_{Ori(dB)} = 4.0 - 0.114 (\phi - 55)$      for $55° \leq \phi < 90°$
$L_{msd(dB)} = L_{bsh} + k_a + k_d \log d + k_f \log f - 9 \log b$
$L_{bsh} = -18 \log (1 + \Delta h_{Base})$      for $h_{Base} > h_{Roof}$
$\Delta h_{Base} = h_{Base} - h_{Roof}$
$k_a = 54$      for $h_{Base} > h_{Roof}$
$k_d = 18$      for $h_{Base} > h_{Roof}$
$k_f = -4 + 1.5 ((f / 925) - 1)$      for metropolitan areas
$f = $ center $f_c = 1910$ MHz

### Substitute values and Calculate:

$k_f = -4 + 1.5 ((f / 925) - 1) = -4 + 1.5 ((1910 / 925) - 1) = -2.4$
$L_{bsh} = -18 \log (1 + \Delta h_{Base}) = -18 \log (1 + 25 - 10) = -21.67$
$L_{msd(dB)} = L_{bsh} + k_a + k_d \log d + k_f \log f - 9 \log b = -21.67 + 54 + 18 \log 1 - 2.4 \log 1910 - 9 \log 30$
$L_{msd(dB)} = 11.17$
$L_{Ori(dB)} = 4.0 - 0.114 (\phi - 55) = 4.0 - 0.114 (90 - 55) = 0.01$
$L_{rts(dB)} = -16.9 - 10 \log w + 10 \log f + 20 \log \Delta h_{Mobile} + L_{Ori} = -16.9 - 10 \log 15 + 10 \log 1910 +$
      $20 \log (10 - 1.5) + 0.01 = 22.75$
$L_{0(dB)} = 32.4 + 20 \log d + 20 \log f = 32.4 + 20 \log 1 + 20 \log 1910 = 98.02$

$L_{b(dB)} = L_0 + L_{rts} + L_{msd} = 98.02 + 22.75 + 11.17 = 131.94$

**Figure 17.4.b**

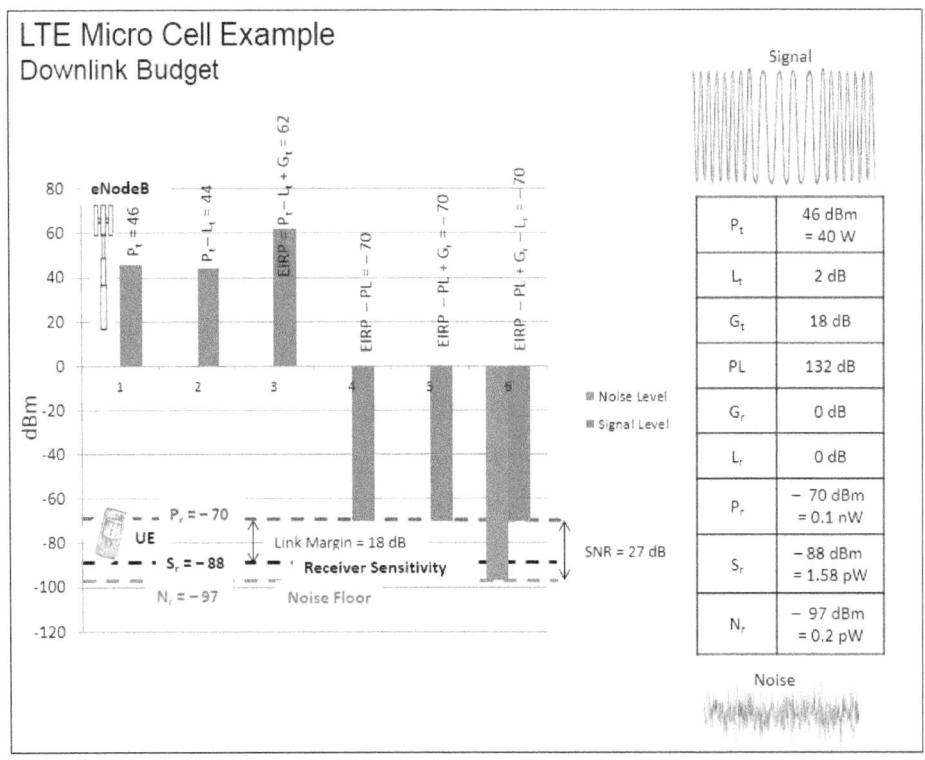

**Figure 17.4.c**

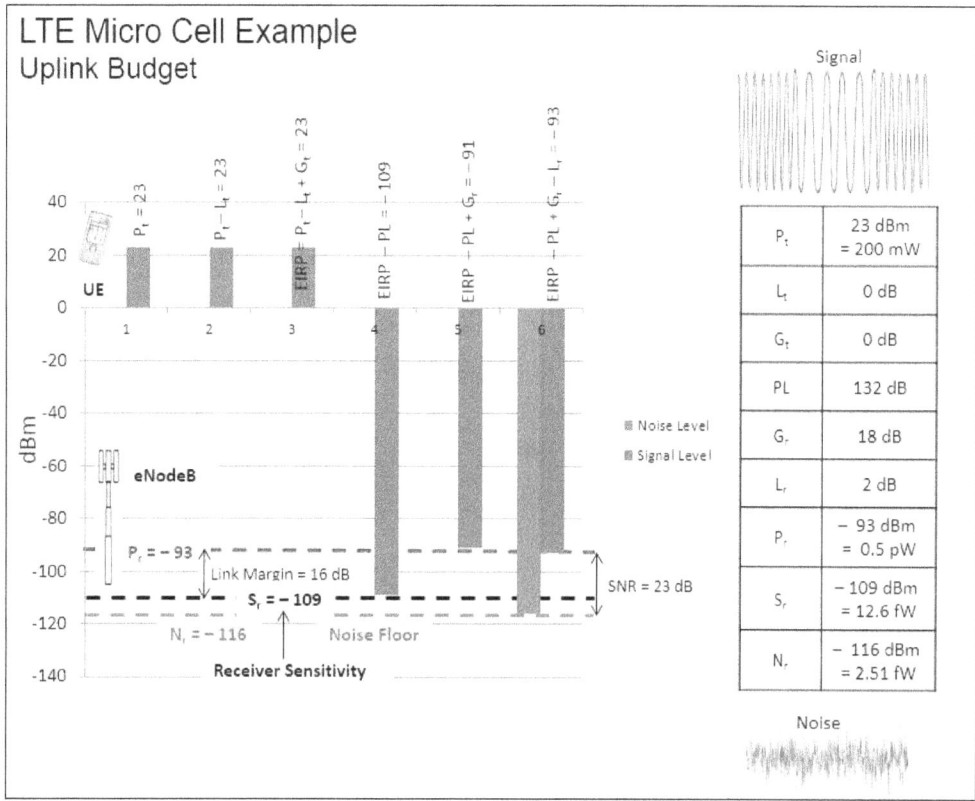

**Figure 17.4.d**

# 17.5 LTE Femto Cell Example

The user equipment (UE) in a LTE femto cell is served by eNodeB. In this example, the area covered by eNodeB is assumed to be the inside of a typical multiwall office. The uplink and downlink frequencies for TDD LTE Band Number 33, the number of penetrated walls are given in **Figure 17.5.a** for a cell radius of 10 m.

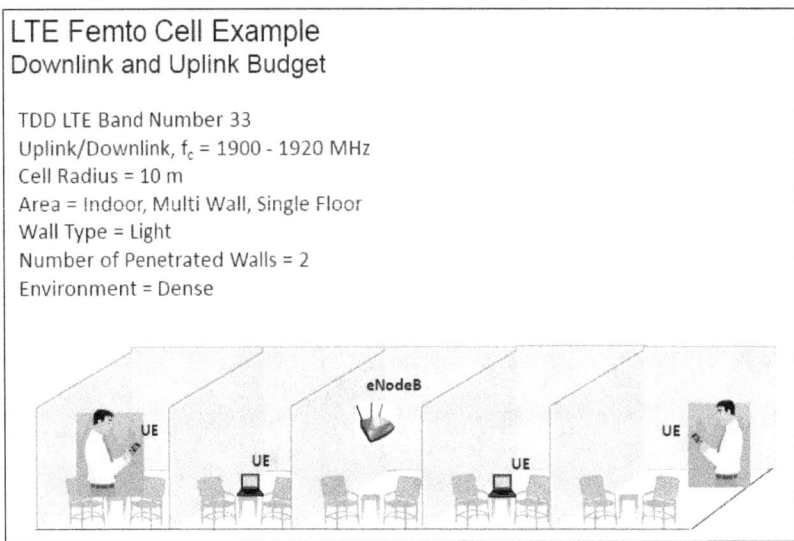

LTE Femto Cell Example
Downlink and Uplink Budget

TDD LTE Band Number 33
Uplink/Downlink, $f_c$ = 1900 - 1920 MHz
Cell Radius = 10 m
Area = Indoor, Multi Wall, Single Floor
Wall Type = Light
Number of Penetrated Walls = 2
Environment = Dense

**Figure 17.5.a**

*Multi Wall Model* for indoor propagation (See **Section 16.11**) is used to calculate the path loss in for uplink path as well as downlink path. Please note that the *uplink path* and downlink path are in the same frequency band of 1900-1920 MHz, because *time division duplexing* (TDD) is used in this band for transmission of data in both directions. The path loss calculations are shown in **Figure 17.5.b**. The downlink and uplink budgets are illustrated in **Figure 17.5.c** and **Figure 17.5.d**. Please review **Chapter 10** to understand the link budget calculations for this example.

Notice that in the *downlink budget* (**Figure 17.5.c**), the *link margin* is calculated to be 39 dB. Also, in *uplink budget* (**Figure 17.4.d**), the *link margin* is 60 dB. This amount of link margin provides an excellent cushion against signal fading.

LTE Femto Cell Example

Use Multi Wall Model (MWM) for Path Loss:

$$L_{dB} = L_{FS} + L_c + \sum_{i=1}^{I} k_{wi} L_{wi} + k_f^{[(kf + 2)/(kf + 1) - b]} \cdot L_f$$

$$L_{FS} = 32.45 + 20 \log f + 20 \log d$$

Substitute Values:

$L_{FS} = 32.45 + 20 \log 1910 + 20 \log (10 \times 10^{-3}) = 78.07$ dB

$L_c = 0$ (negligible)

$k_f = 0$ (single floor)

$k_{wi} = k_{w1} = 2$ (number of penetrated walls)

$L_{wi} = L_{w1} = 3.4$ dB (loss of light wall)

Calculate:

$L_{dB} = 78.07 + 0 + (2 \times 3.4) = 84.87$

**Figure 17.5.b**

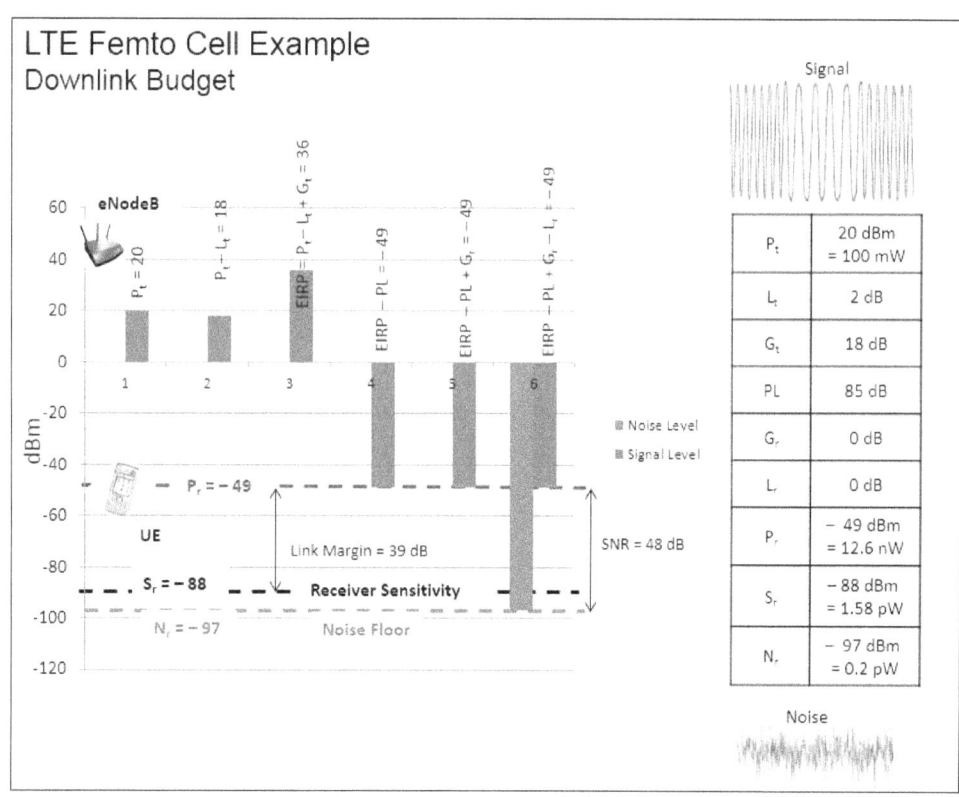

**Figure 17.5.c**

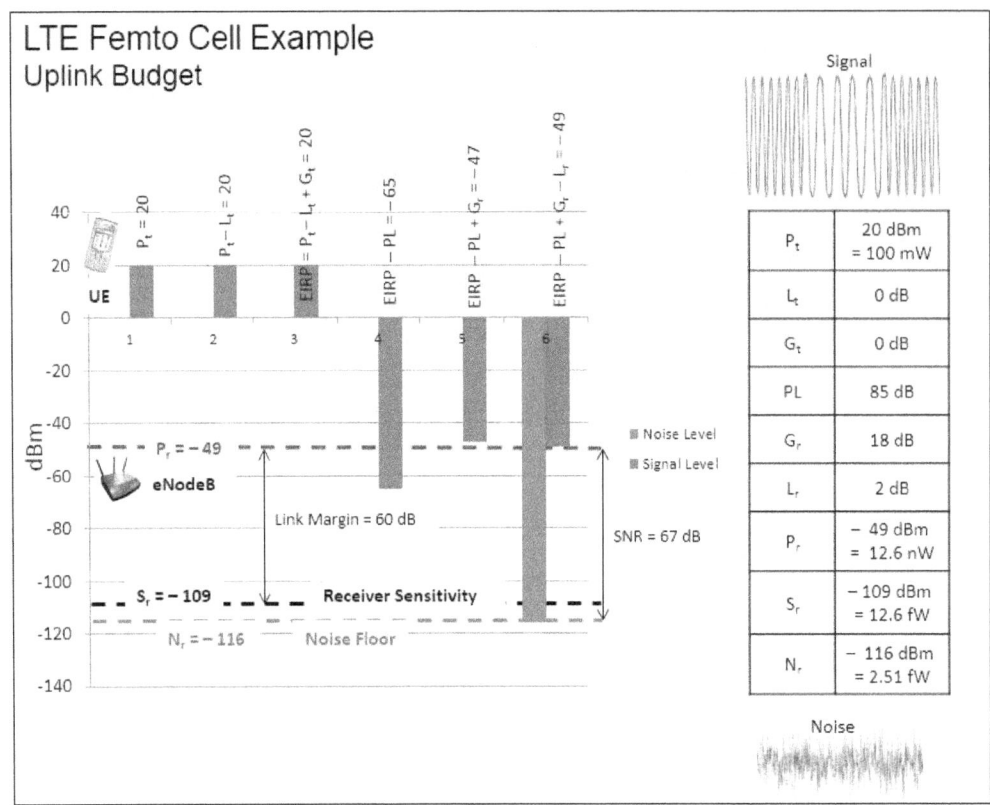

**Figure 17.5.d**

# 18. Wireless Receiver Engineering

In this chapter we will study the defining characteristics of wireless communications receivers, followed by receiver architectures, such as *superheterodyne* and *direct conversion*. We will also study in detail, various digital modulation techniques used in contemporary wireless receivers, such as QAM and OFDM. The chapter will end with a brief introduction to the emerging *software defined radio* (SDR) architecture.

## 18.1 Wireless Communication Systems

A wireless communication system essentially consists of user equipment (UE) served by a BTS, NodeB or eNodeB, commonly called *base station* in cellular systems. For a wireless LAN, it is called an *access point* (AP). Each of these communication systems has an uplink and a downlink, as shown in **Figure 18.1.a**. The uplink represents the radio signal transmission from the UE to the base station or access point. The UE transmits and the base station receives for an uplink. The downlink represents the radio signal transmission from the base station to the UE. In this case, the base station transmits and the UE receives. The UE as well as base station, each has a wireless transmitter as well as a receiver, which in combination is called a *transceiver*. In this chapter we will focus on the engineering of wireless receivers only. The detailed discussion of wireless transmitters is beyond the scope of this book.

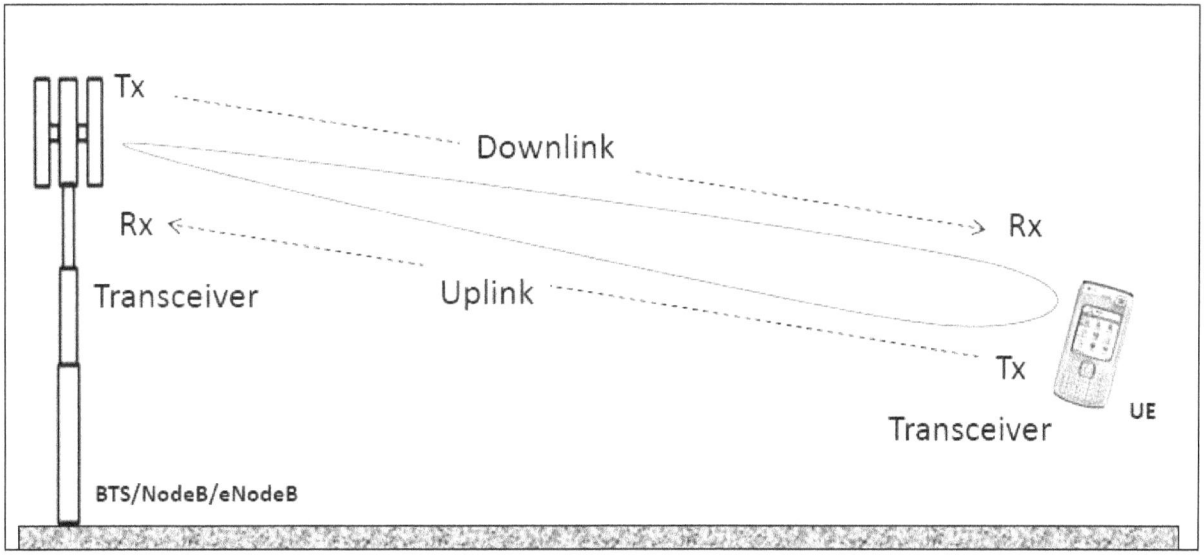

**Figure 18.1.a**

## 18.2 Wireless Receiver Characteristics

For any wireless receiver, irrespective of whether it is used for a base station or a cellphone, there are few defining characteristics, which must be considered to ensure its useful application. These are *sensitivity, noise floor, selectivity, stability, spurious responses, phase noise, dynamic range* and *automatic gain control*. We will now discuss each of these characteristics in a little more detail.

### 18.2.1    Sensitivity

The *sensitivity* of a wireless receiver represents its ability to detect the weakest radio signal. It is usually expressed in dBm. For example, the *sensitivity* of the receiver in mobile hand set shown in **Figure 18.2.1.a** is -88 dBm, which is 1.58 pW in absolute units of power. This means any radio signal weaker than -88 dBm will not be detected. Notice that -88 dBm of *sensitivity* allows a *link margin* of 18 dB in this example. Also, it is well above the receiver *noise floor* of -97 dBm.

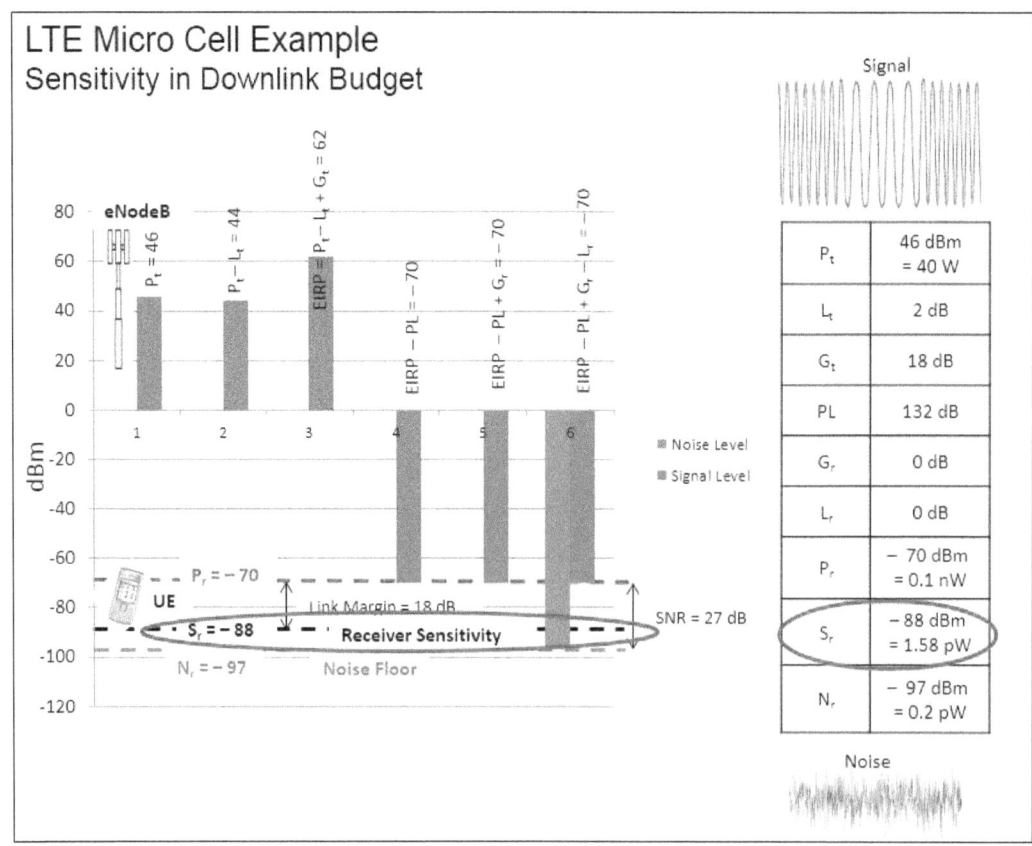

**Figure 18.2.1.a**

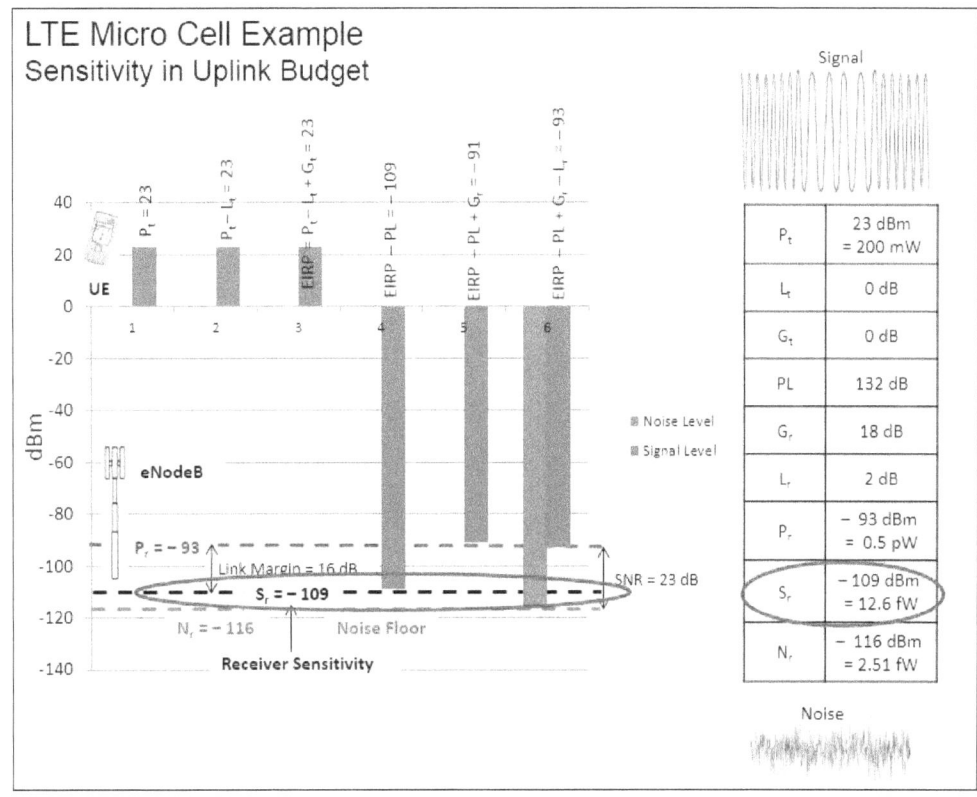

**Figure 18.2.1.b**

The *sensitivity* of receiver in eNodeB of **Figure 18.2.1.b** is -109 dBm, which is better than the sensitivity of -88 dB for the mobile hand set in **Figure 18.2.1.a**. The receivers in eNodeB are designed to be more *sensitive*, so that the weakest signals transmitted from the mobile handsets can be reliably detected.

## 18.2.2    Noise Floor

The *noise floor* of a receiver represents the noise power picked up by the receiving antenna from various natural as well as man-made sources, and also includes the noise generated due to heating inside the receiving equipment. For a detailed understanding of noise power, please see **Section 10.6 – 10.9**.

**Figure 18.2.2a** shows a *noise floor* of -97 dBm for a typical UE. With a received signal power level of -49 dB, the SNR is calculated to be 48 dB. It should be noted that received signal power must be higher than the *noise floor* and *sensitivity* of the receiver for a meaningful detection of the received signal. **Figure 18.2.2b** shows a *noise floor* of -116 dBm for eNodeB and similar analysis can be applied for SNR, received signal power and receiver *sensitivity*.

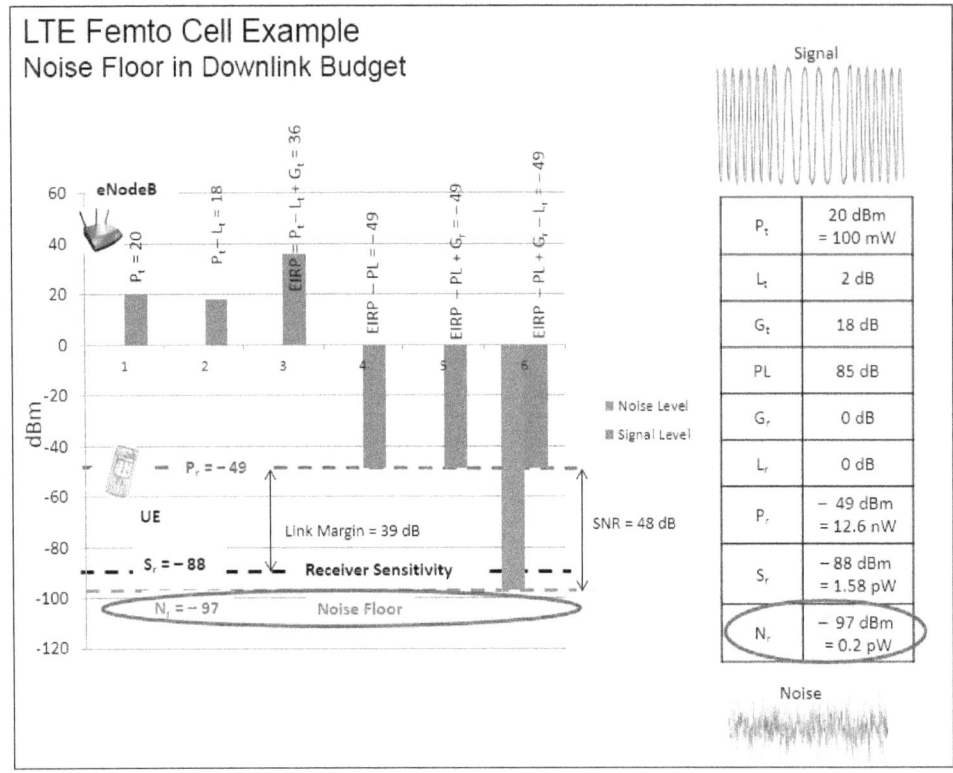

**Figure 18.2.2.a**

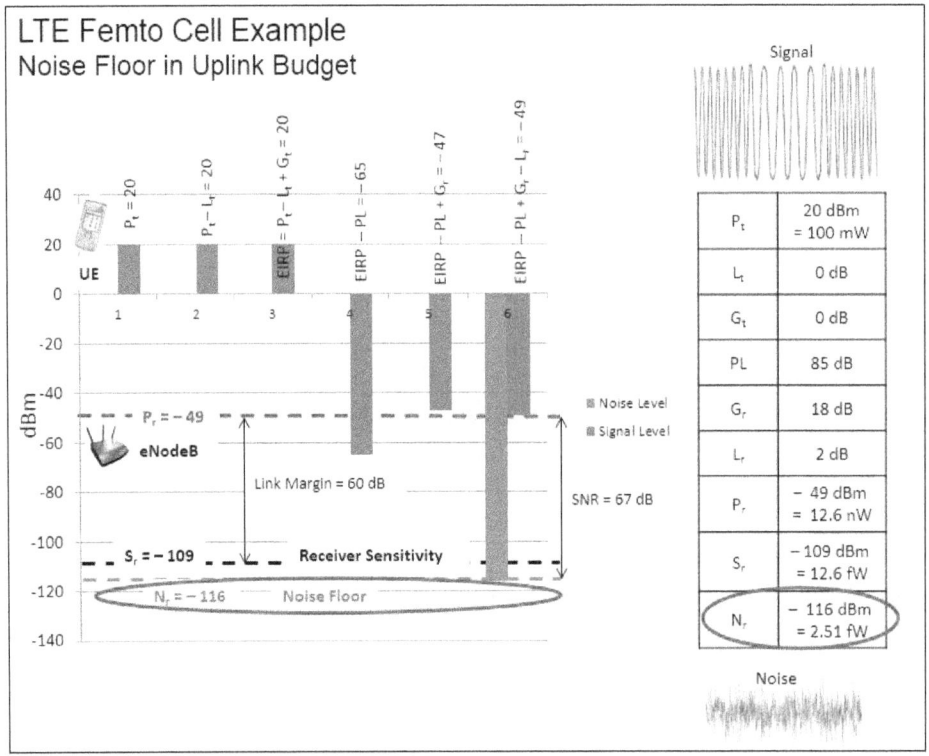

**Figure 18.2.2.b**

## 18.2.3    Selectivity

The *selectivity* of a wireless receiver is its ability to *select* the desired radio signal and reject all undesired signals from other nearby transmitters. **Figure 18.2.3.a** shows a multi band smartphone capable of receiving radio signals at different frequencies and from different generations of radio technologies. In the presence of the multitude of radio signals, the receiver must be *selective* enough to tune into the desired signal, such as 4G LTE 750 MHz and reject all others. This is necessary to avoid interference from adjacent transmitters and even adjacent cells, so as to maximize the *signal to interference and noise* (SINR) ratio.

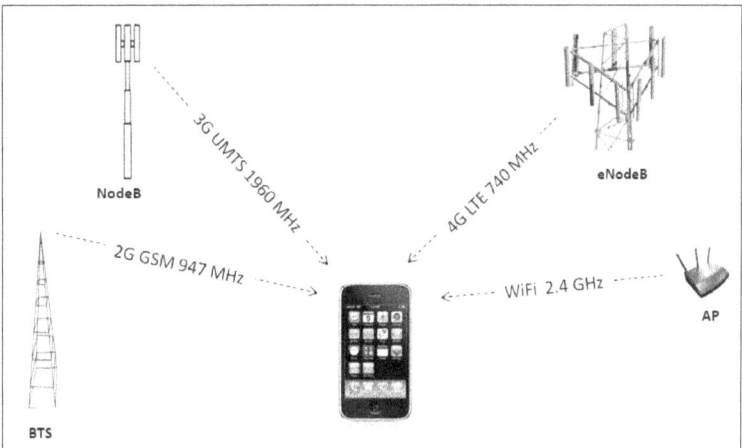

**Figure 18.2.3.a**

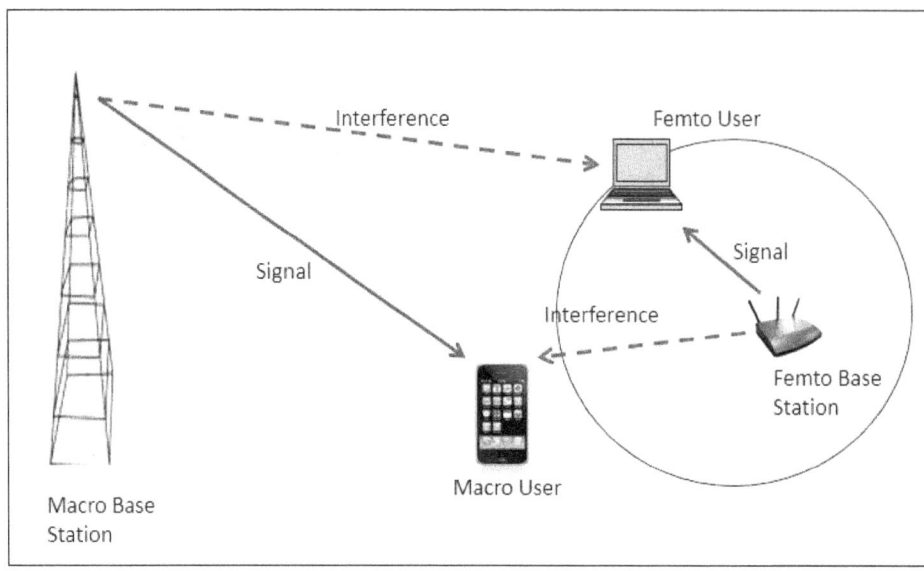

**Figure 18.2.3.b**

**Figure 18.2.3.b** shows another scenario, in which, in the same radio technology, different frequencies are assigned by the operator for macro cell users and femto cell users, in an overlay network of macro cells and femto cells. As shown, for a macro cell user, the signal from macro base station is the valid radio signal; whereas, the one received from the femto base station is interference, so the receiver must be selective enough to select the signal from macro base station. Similarly, the receiver of a femto user must be selective enough to select the signal from femto base station and reject the signal received from macro base station as interference.

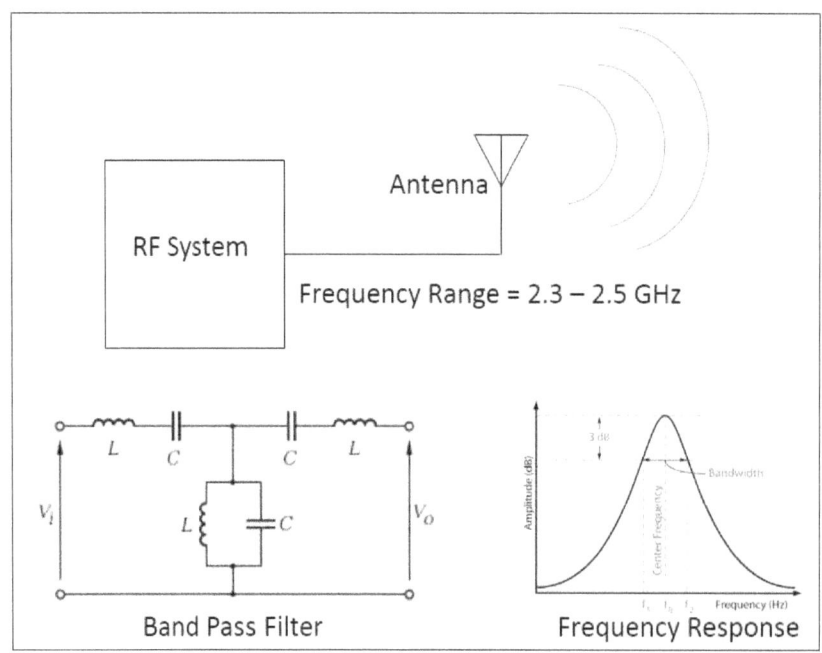

**Figure 18.2.3.c**

The *selectivity* of a receiver is ultimately defined by the frequency response of the band pass filter used in the receiver. **Figure 18.2.3.c** shows a band pass filter frequency response for a WiFi receiver for a frequency range of 2.3 – 2.5 GHz, and a peak at center frequency of 2.4 GHz. If the receiver is used with band pass filter with the given frequency response, it should have high selectivity for WiFi radio signals in the given frequency range.

## 18.2.4    Stability

The *stability* of a wireless receiver is its ability to remain tuned to the desired frequency even in the presence of changes in temperature, environment, mechanical vibration, electrical fluctuations and sources of noise and any other interference, as illustrated in **Figure 18.2.4.a**. The manufacturers of radio equipment ensure *stability* by subjecting the equipment to extreme physical conditions of temperature, humidity, vibration and noise, and testing for any variations in the characteristics of the receiver.

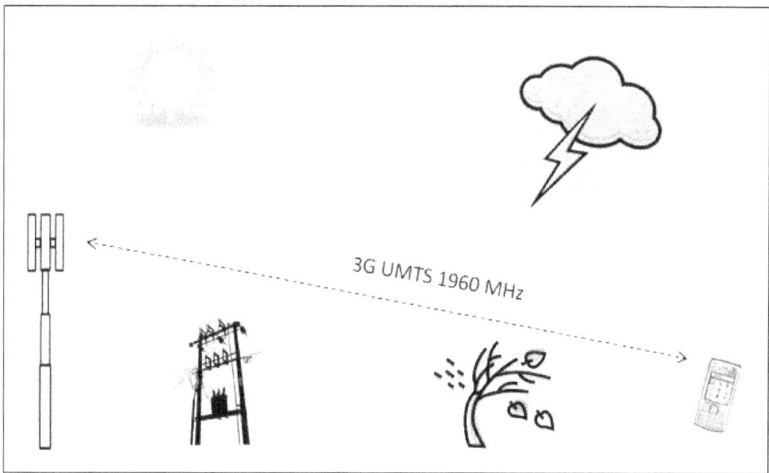

**Figure 18.2.4.a**

## 18.2.5    Spurious Responses

The components and sub modules inside a wireless receiver can generate *spurious responses*. These are undesired frequencies which are caused by non-linearities in the *radio frequency* (RF) and *intermediate frequency* (IF) amplifiers, or by poorly designed and unstable oscillators, or even by the inability of the RF filters to reject the *image frequencies*. An ideal receiver must fully suppress these *spurious responses*, otherwise the desired signal will just be drowned out.

Primarily, *spurious responses* include *harmonic* and *image frequencies*. The *harmonic frequencies* are integer multiples of the original frequency as shown in **Figure 18.2.5.a**, and can be suppressed by designing stable oscillators and good filters that reject the *harmonics.*, The *image frequency* is defined as the one at a distance of twice the *intermediate frequency*, $f_{IF}$, from the carrier frequency, $f_c$, and can be suppressed by a well-designed RF filter at the front end

**Figure 18.2.5.a** illustrates spurious responses in the context of a *superheterodyne receiver*. We will study in detail about *superheterodyne receivers* in **Section 18.3.2**.

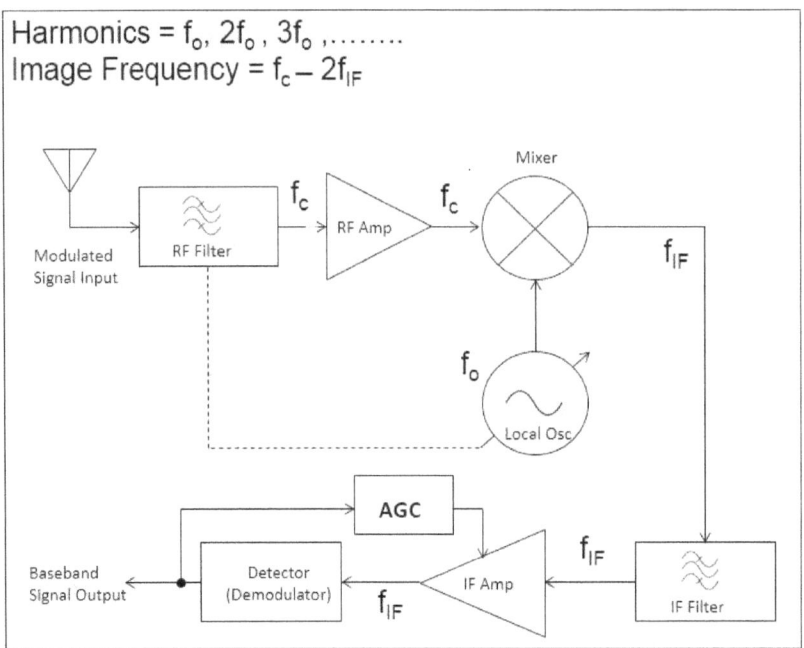

**Figure 18.2.5.a**

# 18.2.6    Phase Noise

The *phase noise* is caused by shifts in phase as the signal travels through the communication channel and various components and modules inside the receiver equipment. The unexpected phase shifts can be especially detrimental to receivers that use phase based modulation techniques, such as *quadrature amplitude modulation* (QAM). All modern digital communication systems use QAM or its variation, so they are susceptible to *phase noise*. These receivers must suppress *phase noise* with accurate phase control as illustrated in the context of a QAM demodulator in **Figure 18.2.6.a**.

We will study more about QAM in **Section 18.4**.

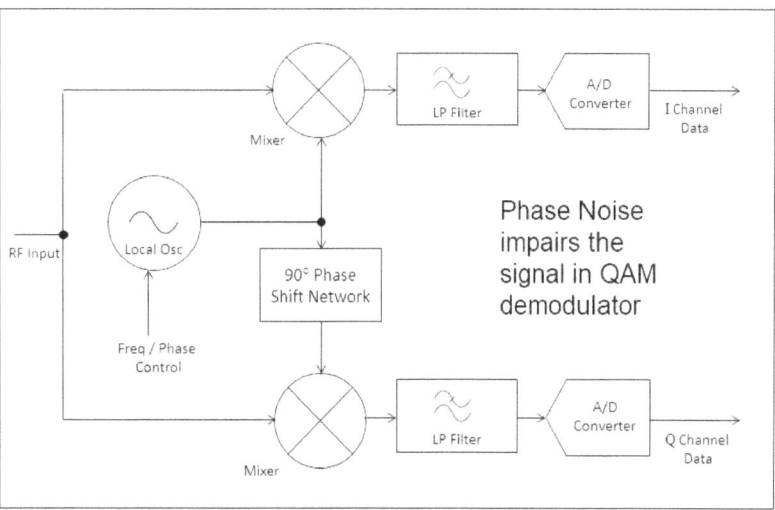

**Figure 18.2.6.a**

## 18.2.7    Spurious Frequency Dynamic Range

The *spurious frequency dynamic range* (SFDR) is defined as the ratio of the strongest signal that causes an overload in the receiver with a given distortion level, to the noise floor of the receiver, as illustrated in **Figure 18.2.7.a**.

As the received signal power is increased, at some point it causes *overload* in the receiver, causing distortion due to generation of third order harmonic frequencies, as the signal level enters the non-linear region of the amplifiers. Therefore, there is a limit to the signal power level that a receiver can tolerate without causing significant distortion. The SFDR measures this tolerance for a receiver.

In the example of **Figure 18.2.7.a**, received signal overload happens at received signal power level of -30 dB, and the SFDR is calculated to be 67 dB at 3% distortion. The measured 3% distortion is indicated for reference.

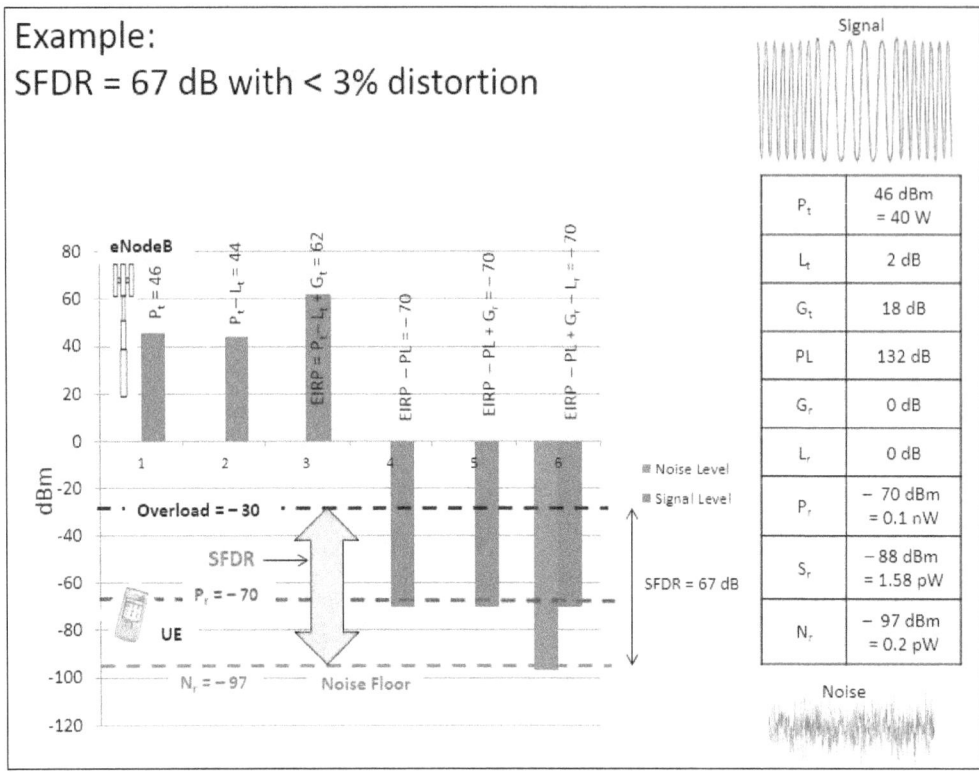

**Figure 18.2.7.a**

## 18.2.8    Automatic Gain Control

The *automatic gain control* (AGC) helps to adjust signal level at the output of an amplifier and prevents the signal level from becoming excessive. A very high level of received signal that goes beyond the range defined by SFDR (See **Section 18.2.7**), can cause distortion, and AGC provides the means to control these signal overshoots.

**Figure 18.2.8.a** illustrates AGC in the context of a superheterodyne receiver. Notice that the AGC block connects the baseband signal output from the *detector* to the IF amplifier. The AGC monitors the level of baseband output signal and adjusts the control input to the IF amplifier. If the input signal to the IF amplifier overshoots, the control input to the IF amplifier from AGC lowers the gain of amplifier, thereby, lowering the level of baseband output signal. In a similar manner, AGC can also be applied to the RF amplifier.

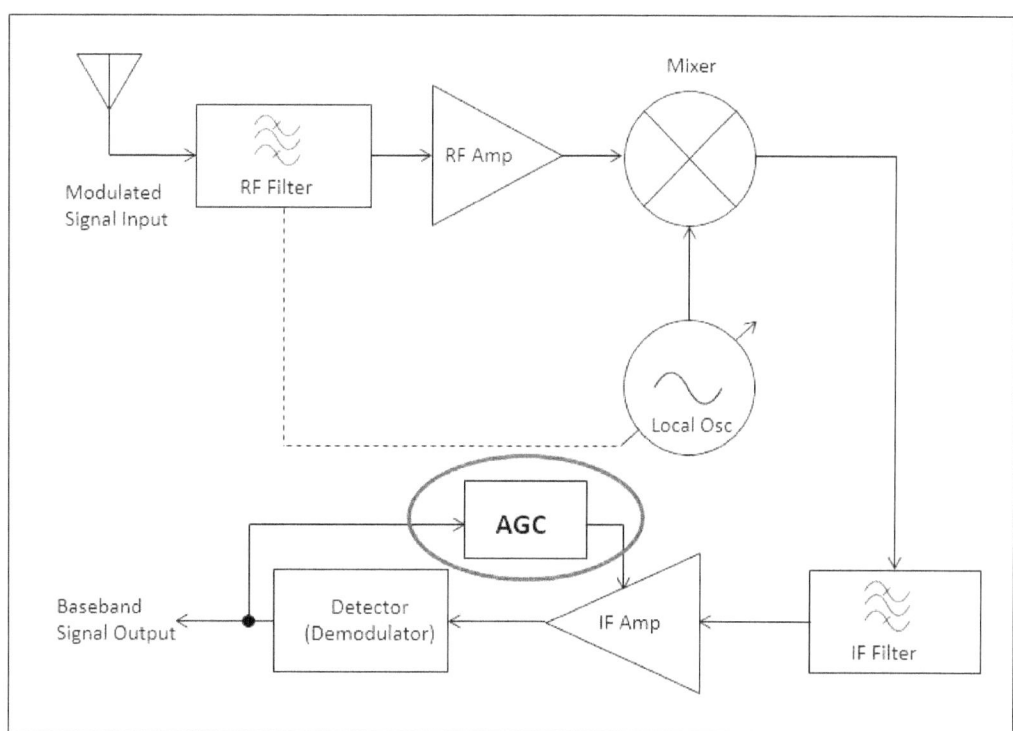

**Figure 18.2.8.a**

## 18.3 Wireless Receiver Topologies

In this section, we will study *superheterodyne* and *direct conversion* receiver topologies, which are commonly used in contemporary wireless communication receivers.

## 18.3.1    Superheterodyne Receivers

In *superheterodyne* receivers, the *radio frequency* (RF) signal is first down converted to an *intermediate frequency* (IF), by *mixing* the RF signal with a stable frequency output from a local oscillator, as shown in **Figure 18.3.1.a**.

The modulated RF signal received at the antenna is first passed through a band pass RF filter to filter out the unwanted frequencies from the adjacent bands, and then passed through an RF amplifier to increase the power level of the received RF signal. The next step is to down convert the RF signal in the mixer by multiplying it with the local oscillator frequency. The output of mixer includes new frequency components which are sum and difference of the original RF and the local oscillator frequencies. It is the difference frequency component which is of interest in a *superheterodyne* receiver, because it is the *intermediate frequency* (IF), and much lower in frequency than the original received RF signal.

The IF signal is filtered out through a band pass filter and passed through an IF amplifier to provide more gain to the signal power level. The next step is to send the IF signal to the demodulator or detector circuit to recover the *baseband signal*.

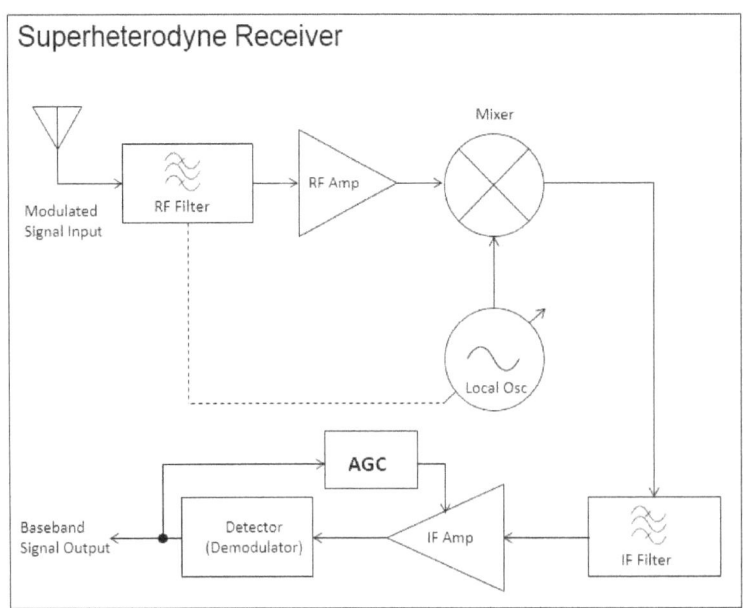

**Figure 18.3.1.a**

The *superheterodyne* receiver topology allows the receiver *gain* or the amplification of received signal to be distributed into two or more stages, of RF and IF amplifiers. This distribution of *gain* provides inherently better stability to a *superheterodyne* receiver, despite higher overall *gain*.

## 18.3.2    Superheterodyne Receiver Examples

**Figure 18.3.2.a** shows an example of a topology for a 902-928 MHz UHF *superheterodyne receiver*. The received radio signal is *frequency modulated* (FM) with a bandwidth of 15 kHz. The *intermediate frequency* (IF) used in the receiver is 28 MHz.

With an IF of 28 MHz, the frequency range of tunable local oscillator must be 874-900 MHz, to keep a difference of 28 MHz with the received frequency band of 902-928 MHz. The received radio signal is passed through a RF band pass filter at the front end, where all frequencies outside the pass band of 902-928 MHz are filtered out. Notice that the image frequency of 872 MHz, which is at a distance of 56 MHz (twice the IF of 28 MHz) from 928 MHz is filtered out. The same is true for the image frequency at 846 MHz at a distance of 56 MHz from the lowest frequency in the pass band of 902 MHz. All image frequencies in the range 846-872 MHz are also filtered out because these frequencies lie outside the pass band of the front end RF filter.

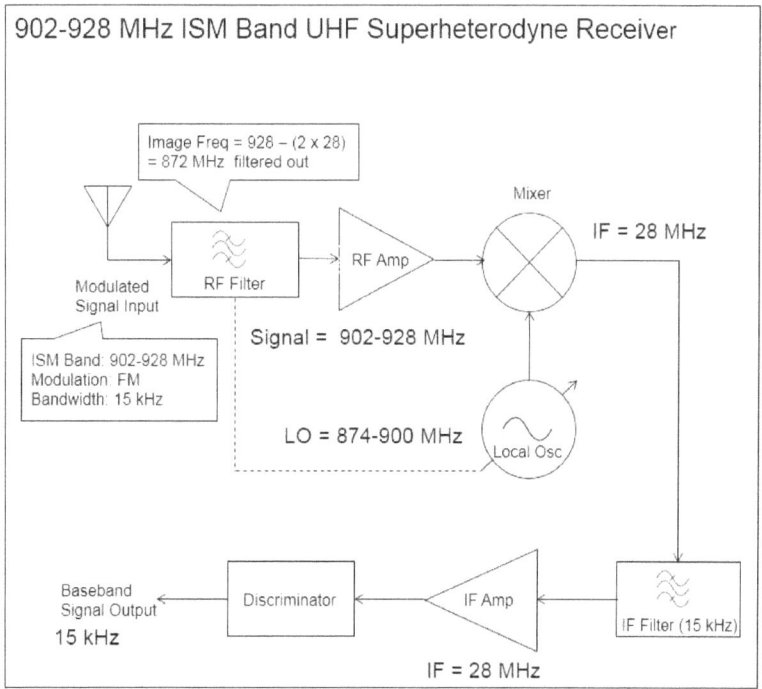

**Figure 18.3.2.a**

The received and filtered RF signal is amplified in RF amplifier and then multiplied with the local oscillator frequency in the *mixer* to produce a *difference* component of 28 MHz as one of the components in the *mixer* output. The *mixer* output also has the *sum* frequencies and the *harmonics*, and may have additional undesired *difference* frequencies.

Notice that if the *image* frequencies are not filtered out before input to the mixer, there will be additional undesired *difference* components at 28 MHz, for example, the one produced by the mixing of image frequency of 872 MHz with the local oscillator frequency of 900 MHz.

The IF filter is a band pass filter centered at 28 MHz with a pass band of 15 kHz around the center frequency. This 15 kHz pass band contains the *baseband signal*, that is the original audio, data or video signal. The IF signal is first amplified in IF amplifier and then fed to a discriminator circuit to demodulate the baseband signal from the 28 MHz FM signal. Notice how the received radio frequencies of 902-928 MHz are down converted to an IF of 28 MHz, before extraction of the base band signal.

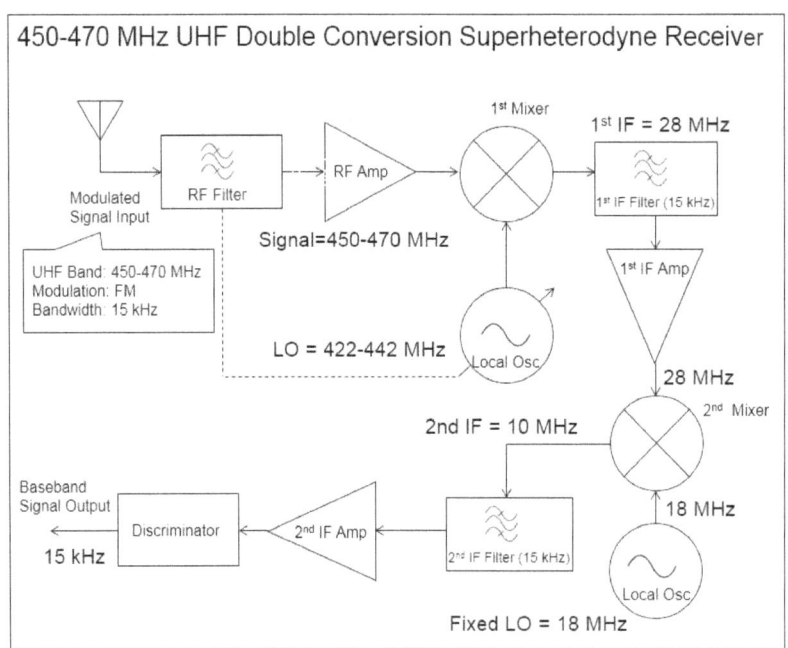

**Figure 18.3.2.b**

**Figure 18.3.2.b** shows the topology for a 450-470 MHz *double conversion superheterodyne* receiver. This receiver topology uses two *intermediate frequencies* (IF), at 28 MHz and 10 MHz. The benefits of having two conversion stages include, higher *gain* and improved *selectivity*. The higher IF of 28 MHz helps reject the image frequencies effectively, and the lower IF of 10 MHz helps improve selectivity by narrowing down the width of pass band.

The analysis of previous example can be applied to this example too for 1st IF of 28 MHz. Thereafter, the 28 MHz signal is fed to a second *mixer*, and is multiplied with a fixed local oscillator frequency of 18 MHz, to produce a *difference* signal component of 10 MHz (2nd IF) in the output of 2nd *mixer*. The 10 MHz output of second IF filter is amplified in the 2nd IF amplifier and then fed to the discriminator circuit to recover the 15 kHz wide baseband signal.

The additional mixer, filter and amplifier stages evidently add more complexity in a double conversion superheterodyne receiver at an additional cost, however the benefits are improved *gain* and *selectivity*. Such receivers, although expensive, do find uses in applications that require highly reliable wireless communications.

### 18.3.3    Direct Conversion Receivers

The *direct conversion* receivers are also called *zero IF* or *homodyne* receivers. These receivers eliminate the need for *intermediate frequency* (IF) and therefore reduce complexity in comparison with *superheterodyne* receivers.

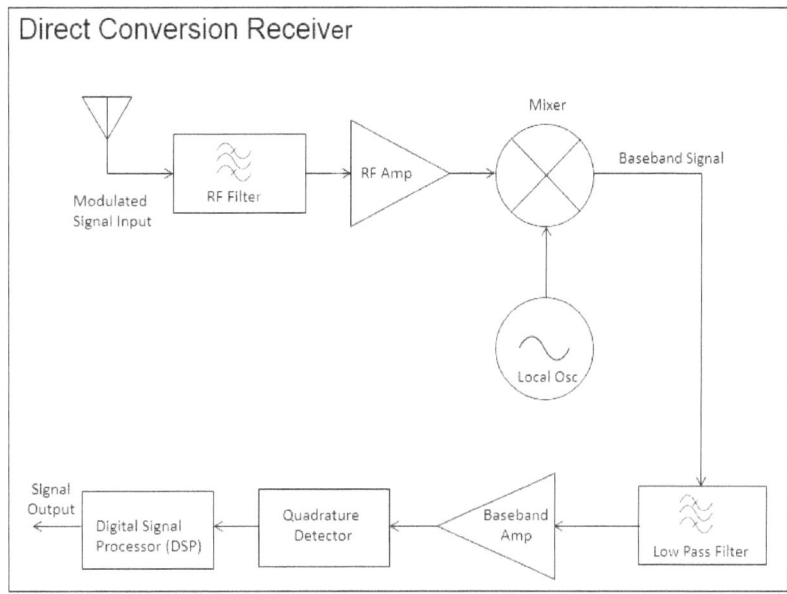

**Figure 18.3.3.a**

As shown in **Figure 18.3.3.a**, the received RF signal is filtered, amplified and then multiplied in the *mixer* with the RF carrier frequency generated by a fixed local oscillator. The *difference* frequency component in the output of *mixer* is the baseband signal. In this way, the baseband signal is directly extracted from the output of mixer without the need for down conversion to IF. Notice that the frequency of the local oscillator is the same as the carrier frequency of the received RF signal. In traditional direct conversion receivers, *phase locked loops* (PLL) were required to lock the frequency of local oscillator to the carrier frequency of incoming RF signal, which were often difficult and expensive to implement.

However, with the advent of low cost and efficient *digital signal processing* (DSP) techniques in modern cellphones and other radio equipment, it is often much easier and cost effective to implement *direct conversion*.

With *direct conversion*, some of the benefits of *superheterodyne* receivers are lost, such as higher *gain* and *selectivity*, and therefore direct conversion receivers are not suitable for *amplitude* (AM) and *frequency* (FM) modulated signals. The loss of *gain* and *selectivity* in direct conversion receivers is compensated by using digital modulation techniques, such as *quadrature amplitude modulation* (QAM), and digitally processing the signals with DSP processors. Please see **Section 18.4** for detailed description of digital modulation techniques. To conclude, for direct conversion receivers, use of DSPs is a must to implement a receiver with the desired *gain*, *selectivity* and other receiver characteristics.

## 18.4 Digital Modulation Techniques

In this section, we will study various digital modulation techniques, such as QAM, FHSS, DSSS and OFDM. These have widespread application in cellular wireless communications technologies, such as UMTS and LTE.

### 18.4.1 Quadrature Amplitude Modulation

The *quadrature amplitude modulation* (QAM) technique comes in many flavors, such as 2-QAM or *binary phase shift keying* (BPSK), 4-QAM or *quadrature phase shift keying* (QPSK), 8-QAM, 16-QAM and so on. The number in front of each QAM denotes the number of states that each QAM symbol represents. It takes 2 states to carry 1 bit of information, which could be 0 or 1. **Table 18.4.1.a** shows the number of states and the number of bits per symbol for each flavor of QAM.

| QAM Flavor | States | Bits per symbol |
|------------|--------|-----------------|
| 2-QAM | 2 | 1 |
| 4-QAM | 4 | 2 |
| 8-QAM | 8 | 3 |
| 16-QAM | 16 | 4 |
| 64-QAM | 64 | 6 |
| 256-QAM | 256 | 8 |
| 1024-QAM | 1024 | 10 |
| 4096-QAM | 4096 | 12 |

**Table 18.4.1.a**

For example, in 4-QAM, the number of possible states are 4 for each symbol and 2 bits can be carried by each possible state of the symbol, that is, 00, 01, 10 and 11. As the order of QAM is increased, more states are possible for each symbol and so more bits can be carried by each symbol. So, to increase data transfer rate, we can use higher order of QAM.

64-QAM is used in LTE and digital cable technologies to enhance the bandwidth of digital communication. In fact LTE uses adaptive modulation techniques that change the flavor of QAM based on available channel capacity.

QAM is essentially a combination of *amplitude* and *phase* modulation, wherein amplitude and phase of a sine and cosine wave are combined in different ways to create multiple states of a symbol.

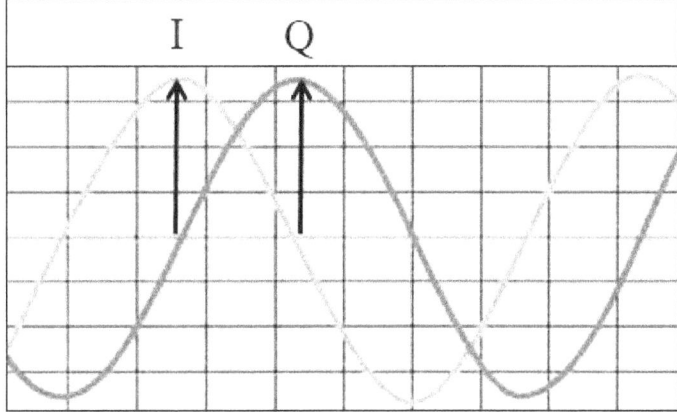

**Figure 18.4.1.b**

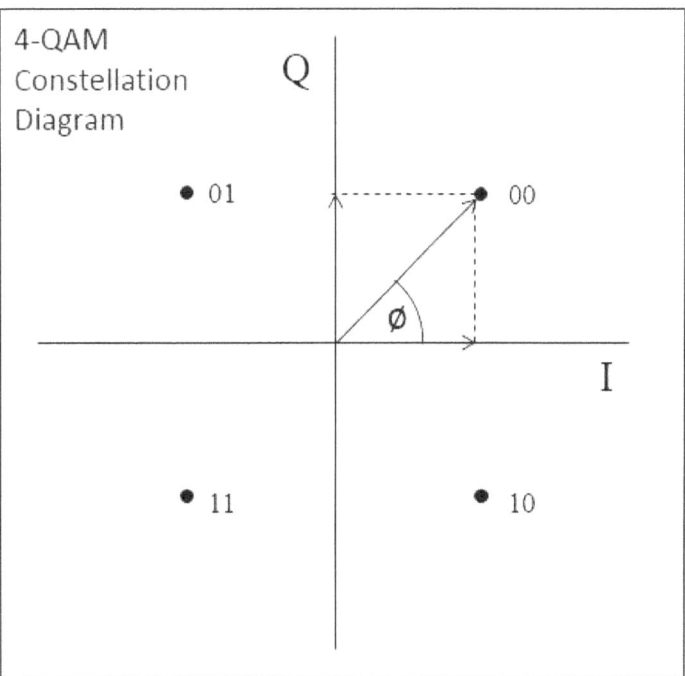

**Figure 18.4.1.c**

**Figure 18.4.1.b** illustrates the in-phase signal component, I and the quadrature signal component, Q, which lags in phase by 90 deg from I. Although the phase difference between I and Q is fixed at 90 deg, their amplitudes are changed to create multiple states of the QAM symbol.

The states of a QAM symbol are represented by a *constellation diagram*, as illustrated for 4-QAM in **Figure 18.4.1.c**. The horizontal axis represents the I component of the signal, and the vertical axis represents the Q component of the signal. Notice that 4-QAM symbol can represent 4 states, 00, 10, 01 and 11. For each of these states, the I and Q components have equal amplitudes, however, the sign of amplitude is determined by the placement of the state in a specific quadrant of the *constellation diagram*. For example, for 00, I and Q have positive amplitudes, however, for 10, the sign of amplitude is reversed for Q; for 01, the sign of amplitude is reversed for I ; and for the state 11, the sign of amplitude is reversed for both I and Q.

The composite *amplitude* of each state is represented by the magnitude of the line joining the placement point (dot) of the state with the *origin*, which is the intersection of I and Q components, as illustrated for the state 00. The *phase* for this state is the angle ϕ between the joining line and the I component. In this way the amplitude and phase of each state can be calculated from the respective amplitudes of I and Q components.

Similar analysis can be done for other flavors of QAM. **Figure 18.4.1.d** illustrates the constellation diagram for 16-QAM, each symbol of which has 16 possible states, and each state represents 4 bits of information.

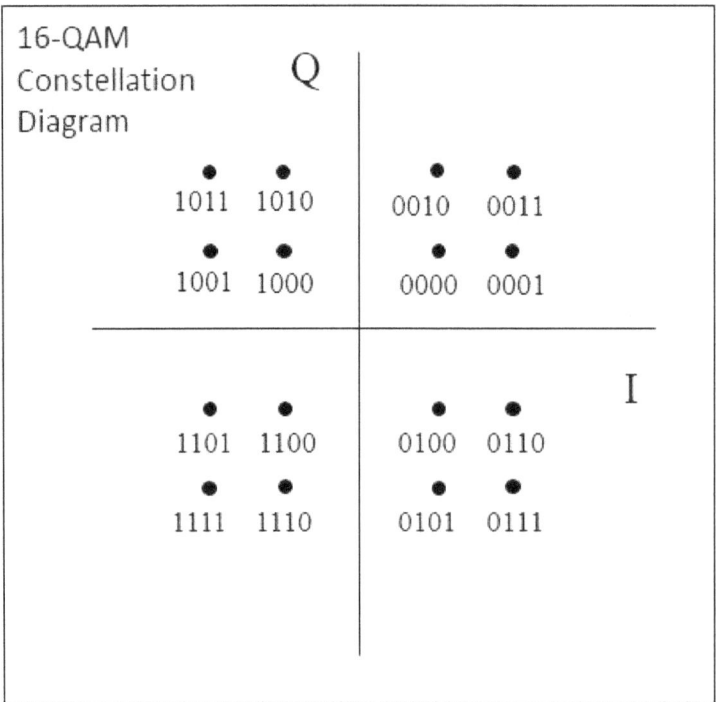

**Figure 18.4.1.d**

In a QAM baseband signal, impairments can be caused due to phase noise, I/Q imbalance, non-linearities in the devices or filter bandwidth. The impairment, if it occurs, results in a shift of the state of a symbol on the constellation diagram from its ideal expected location. For example, in **Figure 18.4.1.e**, the *ideal symbol* location is shown as an opaque black dot, whereas the location of *non-ideal symbol* is shown as a transparent circle. The impairment causes the state of symbol to shift through an angle ф with respect to the origin. The vector joining the ideal symbol to the non-ideal symbol is the *error vector*, and can be used as a measure of impairment in a QAM signal.

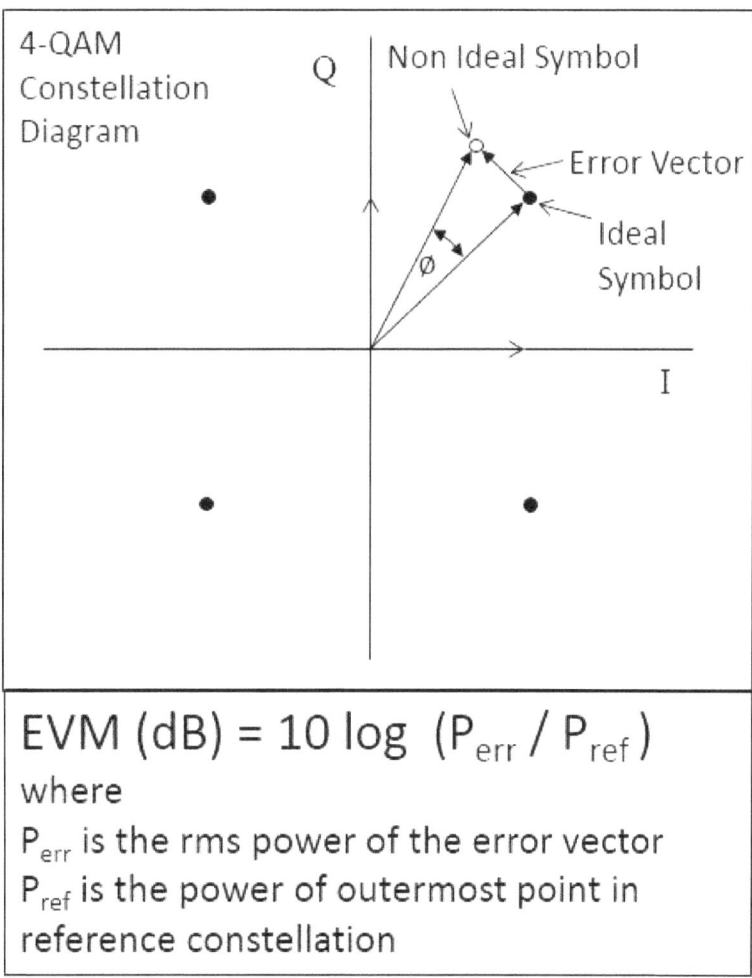

**Figure 18.4.1.e**

The *error vector magnitude* (EVM) is used to numerically measure the impairment of signal in QAM, in the units of dB. See the formula for EVM in **Figure 18.4.1.e**.

**Figure 18.4.1.f** shows the block diagram of a QAM modulator, which is used at the transmitter. Notice that the, I and Q channel data is first fed to their respective *digital to analog* (D/A) converters. The analog I and Q output from the D/A converters is then fed to their respective product *mixers*. The local oscillator frequency is phase shifted by 90 deg and fed to the mixer for Q. This is required to maintain a quadrature (90 deg) phase difference between I and Q components of the signal. The I and Q mixer outputs are then fed to a combiner to output the composite RF signal for transmission.

**Figure 18.4.1.g** shows the block diagram of a QAM demodulator, which is used at the receiver, and is essentially a reverse of what was achieved in the QAM modulator. The RF received signal is first fed to two mixers to separate out the, I and Q components, by using a 90 deg phase shifter in the local oscillator frequency input to the mixer for Q signal. After filtering through *low pass* (LP) filters, the I and Q signals are fed to their respective *analog to digital* (A/D) converters to recover the original I and Q channel data.

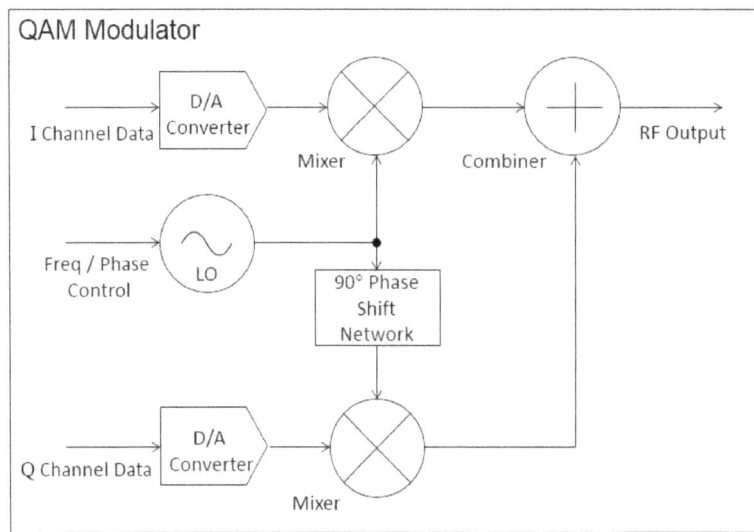

**Figure 18.4.1.f**

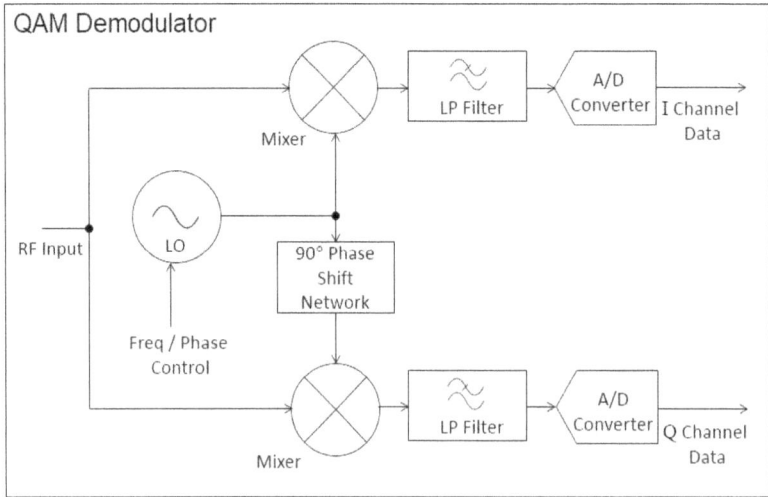

**Figure 18.4.1.g**

## 18.4.2    Spread Spectrum

The *spread spectrum* (SS) technique spreads the bandwidth of a narrow band signal over a wide bandwidth, as illustrated in **Figure 18.4.2.a**, by using a *pseudo-random noise* (PN) code that is independent of the signal to be modulated. The wide bandwidth of the modulated signal makes it resistant to frequency fading effects. Even if some frequencies in the wide signal bandwidth fade, it is possible to recover the signal.

The use of a unique PN code makes the SS signal appear as noise to non-SS receivers, and therefore makes it resistant to interception, interference and anti-jamming effects. A SS receiver can effectively reject other SS signals which have a different PN code than its own. For the same reason, the SS receiver can also effectively reject random noise, and it can even receive signals which are below the noise level. This phenomenon of improved *signal to noise ratio* (SNR) is called *processing gain*, and is measured in dB in terms of the 10 log of the *spreading factor*, which is actually the ratio of *spread bandwidth* to *unspread bandwidth*, as shown in **Figure 18.4.2.a**. However, it may be noted that the improvement due to *processing gain*, compensates for the reduction in SNR, which happens due to spreading of bandwidth from narrow band signal to a wide band signal. Remember, more bandwidth also adds more *noise power* to the channel.

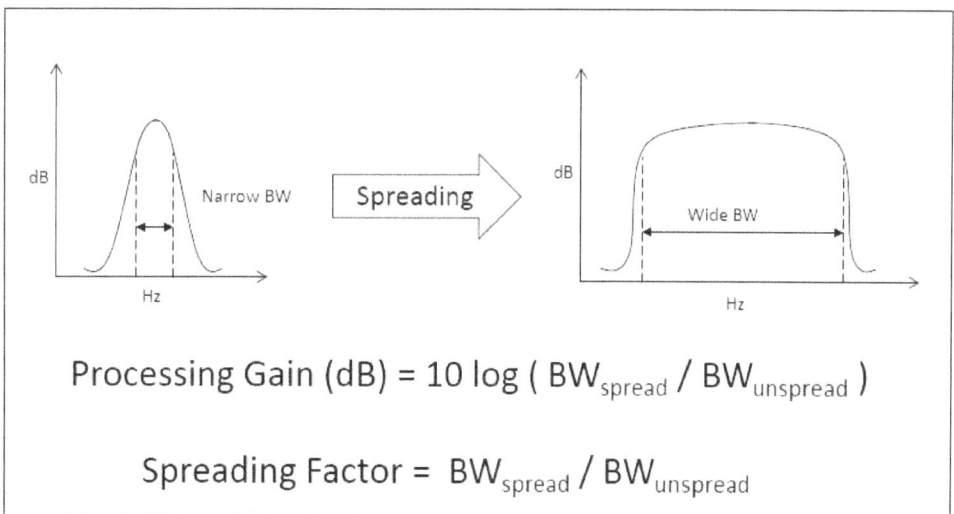

**Figure 18.4.2.a**

There are two ways of spreading the signal bandwidth using a PN code, the *frequency hopping spread spectrum* (FHSS) and a *direct sequence spread sequence* (DSSS).

**Figure 18.4.2.b** shows the block diagram of a FHSS transmitter. The incoming data is *frequency shift keyed* (FSK), to generate the unspread signal, and then fed to a product *mixer*, where it is multiplied with the frequency output of a local oscillator. The local oscillator is essentially a frequency synthesizer, whose frequency rapidly hops back and forth in a pseudo-random fashion over a wide frequency range based on the PN code input from a code generator.

This results in a frequency hopping signal over a wide bandwidth at the output of mixer, which is filtered out through a band pass filter.

One drawback of FHSS is, that frequency synthesizers cannot maintain phase coherence over successive frequency hops, so the unspread signal cannot use phase based modulation techniques, such as QAM. This limits the choice of modulation for unspread signal in FHSS to FSK or MFSK.

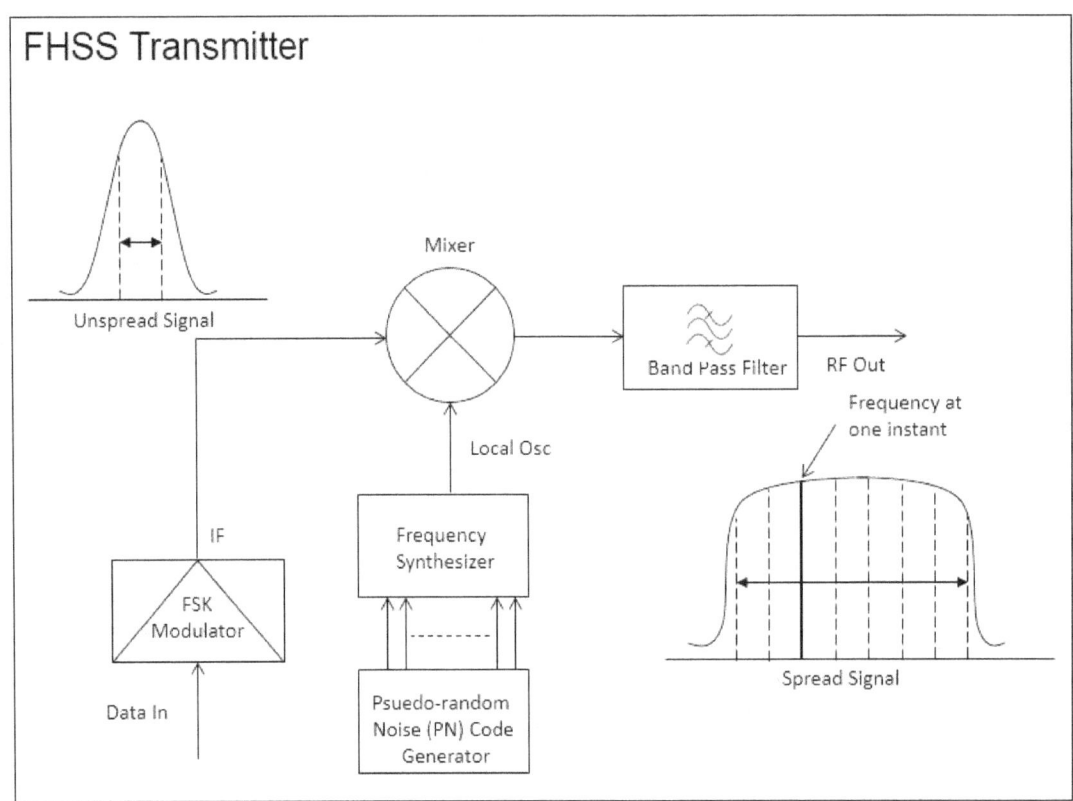

**Figure 18.4.2.b**

**Figure 18.4.2.c** shows the block diagram of a FHSS receiver. The receiver uses the same PN code sequence to hop frequencies back and forth in the frequency synthesizer. This hopping frequency output is multiplied with the incoming wideband spread signal in the mixer, to recover the original narrow band unspread FSK signal at the transmitter. The FSK signal is then demodulated to get the original data out.

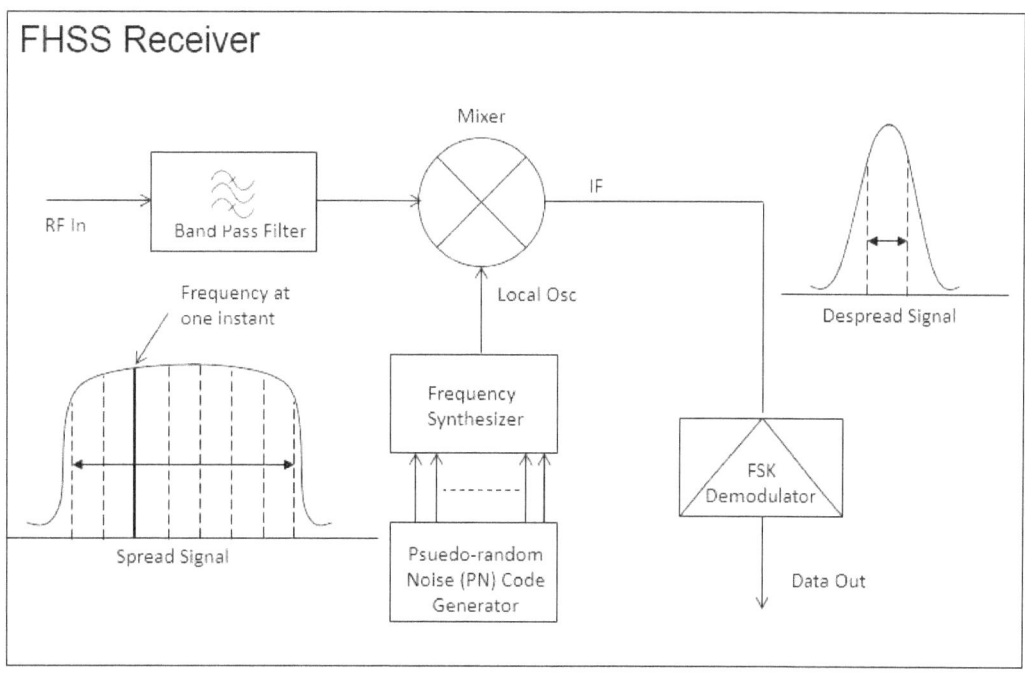

**Figure 18.4.2.c**

In DSSS technique, the PN sequence is applied directly to the incoming data. The PN sequence is applied to data symbols in the form of rectangular pulses of +1 or -1 amplitude as shown in **Figure 18.4.2.d**. Each pulse is called a *chip*. The *chip rate* is the number of *chips* required for each data symbol. In the example shown, there are 24 chips per data symbol.

The data symbols are multiplied with the PN code (sequence of 0s and 1s) chips to generate the wideband DSSS signal, as illustrated in **Figure 18.4.2.d**. The *spreading factor* in this case is the ratio of *chip rate* and the data *symbol rate*. In the example shown, the *spreading factor* is 24 and the *processing gain* is calculated to be 13.8 dB.

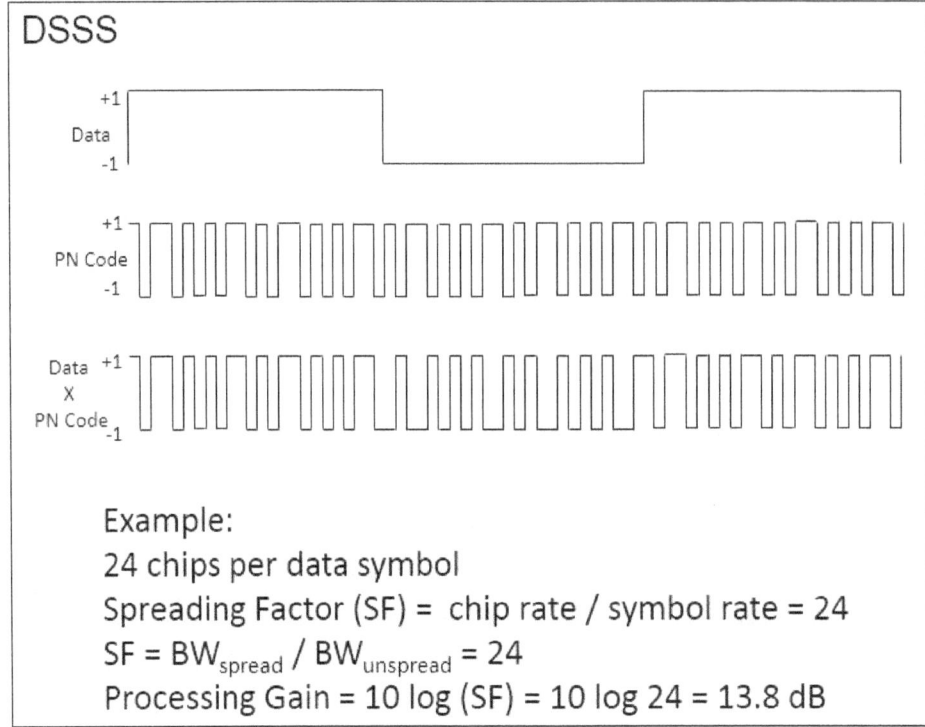

**Figure 18.4.2.d**

**Figure 18.4.2.e** illustrates a DSSS transmitter. The incoming data stream and the *chips* (pulses) from *pseudo random noise* (PN) code generator are fed to a product mixer where they are multiplied to generate a wideband spread signal. This raw spread signal is then fed to a QAM modulator to get the RF signal out for transmission. Notice how the unspread low data rate incoming data is transformed into a spread signal by multiplying it with a high chip rate PN code.

**Figure 18.4.2.f** illustrates a DSSS receiver. The same unique PN code that was used in the transmitter is used in the receiver to recover the original unspread data, by reversing the process. The QAM demodulator first recovers the raw spread signal, which is then multiplied with the PN code to recover the original data.

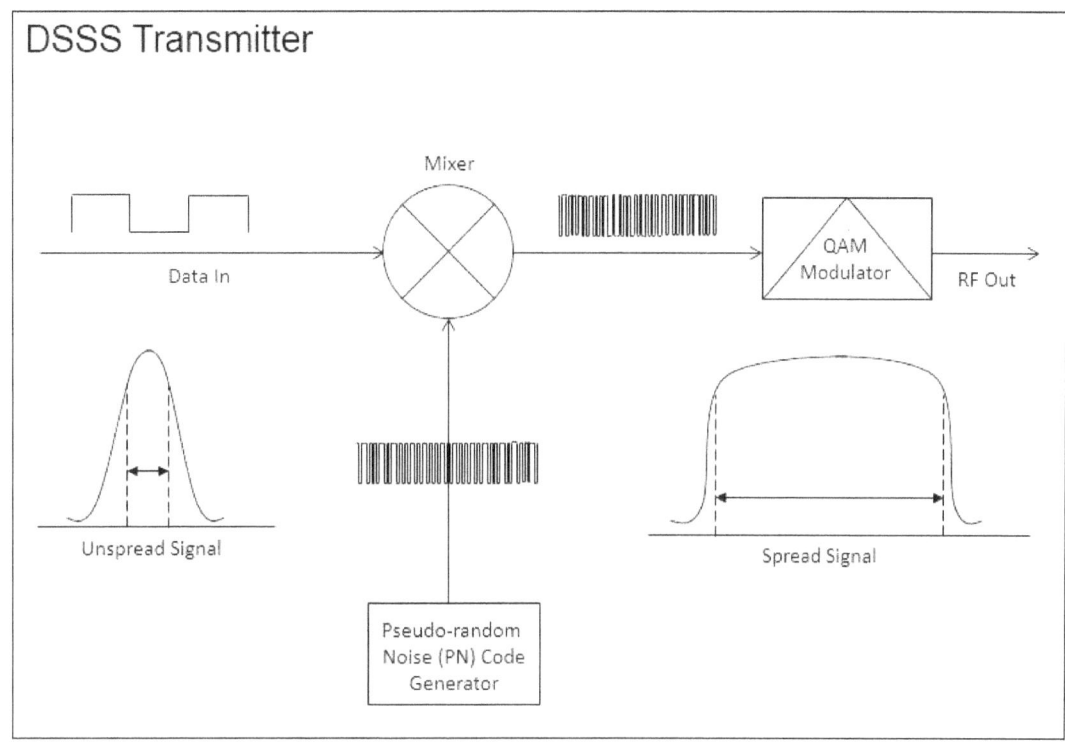

**Figure 18.4.2.e**

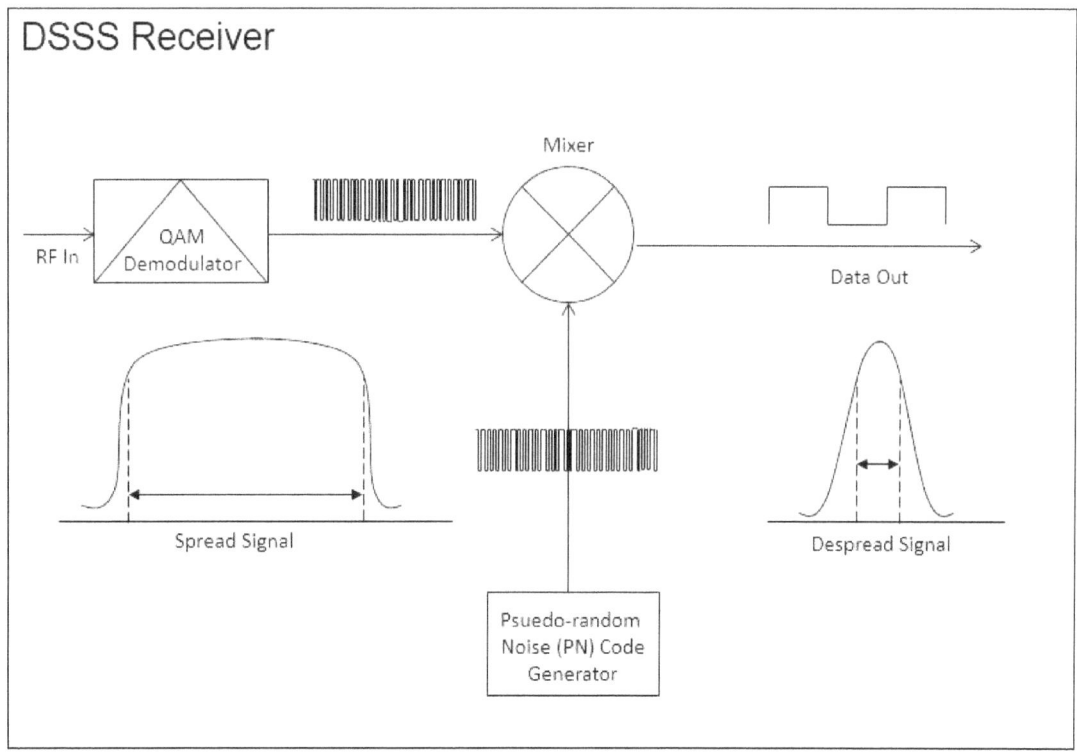

**Figure 18.4.2.f**

## 18.4.3    OFDM

The *orthogonal frequency division modulation* (OFDM) is a multi-carrier modulation technique. Instead of just one carrier frequency, it uses multiple *orthogonal* sub-carriers. The orthogonality or non-interference between sub carriers is maintained by keeping the spacing between sub-carriers in such a way, that the *peak* of a sub carrier coincides with the *null* of adjacent sub-carrier, as shown in **Figure 18.4.3.a**. This allows overlap between adjacent sub-carriers while avoiding interference.

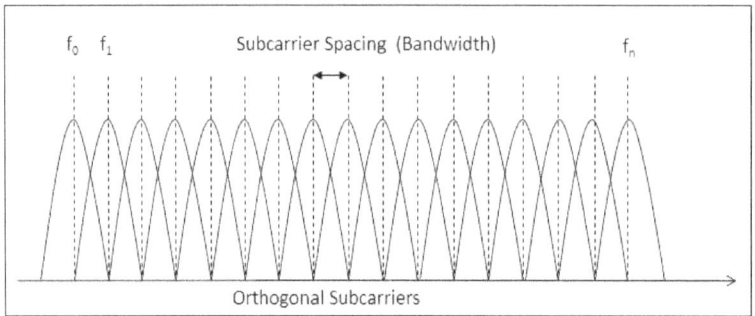

**Figure 18.4.3.a**

Each of the sub-carriers can independently carry a QAM modulated data stream. With multiple sub-carriers, several data streams can be transmitted in parallel. **Figure 18.4.3.b** shows how an incoming high rate data stream is converted into multiple low rate parallel data streams, with each parallel data stream carried by a single sub-carrier. In this way OFDM makes it possible to multiply the total throughput of the wireless channel, and it has high *spectral efficiency* (Mbps/Hz). This is one of the reasons that OFDM is the technique of choice for several wireless broadband technologies, such as IEEE 802.11 (WiFi), IEEE 802.16 (WiMax) and LTE.

The entire channel bandwidth of an OFDM wireless communication channel is divided into several independent sub-bands, with each sub-band centered on the sub-carrier frequency. All sub-bands are of equal width, and frequency response for each sub-band is relatively flat, compared to the frequency response of the entire channel. With a faded OFDM signal, each sub-band will suffer flat fading, to a lesser or greater extent, based on the fading characteristics of the wireless channel. This has practical implications, because it simplifies the design of *equalizer* at the receiver. It is much easier to equalize several flat faded parallel sub carriers at the receiver. If there were no sub-bands, we would have a wide bandwidth channel, with a non-flat fading response, requiring complex equalization at the receiver. See **Figure 18.4.3.c**.

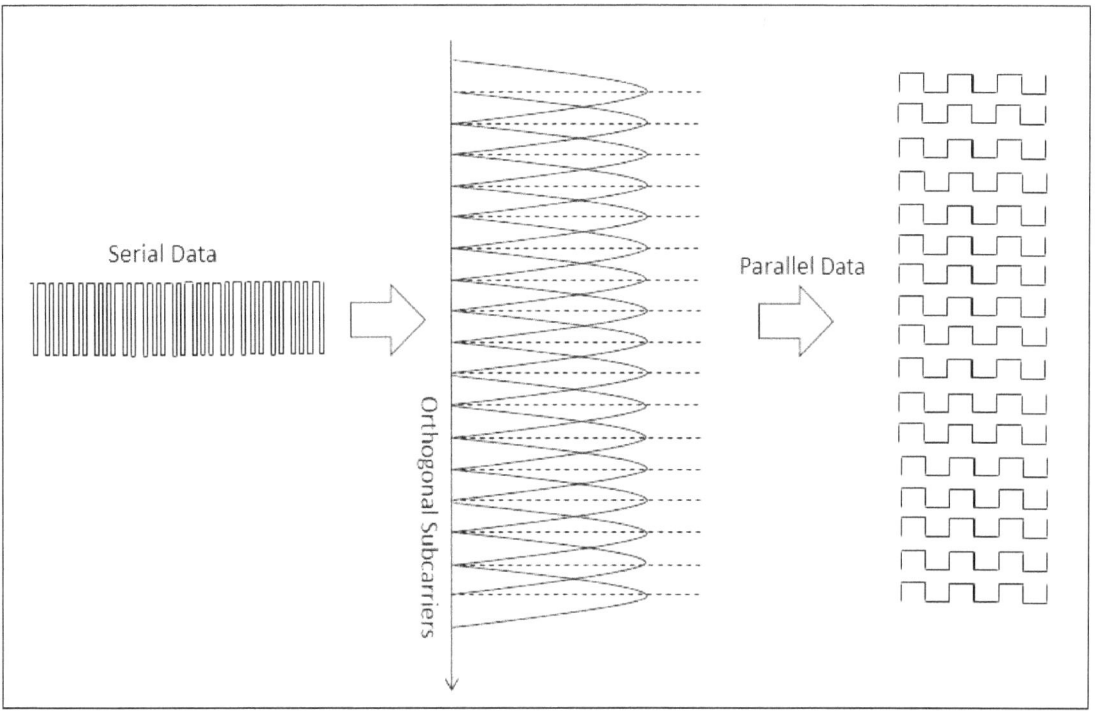

**Figure 18.4.3.b**

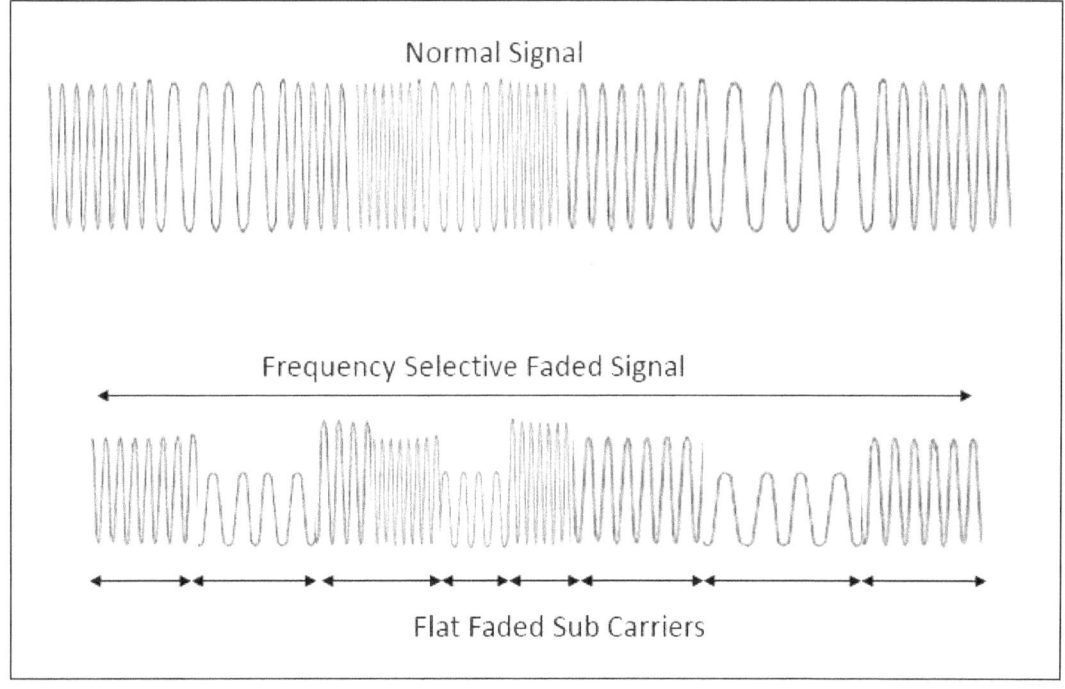

**Figure 18.4.3.c**

Let us for a moment revisit the problem of *inter symbol interference* (ISI) caused by multipath propagation of radio signals, that was earlier discussed in **Section 12.1**. Also, **Figure 18.4.3.d** shows how non-OFDM symbols received from different paths are superimposed and get corrupted, rendering them unintelligible. It is useful to understand the concept of *delay spread* of a wireless channel here, as it relates to different arrival times of symbols over different paths of propagation. The *delay spread* of a wireless channel is defined as the difference in arrival time of the earliest signal component (typ. LOS) and the latest signal component. In **Figure 18.4.3.d**, Path 3 is the shortest and Path 1 is the longest. Therefore, *delay spread* in this case is the difference in arrival time of signal over Path 3 and Path 1. Typical values for *delay spread* can be few hundreds of nanoseconds. For example, if *delay spread* of a channel is 300 ns, it means that multipath propagation causes the transmitted symbols to persist in channel memory for 300 ns.

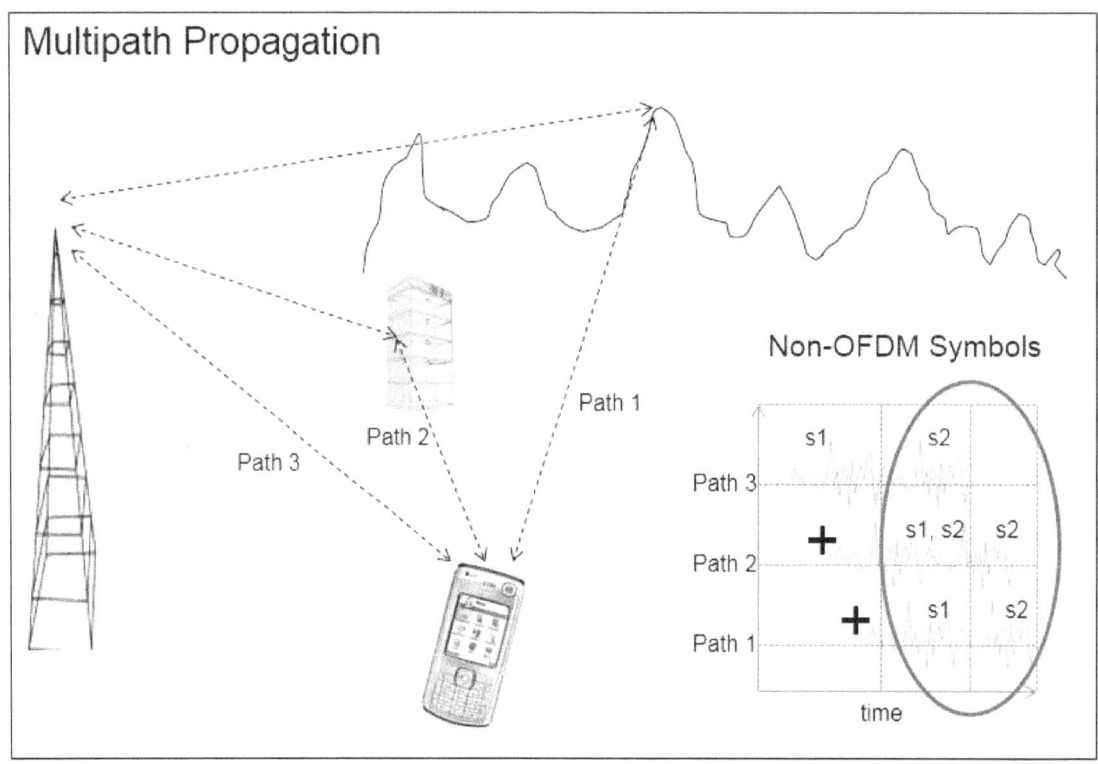

**Figure 18.4.3.d**

The OFDM symbols are transmitted at a comparatively lower symbol rate (because of parallelization of data) and the symbol duration is high. Therefore OFDM symbols can persist in channel memory for longer duration to accommodate the arrival of copies of a symbol over the longest paths of propagation. In other words, the next symbol will begin to be received only after all copies of the previous symbol have been received. In this way, the problem of ISI between consecutive symbols in the stream is avoided.

However, still, copies of symbols received over different paths can overlap causing corruption of the symbol. This is avoided by adding a *cyclic prefix* (CP) in front of each OFDM symbol, as shown in **Figure 18.4.3.e**. The CP is actually a copy of the head or tail of the OFDM symbol, and adds some redundancy to the information carried in the OFDM symbol. The CP provides a kind of *guard interval* between consecutive symbols, and its duration must be longer than the *delay spread* of wireless channel.

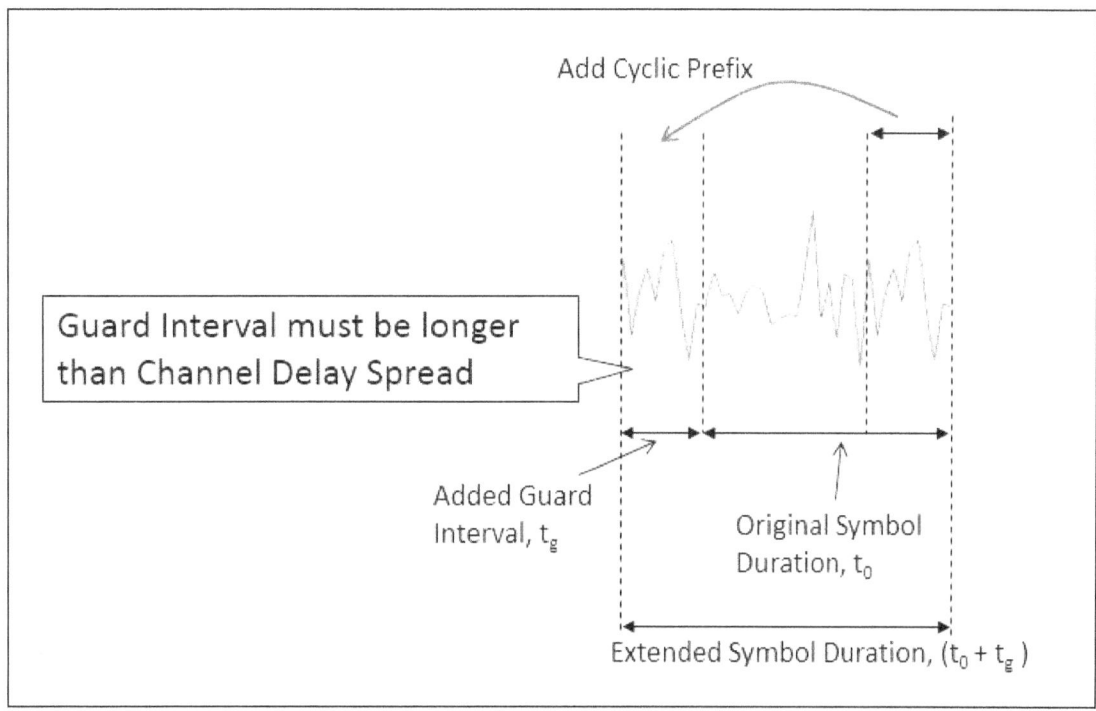

**Figure 18.4.3.e**

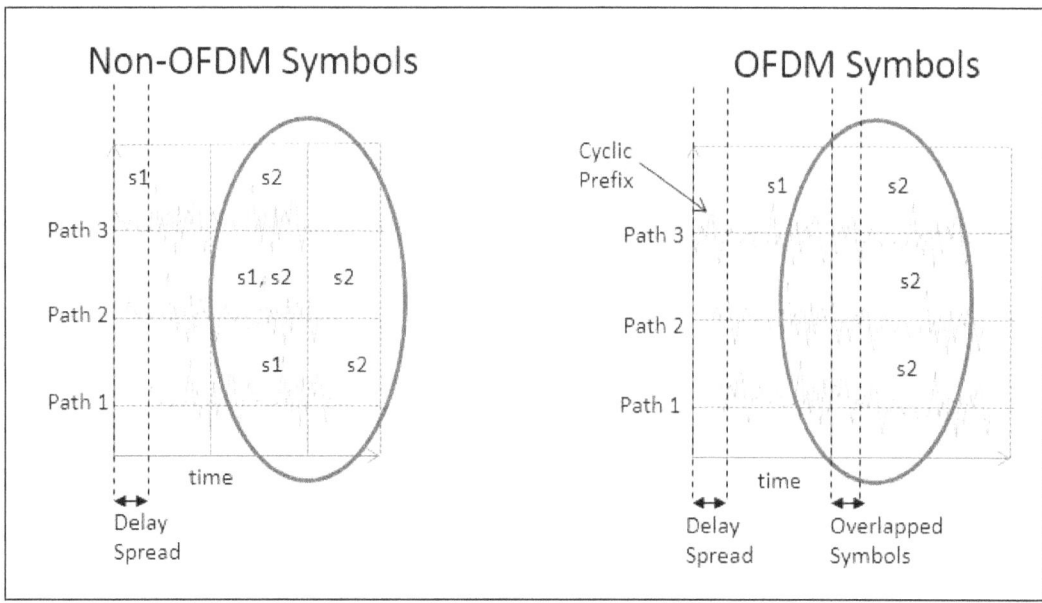

**Figure 18.4.3.f**

**Figure 18.4.3.f** shows a comparison of symbol overlap for non-OFDM and OFDM symbols. The CP in OFDM symbols contains redundant information, therefore, even if copies of symbols received over multiple paths overlap, the symbol corruption is minimized.

In this way, OFDM is actually resistant to effects of *multipath fading* and *inter symbol interference* (ISI).

However, implementing OFDM has its own share of challenges too. It requires tight synchronization between the transmitter and receiver. Loss of synchronization can cause *inter carrier interference* (ICI), causing the sub-carriers to lose orthogonality.

Since QAM is a digital modulation technique, it is highly susceptible to phase noise and modulation quality can be impaired. The challenge is how to maintain *error vector magnitude* (EVM) within an acceptable level.

The OFDM radio signal actually is a composite of hundreds of sub-carriers, and when all these sub-carriers are combined, it can cause high *peak to average power ratios* (PAPR) in RF amplifiers. The peaks in signal power can easily shoot into the non-linear region of the amplifier, causing distortion in the signal. Therefore, mechanisms are needed to limit or cancel the peaks in signal power, making the circuit designs a little more complex.

**Figure 18.4.3.g** shows a block diagram of OFDM transmitter. The incoming high rate data stream is passed through a serial to parallel converter, and converted to n parallel data streams. Each of these parallel streams is passed through a QAM modulator that generates the, I and Q sub-carrier frequency components. These frequency domain components, that contain the amplitude and phase information, are fed to an *inverse fast fourier transform* (IFFT) processor. The IFFT is a *digital signal processing* (DSP) algorithm that is used to convert the incoming frequency components into series of I and Q time samples.

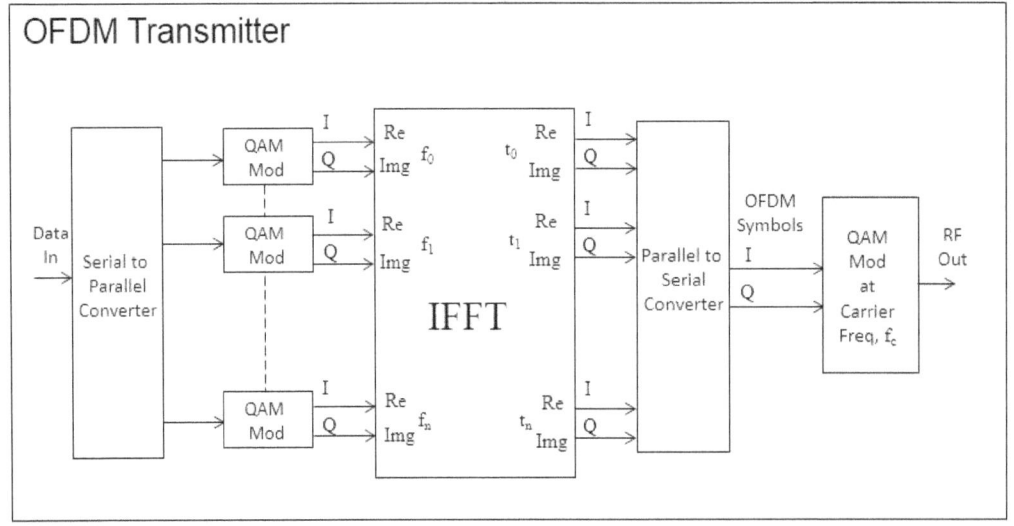

**Figure 18.4.3.g**

The time domain I and Q outputs from IFFT processor are serialized through a parallel to serial converter to generate a composite time domain signal, which is fed to a QAM modulator that operates at the carrier frequency to generate the RF output for transmission.

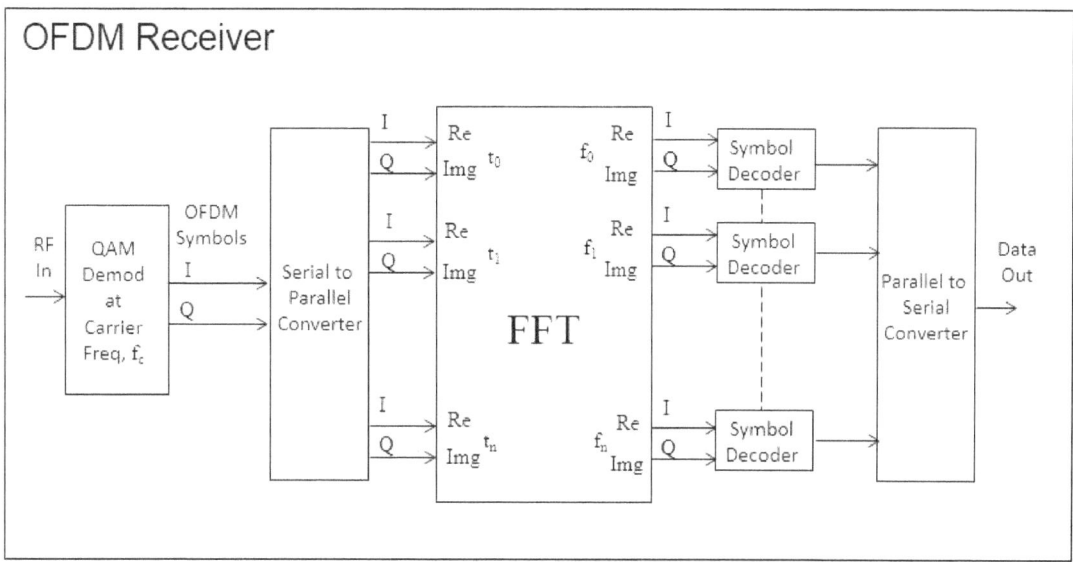

**Figure 18.4.3.h**

**Figure 18.4.3.h** shows the reverse process that is followed in the OFDM Receiver. The received time domain composite OFDM radio signal is first demodulated in a QAM demodulator, to recover the, I and Q components of the time domain serialized OFDM signal. The serialized signal is first passed through a serial to parallel converter to recover the, I and Q time domain components corresponding to each of the original sub-carriers. These time domain I and Q components are fed to a *fast fourier transform* (FFT) processor, which is implemented as a DSP algorithm. The output of the FFT processor are the, I and Q frequency components of the original parallel sub-carriers. The symbols from each parallel stream are decoded and serialized through a parallel to serial converter to recover the original high rate data stream.

# 18.5 Wireless LAN Receiver Example

**Figure 18.5.a** shows the modulation specifications for the IEEE 802.11 a/g WLAN standard. Notice that OFDM is the digital modulation technique used with different flavors of QAM for subcarrier modulation.

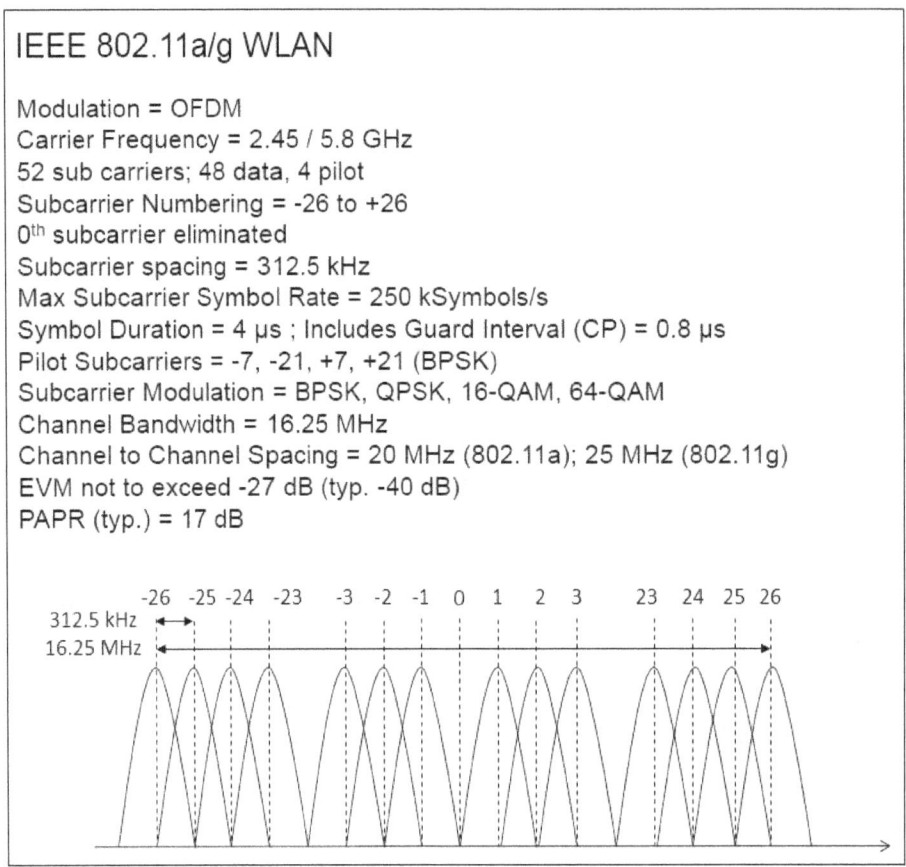

**IEEE 802.11a/g WLAN**

Modulation = OFDM
Carrier Frequency = 2.45 / 5.8 GHz
52 sub carriers; 48 data, 4 pilot
Subcarrier Numbering = -26 to +26
$0^{th}$ subcarrier eliminated
Subcarrier spacing = 312.5 kHz
Max Subcarrier Symbol Rate = 250 kSymbols/s
Symbol Duration = 4 µs ; Includes Guard Interval (CP) = 0.8 µs
Pilot Subcarriers = -7, -21, +7, +21 (BPSK)
Subcarrier Modulation = BPSK, QPSK, 16-QAM, 64-QAM
Channel Bandwidth = 16.25 MHz
Channel to Channel Spacing = 20 MHz (802.11a); 25 MHz (802.11g)
EVM not to exceed -27 dB (typ. -40 dB)
PAPR (typ.) = 17 dB

**Figure 18.5.a**

**Table 18.5.b** shows how with increasing data rates, higher levels of digital modulation are required. The highest data rate of 54 Mbps is achieved with 64-QAM.

The *forward error correction* (FEC) code rate, k/n, means that for every k bits of useful data, n bits are actually transmitted, and (n-k) is the redundant number of bits transmitted These redundant bits are used to correct errors in the received data. For example for 54 Mbps data rate, the code rate is ¾, which means for every 3 bits of data, 4 bits are actually transmitted with 1 redundant bit.

**Figure 18.5.c** shows an example to calculate the data rate for a given flavor of QAM used for subcarrier modulation.

In the next few sections of this chapter, we will study few examples of WLAN receiver architectures and explain the relative merits and demerits of each architecture.

| Data Rate (Mbps) | Modulation | FEC Code Rate (k / n) | Sensitivity Requirement (dBm) |
|---|---|---|---|
| 6 | BPSK | 1 / 2 | − 82 |
| 9 | BPSK | 3 /4 | − 81 |
| 12 | QPSK | 1 / 2 | − 79 |
| 18 | QPSK | 3 / 4 | − 77 |
| 24 | 16-QAM | 1 / 2 | − 74 |
| 36 | 16-QAM | 3 / 4 | − 70 |
| 48 | 16-QAM | 2 / 3 | − 66 |
| 54 | 64-QAM | 3 / 4 | − 65 |

**Table 18.5.b**

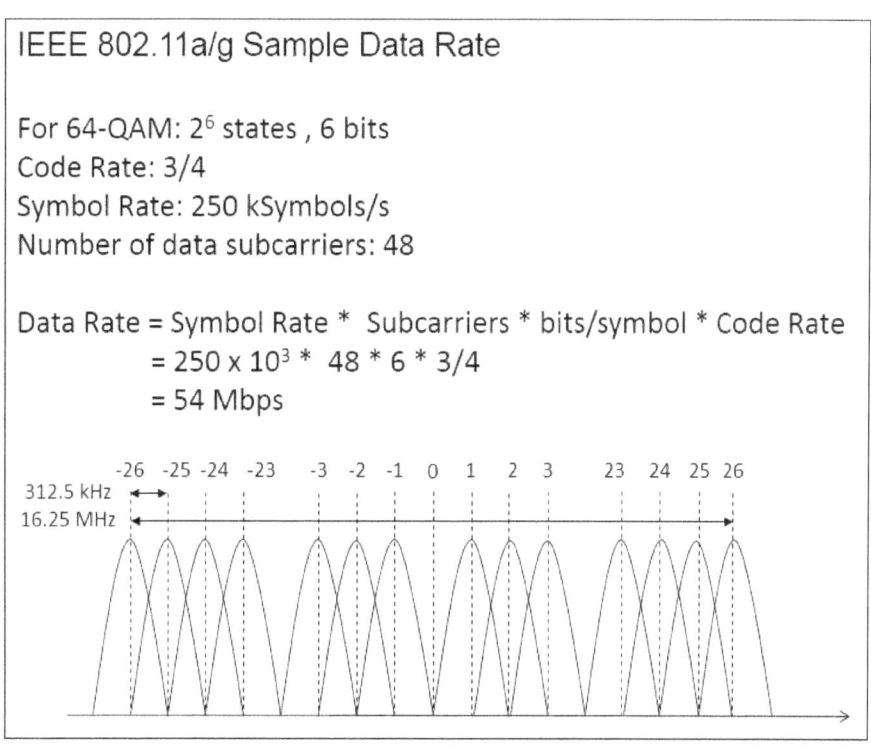

**Figure 18.5.c**

## 18.5.1    Superheterodyne Receiver

**Figure 18.5.1.a** shows a *superheterodyne* receiver architecture for WLAN receiver. For the basics of *superheterodyne* receivers, please refer Section **18.3.1**. Notice that the band pass filter at the front end is a RF ceramic filter, which offers space and cost saving features, as well as superior performance, compared to air-cavity filters. The *low noise amplifier* (LNA) amplifies the signal at carrier frequency before it is fed to the *image reject* filter. The local oscillators for the RF as well as the IF use *phase locked loop* (PLL), to generate stable frequency output. The RF PLL output is fed to the mixer along with the incoming band of carrier frequency. The product mixer generates an output of *intermediate frequency* (IF), that is passed through a band pass IF *surface acoustic wave* (SAW) filter. The SAW filters have many advantages, such as, reduced size and weight, high reliability and ruggedness, no need for tuning or readjustment, and capable of mass production.

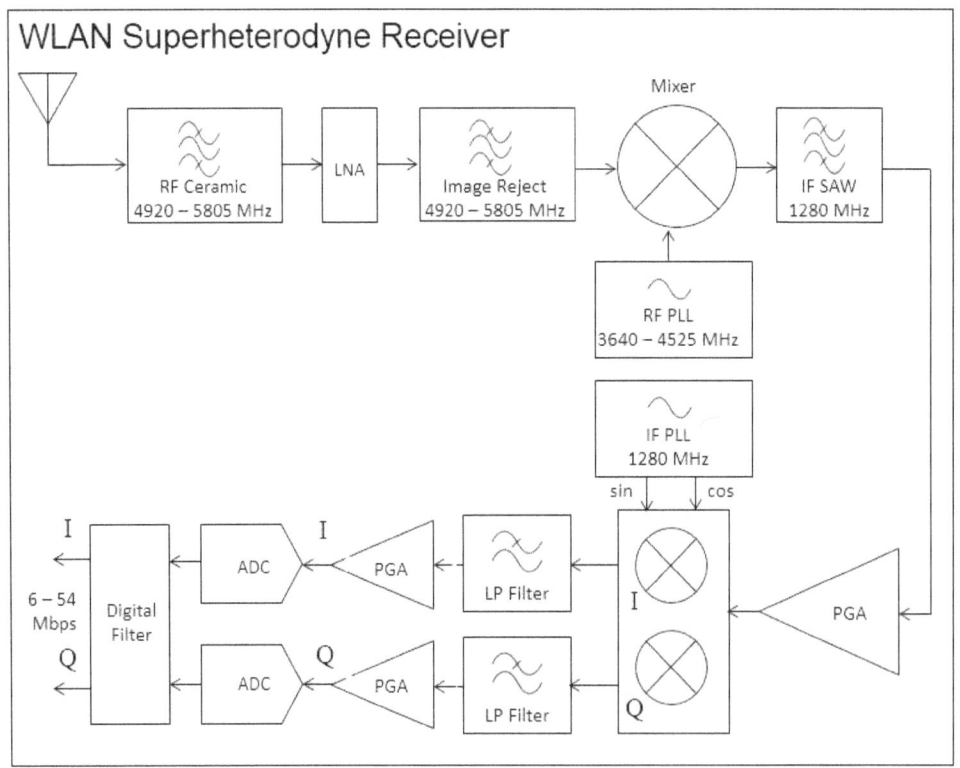

**Figure 18.5.1.a**

The output from IF SAW filter is fed to a *programmable gain amplifier* (PGA). The gain of a PGA can be externally controlled by a digital or analog signal. The output of IF PGA is then fed to the QAM I/Q demodulator to recover the original time domain OFDM composite signal, which consists of time domain samples of amplitude and phase components of 52 sub carriers. The IF PLL oscillator output is used to demodulate the baseband I and Q signal components. After passing through the low pass (LP) filter, the I and Q baseband signal components are first amplified through PGAs and then converted through *analog to digital* (ADC) converter.

Thereafter a digital filter provides the output data stream for I and Q baseband components, which is ready for feeding into the OFDM receiver, in order to recover the parallel data streams from each of the sub carriers. (See **Section 18.4.3**).

The *superheterodyne* receiver architecture for WLAN provides high gain with improved stability. There are no DC offsets and there is better I/Q matching at the down converted *intermediate frequency* (IF). However, *superheterodyne* WLAN receivers tend to be large, expensive, require many external discrete components and require multiple IF filters for *multimode* applications, for example a *multimode* handset is required to work on WLAN, GSM as well as CDMA technologies.

# 18.5.2    Low IF Receiver

**Figure 18.5.2.a** shows a *low IF* receiver architecture. In this architecture, the IF is reduced to 10 MHz and the I and Q carrier components are separated out at the front end itself. This eliminates the need for IF SAW filter and IF PLL. There are no DC offsets too. This architecture has found use in narrowband applications, such as Bluetooth and GSM.

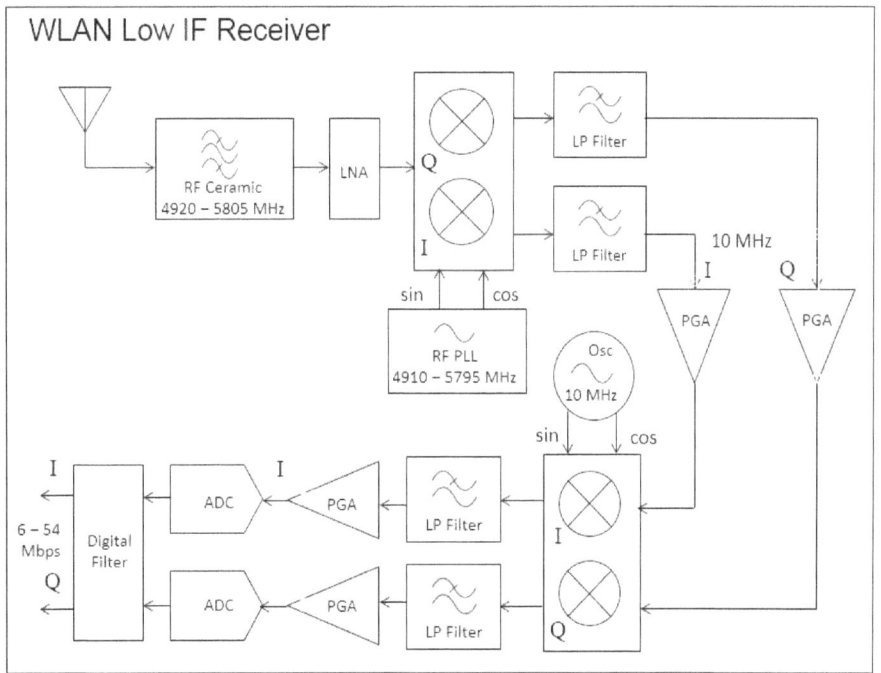

**Figure 18.5.2.a**

On the down side, it is difficult to implement I/Q RF down converters at the front end, especially at 2.4/5.8 GHz, and a second I/Q converter at 10 MHz IF needs to be added. The architecture requires several discrete external components and is not ideal for broadband applications, such as WLAN.

## 18.5.3    Direct Conversion Receiver

**Figure 18.5.3.a** shows a *direct conversion* receiver for WLAN. For the basics of direct conversion, see **Section 18.3.3**.

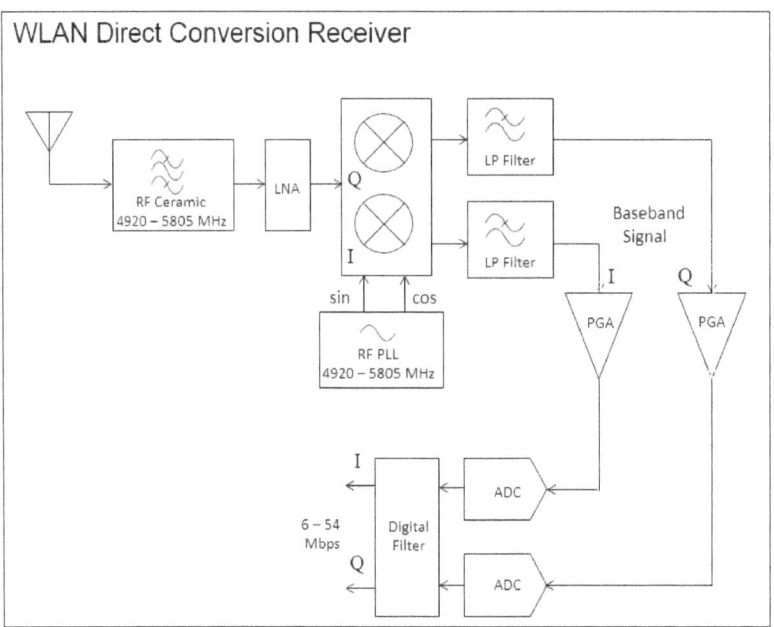

**Figure 18.5.3.a**

As shown in the figure, there is no conversion of the incoming RF signal to intermediate frequency (IF), so the architecture of the receiver is greatly simplified. The frequency of the RF PLL local oscillator is exactly the same as that of the incoming RF carrier, therefore, the I and Q baseband signal components can be directly recovered at the output of I/Q down converter at the front end. Thereafter the I and Q baseband signal components are amplified and converted from analog to digital, before sending them to the OFDM receiver in order to recover the parallel data streams from each of the sub carriers. (See **Section 18.4.3**).

The implementation of *direct conversion* receivers has its own challenges. The I/Q RF down converter is difficult to implement at 2.4/5.8 GHz. Also, the local oscillator frequency is the same as the incoming RF carrier, so it can easily couple with the antenna input, enter the mixer and self-mix with itself creating a DC offset signal that could be large enough to overload the baseband amplifiers and therefore distort the baseband signal. However, despite these challenges, the benefits of lower cost, simple architecture, fewer components and low power consumption, far outweigh the trouble it takes to design and implement *direct conversion* receivers.

# 18.6 Software Defined Radio

In this section, we will introduce ourselves to the *software defined radio* (SDR). In a typical SDR based radio receiver, some or all functions are implemented in software by using DSP algorithms. The radio functions which can be implemented in software are, *digital filters* (FIR, IIR), *automatic notch filters, adaptive equalizers, noise cancellers, AGC, digital modulators* and *demodulators, direct digital synthesizers* (DDS) or *local oscillators* (LO).

The hardware design of a typical SDR is partitioned between *digital signal processors* (DSP), *application specific integrated circuits* (ASICS) and *field programmable gate arrays* (FPGA).

A full featured SDR based radio receiver lends itself to reconfigure its software to implement new features and functionality. The software in a SDR based receiver can be reconfigured using a smart card, or with *over the air* (OTA) reconfiguration, or with dynamic OTA reconfiguration during a voice or video call or a user data session.

However, there are five distinct levels of evolution from the conventional radio receiver to the full featured SDR based radio receiver. These are Level 0 to Level 4. We will now study each of these levels of SDR evolution.

## 18.6.1    Level 0

The *Level 0* corresponds to the conventional fully hardware based radio receiver as shown in **Figure 18.6.1.a**.

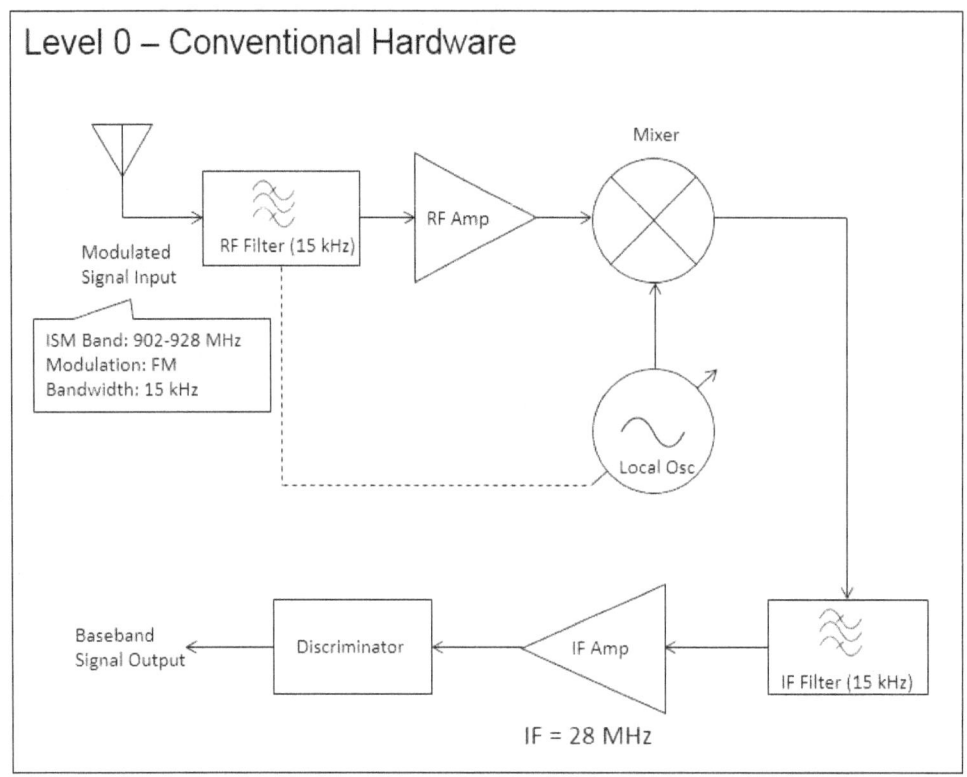

Level 0 – Conventional Hardware

Mixer

Modulated
Signal Input

RF Filter (15 kHz)

RF Amp

Local Osc

ISM Band: 902-928 MHz
Modulation: FM
Bandwidth: 15 kHz

Baseband
Signal Output

Discriminator

IF Amp

IF Filter (15 kHz)

IF = 28 MHz

**Figure 18.6.1.a**

## 18.6.2    Level 1

In *Level 1*, the radio functions are still implemented in hardware, however, the man machine or user interface and the control functions are implemented in software, as shown in **Figure 18.6.2.a**. Therefore, this design is also called *software controlled radio* (SCR).

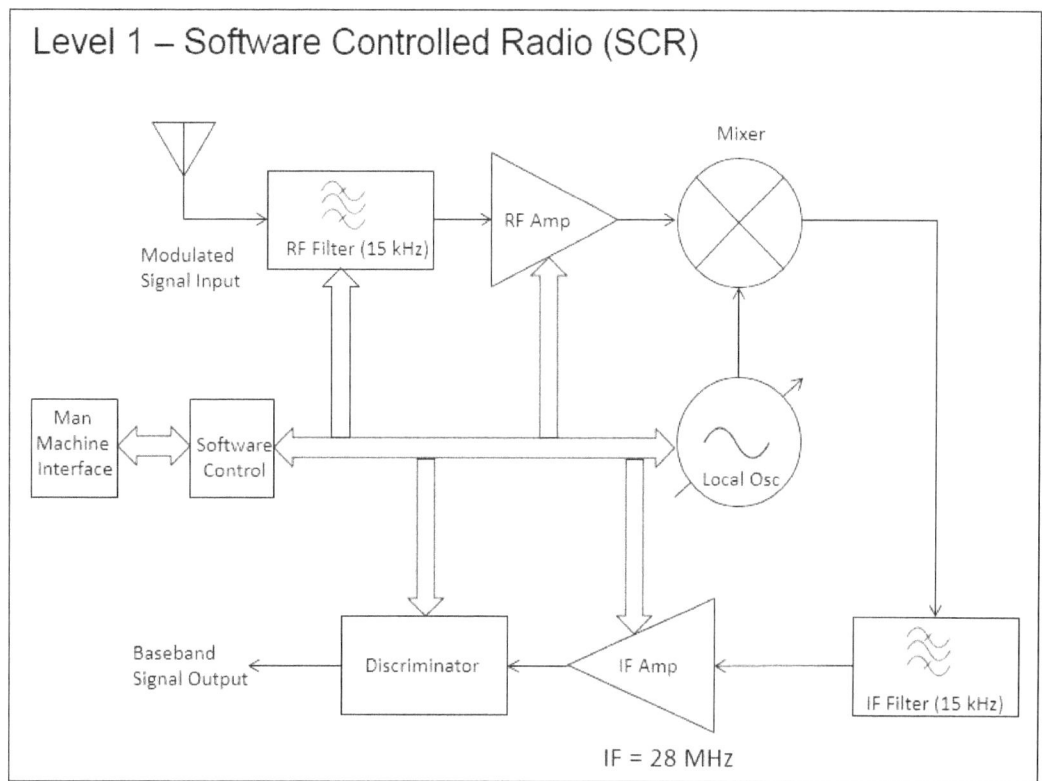

**Figure 18.6.2.a**

## 18.6.3    Level 2

In *Level 2*, the radio functions are still in hardware, however, the baseband signal processing, demodulation, user interface and control are implemented in software, as shown in **Figure 18.6.3.a**. This design is truly called a *software defined radio* (SDR).

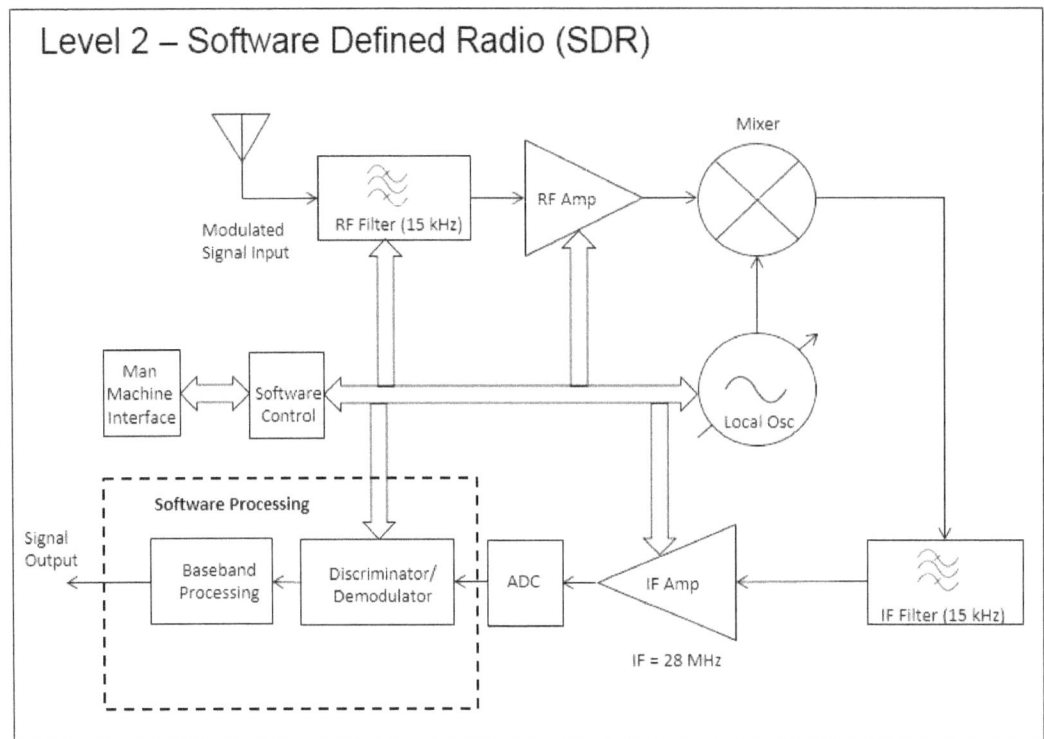

**Figure 18.6.3.a**

## 18.6.4   Level 3

In *Level 3*, except for the front end *analog to digital converter* (ADC), all other functions are implemented in software, as shown in **Figure 18.6.4.a**. This design is called *ideal software radio* (ISR).

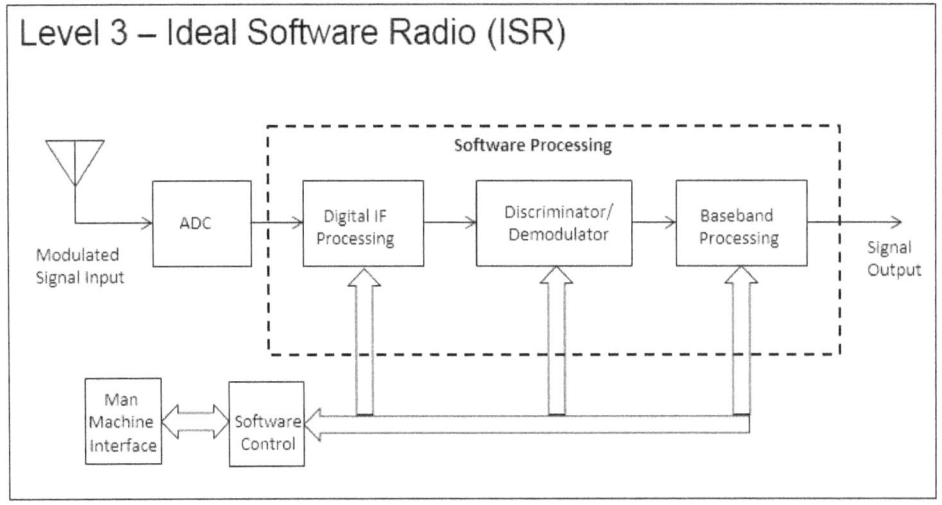

**Figure 18.6.4.a**

## 18.6.5 Level 4

The *Level 4* is similar to *Level 3* for the fact that except for front end ADC, all other functions are implemented in software. However, the similarity ends there. The *Level 4* design allows the software to support multiple wireless communication standards, such as CDMA, GSM, LTE or WiFi, or any new emerging standard. This is the ideal design for a multimode smartphone, and therefore this design is called the *ultimate software radio* (USR).

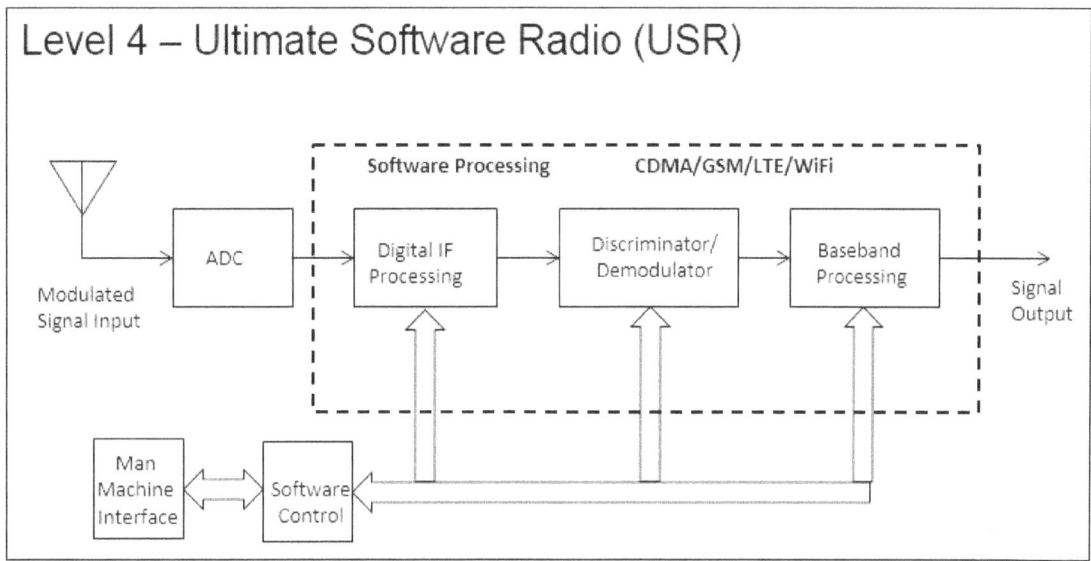

**Figure 18.6.5.a**

## 18.6.6 SDR Benefits and Challenges

The SDRs offer several benefits. The software can be easily upgraded without changing hardware to implement new functions or new communication protocols. It is much easier to customize the software in a SDR for a specific application or customer. Moreover SDRs offer the ability to dynamically reconfigure *over the air* (OTA).

The reconfiguration capability in SDRs can be used to bypass failed hardware and build hardware redundancy in the receiver to eliminate single points of failure. The access network can be remotely reconfigured to suit changed network traffic conditions, by dynamically reconfiguring the software in base stations, access points, NodeBs or eNodeBs. This helps to reduce physical visits to cell sites and therefore increase cost efficiency.

Another important benefit of SDRs is, that any modulation or demodulation scheme can be implemented in software, which includes, *amplitude modulation* (AM), *frequency modulation* (FM) and *quadrature amplitude modulation* (QAM), or any new modulation technique that may be invented in future.

However, SDRs are not free from challenges. The biggest challenge is secure download and reconfiguration of software. It is of utmost importance to secure SDR based receivers from malicious software that might be inadvertently downloaded to impair their functionality.

Another challenge is processing power available on the radio receivers. New communication algorithms and increasing demand for features on multimode smartphones can easily stretch the demand for processing power, therefore there is plenty of scope to invent novel methods to optimize the use of available processing power in a wireless receiver.

# Acronyms

| | |
|---|---|
| 1SM | 1 Slope Model |
| 3GPP | 3$^{rd}$ Generation Partnership Project |
| A/D | Analog to Digital |
| ADC | Analog to Digital Converter |
| AGC | Automatic Gain Control |
| AI | Aperture Integration |
| AM | Amplitude Modulation |
| AP | Access Point |
| AR | Axial Ratio |
| ASIC | Application Specific Integrated Circuit |
| AUT | Antenna Under Test |
| AWGN | Additive White Gaussian Noise |
| Balun | Balanced Unbalanced |
| BER | Bit Error Rate |
| BPSK | Binary Phase Shift Keying |
| BS | Base Station |
| BTS | Base Transceiver Station |
| CCI | Co Channel Interference |
| CDMA | Code Division Multiple Access |
| CGM | Conjugate Gradient Method |
| CIR | Carrier to Interference Ratio |
| CMA | Constant Modulus Algorithm |
| COST | European Cooperation in Science and Technology |
| CP | Circular Polarization, Cyclic Prefix |
| CQI | Channel Quality Indicator |
| D/A | Digital to Analog |
| DAC | Digital to Analog Converter |
| DDS | Direct Digital Synthesizer |

| | |
|---|---|
| DSP | Digital Signal Processor |
| DSSS | Direct Sequence Spread Spectrum |
| EHF | Extremely High Frequency |
| EIRP | Effective Isotropic Radiated Power |
| ELF | Extremely Low Frequency |
| EM | Electro Magnetic |
| EMI | Electro Magnetic Interference |
| EVM | Error Vector Magnitude |
| eNodeB | Evolved Node B |
| ERP | Effective Radiated Power |
| FDD | Frequency Division Duplexing |
| FDM | Frequency Division Multiplexing |
| FDTD | Finite Difference Time Domain |
| FEC | Forward Error Correction |
| FEM | Finite Element Method |
| FFR | Fractional Frequency Reuse |
| FFT | Fast Fourier Transform |
| FHSS | Frequency Hopping Spread Spectrum |
| FIR | Finite Impulse Response |
| FM | Frequency Modulation |
| FPGA | Field Programmable Gate Array |
| FRF | Frequency Reuse Factor |
| FSK | Frequency Shift Keying |
| GSM | Global System for Mobile Communications |
| GTD | Geometrical Theory of Diffraction |
| HF | High Frequency |
| HPBW | Half Power Beam Width |
| Hz | Hertz |
| ICEPAC | Ionosphere Communications Extrapolation of Polar Absorption Condition |
| ICI | Inter Carrier or Cell Interference |

| | |
|---|---|
| ICIC | Inter Cell Interference Coordination |
| IEEE | Institute for Electrical and Electronics Engineers |
| IF | Intermediate Frequency |
| IFFT | Inverse Fast Fourier Transform |
| IIR | Infinite Impulse Response |
| INTELSAT | International Telecommunications Satellite Organization |
| ISI | Inter Symbol Interference |
| ISR | Ideal Software Radio |
| ITU | International Telecommunication Union |
| ITU | International Telecommunication Union |
| LAM | Linear Attenuation Model |
| LF | Low Frequency |
| LHCP | Left Hand Circular Polarization |
| LHEP | Left Hand Elliptical Polarization |
| LMS | Least Mean Squares |
| LNA | Low Noise Amplifier |
| LO | Local Oscillator |
| LOS | Line of Sight |
| LP | Linear Polarization, Low Pass |
| LPDA | Log Periodic Dipole Array |
| LTE | Long Term Evolution |
| LTE-A | LTE_Advanced |
| LUF | Lowest Usable Frequency |
| MIMO | Multiple Input Multiple Output |
| MISO | Multiple Input Single Output |
| MF | Medium Frequency |
| MFSK | Multilevel FSK |
| MoM | Method of Moments |
| MUF | Maximum Usable Frequency |
| MU-MIMO | Multi User-MIMO |

| | |
|---|---|
| MWM | Multi Wall Model |
| NF | Noise Figure |
| NLOS | Near Line of Sight |
| OATS | Open Area Test Site |
| OF | Operating Frequency |
| OFDM | Orthogonal Frequency Division Multiplexing |
| OTA | Over The Air |
| PCB | Printed Circuit Board |
| PCS | Personal Communication Services |
| PGA | Programmable Gain Amplifier |
| PI | Position Information |
| PIFA | Planar Inverted F Antenna |
| PL | Path Loss |
| PLL | Phase Locked Loop |
| PM | Phase Modulation |
| PN | Pseudo random Noise |
| PO | Physical Optics |
| QAM | Quadrature Amplitude Modulation |
| QoS | Quality of Service |
| QPSK | Quadrature Phase Shift Keying |
| RADAR | Radio Detection And Ranging |
| REC | RECommendation |
| RET | Remote Electrical Tilt |
| RF | Radio Frequency |
| RHCP | Right Hand Circular Polarization |
| RHEP | Right Hand Elliptical Polarization |
| RL | Return Loss |
| RLS | Recursive Least Squares |
| Rx | Receive |
| SAW | Surface Acoustic Wave |

| | |
|---|---|
| SDM | Space Division Multiplexing |
| SCR | Software Controlled Radio |
| SDR | Software Defined Radio |
| SFDR | Spurious Frequency Dynamic Range |
| SFR | Soft Frequency Reuse |
| SHF | Super High Frequency |
| SID | Sudden Ionosphere Disturbance |
| SINR | Signal to Interference and Noise Ratio |
| SIR | Signal to Interference Ratio |
| SIMO | Single Input Multiple Output |
| SISO | Single Input Single Output |
| SMI | Sample Matrix Inversion |
| SNR | Signal to Noise Ratio |
| SU-MIMO | Single User-MIMO |
| SWAP | Size, Weight and Power |
| TDD | Time Division Duplexing |
| TDM | Time Division Multiplexing |
| THz | Tera Hertz |
| TV | Tele Vision |
| Tx | Transmit |
| UE | User Equipment |
| UHF | Ultra High Frequency |
| ULF | Ultra Low Frequency |
| UMTS | Universal Mobile Telecommunication System |
| USR | Ultimate Software Radio |
| VHF | Very High Frequency |
| VLF | Very Low Frequency |
| VNA | Vector Network Analyzer |
| VOACAP | Voice of America Coverage Analysis Program |
| WCP | Wireless Communications Professional |

| | |
|---|---|
| WCET | Wireless Communication Engineering Technologies |
| WEBOK | Wireless Engineering Body OF Knowledge |
| WiFi | Wireless Fidelity |
| WiGig | Wireless Gigabit Alliance |
| WiMax | Worldwide Interoperability for Microwave Access |
| WLAN | Wireless Local Area Network |

# Bibliography

1. The ARRL Handbook for Radio Communications, 2010
2. The ARRL Antenna Book, 21st Edition
3. Reference Manual for Telecommunications Engineering- Freeman-Wiley-Interscience
4. A Guide to the Wireless Engineering Body of Knowledge (WEBOK) - IEEE
5. Electronics Engineers Handbook- Fink and Christiansen – McGraw Hill
6. Electrical Engineering Reference Manual – Camara – PPI
7. Telecommunication System Engineering – Freeman – Wiley-Interscience
8. Telecommunication Transmission Handbook-Freeman-Wiley Interscience
9. Monopulse Radar -Leonov-Fomichev-Artech House, Inc.
10. Helical and Spiral Antennas-A Numerical Approach-Research Studies Press
11. Modern Radar Techniques-Scanlan-McMillan
12. Telecommunications Transmission Engineering- Volume 2- Facilities- Bellcore
13. Mobile Radio Communications-Steele-IEEE Press
14. Introduction to Radar Systems-Skolink-McGraw Hill
15. Fractal Element Antennas : A compilation of configuration with novel characteristics-Gianvittorio and Rahmat –Samii-IEEE
16. Properties of Phased Arrays-Aulock-Proceedings of the IRE
17. A Plane Polar Approach for Far-Field Construction from Near-Field Measurements-Rahmat-Samii & Galindo-Israel- IEEE Transactions on Antennas and Propagation, Vol. AP- 28, NO.2, March 1980
18. Aperture Integration and GTD Techniques used in the NEC Reflector Antenna Code-Lee-IEEE Transactions on Antennas and Propagations, Vol. AP-33, NO.2, Feb 1985
19. The Art and Engineering of Modern Antenna Near-Field Measurements-Rahmat-Samii-Farhad-Razavi-
20. International Microwave and Opto Electronics Conference-2009
21. Improved Closed Form Solution For Center Fed Straight Wire Antennas-2010-14th International Symposium on Antenna Technology and Applied Electromagnetics (ANTEM) and the American Electromagnetics Conference (AMEREM)
22. Pseudo Closed Form Solution for Center fed Monopole and Dipole Antenna-Shams, Sharaf, Nafe & - 28th National Radio Science Conference- (NRSC 2011)
23. Two-Branch Space and Polarization Diversity Schemes For Dipoles-Kar & Wahid-IEEE-2001
24. Space, Polarization and Pattern Diversity for Wireless Hand-held Terminals-Dietrich, Dietze, Niely & Stutzman-IEEE Transactions on Antennas and Propagation, Vol. 49, NO.9, Sept 2001
25. Time and Antenna Diversity in Wireless Sensor and Actuator Networks-Nethi, Jantti & Nassi-Proceedings of the 2009 IEEE 9th Malaysian Conference on Communications-December 2009
26. Characterization of NLOS Wireless Propagation Channels with a Proper Coherence Time Value in a Continuous Mode Stirred Reverberating Chamber-Sorrentino,Ferrara & Migliaccio-Proceedings of the 2nd European Wireless Technology Conference
27. The effective coherence time of common channel models-Bergel & Benedetto-SPAWC 2010

28. A Polarization Diversity Antenna by Printed Dipole and Patch with a Hole-Michishita-IEEE 2001

29. Polarization Diversity for Indoor Cellular and PCS CDMA Reception-Vargas,Victor & Baker-IEEE 1997

30. Electromagnetic Analysis of Effective and Apparent Diversity Gain of Two Parallel Dipoles-Kildal & Rosengren-IEEE Antennas and Wireless Propagation Letters, Vol 2, 2003

31. An Efficient Frequency Reuse Scheme by Cell Sectorization in OFDMA Based Wireless Networks-Boustani,Khorsandi,Danesfahani & MirMotahhary- 2009 4th International Conference on Computer Sciences and Convergence Information Technology

32. A Frequency Reuse Scheme for GPRS Network-Gitlits, Ghanaati-Nazari & Yin- 2005 Asia-Pacific Conference on Communications, 2005

33. A Universal Frequency Reuse System in a Mobile Cellular Environment-Kim & Oh-IEEE 2007

34. Performance Analysis of a CDMA/FDMA Cellular Communication System with Cell Splitting-Hamidian & Payne-IEEE 1997

35. Co-channel Interference Reduction on the Forward Channel of a Wideband CDMA Cellular System-Mayer,Robertson & Ha- IEEE

36. Frequency Reuse in Mobile WiMAX and LTE Networks with Sectored Cells-Sari,Sezginer & Vivier-2009 IEEE Mobile WiMAX Symposium

37. Inter-cell Interference Coordination Through Adaptive Soft Frequency Reuse in LTE Networks- Qian,Hardjawana, Li, Vucetic, Shi & Yang- 2012 IEEE Wireless Communications and Networking Conference: MAC and Cross-Layer Design

38. Generalized Frequency Reuse Schemes for OFDMA Networks: Optimization and Comparison-Chen & Yuan-IEEE

39. Femtocell Frequency Planning Scheme in Cellular Networks Based on Soft Frequency Reuse-Jeong,Lee, Chung, Lee & Choo-2010 International Conference on Cyber-Enabled Computing and Knowledge Discovery

40. An Introduction to MU-MIMO Downlink - IEEE Communications Magazine, October 2004

41. MIMO-OFDM based air interface- IEEE Communications Magazine, January 2005

42. Downlink MIMO in LTE-A- IEEE Communications Magazine, February 2012

43. Understanding IEEE 802.11n amendment- IEEE Circuits and Systems Magazine 1Q 2008

44. Advancement of MIMO in WiMax- IEEE Communications Magazine June 2009

45. MIMO in WiMax and LTE- IEEE Communications Magazine May 2010

46. MIMO-OFDM Wireless Systems- IEEE Wireless Communications August 2006

47. Antennas for WiFi Connectivity- Proceedings of the IEEE July 2012

48. Overview of Mobile WiMax – Technology and Evolution- IEEE Communications Magazine October 2008

49. Radio Network Planning and Optimization for UMTS-Second Edition-Jaana Laiho-Achim Wacker-Tomas Novosad

50. The Design of a Radio Link for Indoor Wireless Communications at 29 GHz- G. Kalivas, M El-Tanany, S. Mahmoud-IEEE

51. Efficient Design Techniques for Wireless Backhaul Networks- Pakorn Leesutthipornchai, Naruemon Wattanapongsakorn+ and Chalermpol Charnsripinyo- 2008-IEEE

52. Microwave Link Budget Analysis-S. Loyka, A. Kouki-IEEE Antennas and Propagation Magazine. Vol. 43. No. 5. October 2001

53. Transactions on Vehicular Technology, Vol. 47, No. 4, November 1998

54. RF Link Budget Analysis in Urban Propagation Microcell Environment for Mobile Radio Communication Systems Link Planning-Ardavan Rahimian, Farhad Mehran-2011 IEEE

55. Statistical Characteristics of Microwave Signals Scattered from a Randomly Rough Surface-Saba Mudaliar and Freeman Lin-2007 IEEE

56. System Level Analysis of Mobile Cellular Networks considering Link Unreliability-Carmen B. Rodríguez-Estrello, Genaro Hernández-Valdez, Felipe A. Cruz-Pérez- IEEE Transactions on Vehicular Technology, Vol. 58, No. 2, February 2009

57. IEEE WLANs: 802.11, 802.11e MAC and 802.11a, 802.11b, 802.11g PHY Cross Layer Link Budget Model for Cell Coverage Estimation-Roger PierreFabris Hoefel-IEEE

58. Digital Mobile Radio Towards Future Generation Systems

59. COST 231 Final Report – COST Telecom Secretariat

60. Link Budget Analysis in Mobile Communication Systems-Haotian DAI, Mike Cherniakov, JunHong – IEEE

61. Networking at 60 GHz: The Emergence of MultiGigabit Wireless-Upamanyu Madhow-IEEE

62. Performance Analysis of Cellular CDMA Networks over Frequency Selective Fading Channel-Jeich Mar and Hung-Yi Chen-IEEE

63. Transactions on Vehicular Technology, Vol. 47, No. 4, November 1998

64. Farhad Mehran-2011 IEEE Statistical Characteristics of Microwave Signals Scattered from a Randomly Rough Surface-Saba Mudaliar and Freeman Lin-2007 IEEE

65. Analysis of Designing Software Defined Radio, Duan Lian, International Conference of Control and Engineering, 2010

66. SDR for Broadband OFDM Protocols, Brian Kelly, IEEE International Conference on Systems, Man and Cybernetics, 2009

67. OFDMA for 4G, Gholamreza Parsaee & Abdulrahman Yarali, IEEE 2004

68. Principles of OFDM, Ashish Pandharipande, IEEE 2002

69. Wireless LAN Radios: System Definition to Transistor Design, Arya Behzad

# Web Sites

1. Institute of Electrical and Electronics Engineers- www.ieee.org
2. IEEE Communications Society- www.comsoc.org
3. American Radio Relay League- www.arrl.org
4. International Telecommunications Union- www.itu.int
5. International Telecommunications Satellite Organization – www.itso.int
6. Third Generation Partnership Project- www.3gpp.org
7. UMTS Forum – www.umts-forum.org
8. Wi-Fi Alliance – www.wi-fi.org
9. US Purtek LLC – www.uspurtek.com

www.ingramcontent.com/pod-product-compliance
Lightning Source LLC
Chambersburg PA
CBHW081720170526
45167CB00009B/3640